MODERN PRACTICAL MASONRY

Edmund George Warland

With an introduction by Christopher Weeks

Routledge
Taylor & Francis Group
LONDON AND NEW YORK

First Edition published in 1929 by B.T Batsford, London

Published by Donhead Publishing Ltd 2006

Published 2015 by Routledge
2 Park Square, Milton Park, Abingdon, Oxon OX14 4RN
711 Third Avenue, New York, NY 10017, USA

Routledge is an imprint of the Taylor & Francis Group, an informa business

ISBN 13: 978-1-873394-76-2 (hbk)

A CIP catalogue for this book is available from the British Library

Warland, Edmund George
Modern practical masonry
1.Stonemasonry
I.Title
693.1
ISBN-13: 9781873394762
ISBN-10: 1873394764

Introduction to the 2006 edition

The 'modern' masonry to which E. G. Warland refers in his title was that practised by a thriving industry adapting rapidly to the widespread introduction of steel-framed construction. Today's stone and glass-clad structures bear small relation to buildings of the 1920s and thirties (the example which frequently furnishes plates for *Modern Practical Masonry* is that of Bush House in London). The decorative vocabulary was predominantly neo-Classical, creating demand for large quantities of classical moulding and other ornament. Indeed, the masonry 'cladding' was really only a pared-down version of load-bearing masonry. Nevertheless, the associated changes in masonry practice, including the use of automated cutting and finishing machinery, led Warland to set his volume in clear opposition to earlier works such as *Practical Masonry* by W. R. Purchase (Crosby Lockwood & Son, 1896).

Modern Practical Masonry was Warland's first major published work. It would appear to have been well received, and there followed *The Fabric of Modern Building* (Pitman, 1937) *Constructional Masonry* (Pitman, 1946), and *The Technique of Building* (Hodder and Stoughton Ltd., 1949). All of these titles display a similar format and style, and all have a straightforward didactic purpose rather than anything approaching the discursive. During this period, Warland's published works included a number of syllabus-specific textbooks aimed at students studying for examinations with the *City & Guilds Institute* and the *Institute of Builders*. He also wrote in similar vein for younger students in technical schools. His publications form part of a drive to modernize and standardize building quality and training in construction. The foundation of the Building Research Station (now the Building Research Establishment) after the First World War and the publication of seminal works such as Schaffer's *The Weathering of Natural Building Stones* (HMSO, 1932, reprinted Donhead, 2004) can also be seen in this context.

Very little biographical material on Edmund George Warland is available in the public domain. He was, above all, a teacher, for many years Chief Lecturer on Masonry and Geometry for Masons at the London County Council School of Building in Brixton (founded in 1904 and now part of London South Bank University). A refreshing sense of that humility proper to his profession occasionally emerges in his writing, and may account to some degree for his relative obscurity today.

From 1928 onward, he was an active member of the Institution of Structural Engineers, particularly during the 1930s, and was for a time chair of the Institution's Lancashire and Cheshire Branch. Some impression of his character can be gained through transcripts of his commentary on the published work of others that appear in the journal *The Structural Engineer,* where he appears forthright, knowledgeable and above all sensible. He was resident in Liverpool when he died during the summer of 1964.

Modern Practical Masonry is an encyclopaedic guide to the subject, based heavily on meticulous and beautiful diagrams. At its heart is a treatise on geometry for setting-out, though it contains a number of other valuable chapters. The structure is logical and straightforward, and although the entries are never organized alphabetically, it remains an object lesson in clarity and simplicity. In three sections, the bulk of the material appears in those chapters detailing specific constructions; methods of working particular structural shapes; basic, and finally advanced geometry and setting out. The 1929 edition, which this volume reproduces, is beautifully finished with sharp drawings, a number of which are over a double page spread. The photographic plates are of good quality, well chosen and clearly reproduced. There is a reasonable glossary at the back, and suggestions for further reading appear at the end of pertinent chapters.

In the second edition (Pitman, 1953), some regrettable omissions were made. Gone are the useful, if somewhat quaint sections on mason's tools and masonry machinery. The third and final section, 'Costing and Estimating', was also omitted. This is a pity, because its forms and tables constitute a very carefully thought out and exhaustive set of templates to aid in estimating work, that are as relevant today as they were in the 1920s (with very little adaptation they would furnish an excellent estimation spreadsheet for a computer). The table of contents was condensed and the photographic plates were collated in series. Both changes detract from the clarity and logic of the layout. The size of the illustrations and their quality, as well as that of the photographic plates was somewhat reduced. This is a shame because they are the crowning glory of the book. The frontispiece of Bush House was substituted with one showing the construction of Rylands Building in Manchester (1931). The addition of a list of plates and a quotation from Longfellow's 'The Builders' is but poor mitigation.

In their book, *Practical Stone Masonry* (Donhead, 1995) Peter Hill and John David present a well-reasoned update to *Modern Practical Masonry*. Their excellent work is nevertheless explicitly defined in relation to Warland's book, speaking eloquently of the relevance of the older book to today's practitioners. Nonetheless, any handbook in current use will be showing signs of age in its eighth decade, and *Practical Modern Masonry* is no exception. Some sections, particularly 'Description of Machines in Use', are no longer of direct practical utility. Readers will note dated terminology ('alabaster' for 'onyx', etc.) and factual content such as listings for quarries now closed or worked out. They might shudder at the wooden scaffolds and health and safety regime apparent in the photographs taken on site. The brief sections on stone decay and restorative treatments betray their times, as do descriptions of the use of iron and steel fixings and ordinary Portland cement. The attentive reader will view such matters in context and cross reference with more modern sources.

Apart from students of masonry and carving, teachers, architects, surveyors and conservation officers, this is a book for anyone with an interest in historic buildings, be it technical or aesthetic. It is, too, pleasurable simply to browse the book for the beauty of the geometrical diagrams and the rich vocabulary evocative of working mens' lives past.

E. G. Warland retired from the profession in 1956, at a time when computer programmes were beginning to be employed in structural engineering, and some of the techniques he made a special study of had largely been superseded. Today, the industry to which he dedicated his professional energies has transformed beyond recognition, and stone working of the kind that concerned him is now the province of restoration. Warland was clearly a practical man with an obvious regard for craft. He was at pains to stress his view that the use of automated machinery would never obviate the involvement of true craftsmen. It is to be hoped the validity of this view will be underscored by the re-edition of Warland's work at a time when the advance of slick automation in the stone industry is so overwhelming.

Christopher Weeks
Christopher Weeks has practised as a conservator in France, Italy, Eire and the UK, largely on architectural sculpture. His particular interest is the conservation of painted stone, and he now runs an independent conservation practice in Hertfordshire.

FRONTISPIECE

FAÇADE TO BUSH HOUSE, FACING " KINGSWAY," SHOWING COFFERED DOME.

Helmle & Corbett, Architects.
New York, U.S.A.

MODERN PRACTICAL MASONRY

A COMPREHENSIVE TREATISE ON

THE PRACTICE OF THE VARIOUS BRANCHES OF THE CRAFT

GENERAL, MONUMENTAL, MARBLE, AND GRANITE

FOR THE USE OF

CRAFTSMEN, STUDENTS, TEACHERS, BUILDERS, & ARCHITECTS

CONTAINING A FULL DESCRIPTION OF

TOOLS AND MACHINES; DETAILS AND CONSTRUCTION; HOISTING AND SETTING OF STONEWORK; PRINCIPLES OF STONE CUTTING; BUILDING STONES AND THEIR CLASSIFICATION; PRESERVATION OF STONE; GEOMETRY AND SETTING OUT; COSTING AND ESTIMATING

WITH A FULL GLOSSARY AND INDEX

BY

E. G. WARLAND, A.I.STRUCT.E.

Chief Lecturer on Masonry and Geometry for Masons, L.C.C. School of Building, Brixton, London, S.W.

With upwards of 600 Illustrations and 18 double-page Plates from Drawings by the Author and 59 Reproductions from Photographs

LONDON

B. T. BATSFORD LTD., 94 HIGH HOLBORN

DEDICATED

TO THE MEMORY OF

PRINCIPAL H. W. RICHARDS

but for whom this book would not have been written

FOREWORD

CONSTRUCTION in stone is an outstanding section of building construction, because it is a principal medium for architectural expression in monumental and commercial buildings, churches, and public institutions. Modern construction, as portrayed in the large steel-framed buildings of our principal cities, has not destroyed, but rather enhanced the value of masonry by its association with framed construction.

As in the past, careful selection of materials and wise arrangement in the details of masonry construction, combined with good craftsmanship, determines, in a large measure, the durability and life of the structure in which it is employed.

As a subject for specialised study, masonry, therefore, takes a high place, and the purpose of this volume is to provide, for the craftsman and builder, a reasonably comprehensive treatise on modern mason's work, in which all the sections of the work of the craft receive due attention. It is recognised also that good masonry has a definite relation to good architecture, and that the architect will have a legitimate interest in this work. Care has therefore been taken to select for discussion examples of modern masonry construction which are acceptable and appropriate, and selected chiefly from important buildings of recent construction. Readers interested in design and construction from the point of view of professional practice should, therefore, find much in this work which is of personal interest, and likely also to be helpful to them in the solution of constructional and geometrical problems which may arise in relation to their private practice and personal work.

The volume deals with tools, materials, machines, principles of construction, the geometry of masonry, setting out and stone cutting, hoisting and setting, costing and estimating, and terminates with a glossary of terms used in masonry.

The most important sections of this volume are probably those in which the applications of geometry to the setting out and shaping of stonework are developed. This branch of the mason's art has an absorbing interest for a keen student of masonry and also for the general student of building construction who has a natural liking for geometrical problems. For successful study and ultimate confidence in the application to unusual problems a thorough knowledge of the main principles of plane and solid geometry is necessary, and readers who are desirous of becoming experts in geometrical masonry should not shirk the concentrated study of the principles of geometry in the early stages of their training. With such a foundation the study of this volume will be a stimulating pleasure, as well as a thorough preparation for an important position in the building industry.

The ambitious craftsman, having become a master of his craft, will probably aspire to some position involving a wider knowledge, more responsibility, and better remuneration. He will realise that the cost of structural work is a matter of prime importance, and will desire knowledge of costing and estimating. Such knowledge will enable him better to appreciate the responsibility undertaken by the estimator and contractor in tendering for first-class masonry work, and will also enable him to relate his efforts in the time taken to accomplish a task, to the ultimate cost of a completed unit of work.

A very helpful chapter on methods of arriving at costs and the uses of established costs in preparing estimates has been added, and many examples of tabulation given for the purpose of developing systematised clerical work.

The writer expresses the hope that the thoroughness displayed by the author in his selection and preparation of matter, in the care bestowed on the arrangement of the examples, and in the accurate and well-drawn diagrams and plates, will meet with appreciation, and that the volume will be used, as the author would desire, for the increase of craft knowledge amongst masonry apprentices, and for the assistance of all who are actively connected with architectural masonry construction.

F. E. DRURY.

PREFACE

THE changing requirements during recent years, both of modern building construction and the methods of manufacture of wrought stonework, would appear to call for a book dealing with modern masonry.

This work is the result of the author's practical experience in the various branches of the craft, combined with several years' comprehensive teaching of the subject. The writer has endeavoured to explain in a direct and simple manner the art of modern practical masonry, thereby bringing before his readers the technicalities of the craft with a view to meeting the need of the young student who is unable to utilise the facilities offered by those technical institutes which specially include masonry among their courses. It is hoped that the information contained in this work will be also of service to teachers of building subjects. To this end it has been arranged to cover a course of instruction in masonry extending over four years, and should meet the requirements of students entering for the final certificate in masonry of the City and Guilds of London Institute and the masonry section of the examination of the Institute of Builders. The writer is glad to acknowledge his indebtedness to Mr J. P. C. Bowden for his help during the preparation of the notes on Costing and Estimating, together with helpful criticisms of the illustrations, and to Mr F. W. Hill, B.Sc., A.I.Struct.E., for correcting the MSS., and to Mr A. R. H. Jackson, A.R.C.A., for shading the plates, and for his kindly criticisms of the details. He is also desirous of expressing his thanks to Mr F. E. Drury, M.Sc., F.I.S.E., M.I.Struct.E., Principal of the L.C.C. School of Building, Brixton, London, S.W., for his kindly criticisms, and Mr E. J. Burlington for reading the proofs. The writer wishes to thank those who have kindly assisted by loaning and permitting photographs to be taken.

CONTENTS

INTRODUCTION

MASONRY is divided into several branches, each of which is subdivided into several groups. The chief branches are—

General Masonry.	Marble Masonry.
Monumental Masonry.	Granite Masonry.

The craftsmen connected with these branches are broadly divided into groups, the craftsman in each section being thoroughly conversant with the technique of the branch in which he works. There are the Banker Mason or Stone Cutter, Machinist, Fixer, Waller, Mould Cutter, and Setter-out or Draughtsman.

These are again subdivided into groups, such as those working on—

Soft stone.	Marble.
Portland stone.	Granite.
Hardstone.	

At one time it was considered unnecessary to have a skilled craftsman to operate the planing and moulding machines, but experience has taught all concerned that the most efficient operators are those who are skilled in the working of stone, and who are acquainted with lines and the application of moulds. We find, therefore, a large number of masons engaged in the engineering branch of masonry, which requires a certain amount of specialised knowledge.

On the constructional side of masonry there are two kinds of craftsmen: the wallers and the fixers. The wallers prepare and set rubble walls, the stones being supplied to them in a partially prepared condition on the site. Their task is to dress them roughly to shape, and to build the structure in the manner required by the architect. Much depends on the skill and the artistic temperament of these craftsmen, for the satisfactory execution of any particular piece of work.

The duty of the fixer is to set worked stones into their correct position in the building. An efficient fixer must have a knowledge of general building construction. He should be acquainted with all types of scaffolding and appliances for hoisting stones of various weights, and understand their construction and erection. He should be familiar with the preparation and uses of limes and cements, and their combination with a matrix into suitable mortars for particular purposes. The masonry draughtsman, although mentioned last in this list, is by no means the last in order of merit. The draughtsman, or setter-out as he is termed, plays an important part, and upon him rests a large share of responsibility in interpreting the drawings received from the architect, which are often only in the form of rough sketches.

To become an efficient setter-out, it is necessary to become well acquainted with the principles of geometry, and to be able to apply these principles to the drawings, so that the individual stones in any particular piece of work may be worked to the pattern cut to the setting out. Good setting out is indispensable for good results.

A thorough knowledge of building construction is required by the setter-out, so that he may be in a position to advise on the general arrangement of the beds and joints for the stones throughout the building. It is his duty to prepare scale lay-outs of the elevation and to indicate on these lay-outs every stone in the building.

To-day it is necessary that the setter-out should have a knowledge of steel construction, and should be in a position to interpret steel work drawings in order that he may fully appreciate the cutting and construction necessary where steelwork and stonework are in combination. In the larger establishments mould cutters are employed in association with the setter-out. It is their duty to transfer, from the setting out, all the lines and particulars necessary for the working of the stones on to zinc templets.

The mould cutter is usually in a transition stage, from the position of banker mason to that of setter-out or draughtsman; if he desires advancement it is a vital necessity for him to become thoroughly acquainted with all the requirements and attainments of the draughtsman.

Although there are so many branches in masonry, the tendency is towards a closer combination of these branches into one craft. Stone masons are now working in marble masonry works and *vice versa* with excellent results. This migration of the various kinds of masons into different sections of the craft implies a more extended knowledge of the whole technique of masonry, and widens the outlook of the draughtsman. The technique of masonry calls for a knowledge of the principles of geometry far beyond the imagination of those who are not closely connected with the craft, and no branch of the building industry requires a more thorough grounding in the principles of sound construction.

The line of separation between the work of the carver and that of the mason is very indefinite, so that it is essential for the young craftsman in masonry to be encouraged to study the arts closely allied to his own work, in order that he may obtain that freedom of mind and hand which is necessary for the execution of any piece of enrichment, in the material upon which he labours.

Technical classes in masonry are being formed in various parts of the country, in which facilities for studies in the practical technique of the craft in its entirety are available to the young craftsman. Closely allied to these, in the form of group classes, it is possible for him to study the principles of geometry, construction, and free-hand drawing, all of these being essential to his training as a competent craftsman.

Although the craftsman is not expected to be a scientist, he should acquire a knowledge of the physical and chemical properties of the various stones used for constructional and decorative purposes in building, and he may become an expert in his knowledge of the behaviour of stones and the causes of decay by a careful study of practical geology, combined with the craft-experience gained by actual contact with the stone and keen observation on his own part.

As accurate estimating can only be based upon a sound knowledge of the principles of masonry, and the properties of the stone used, then it is important that the *Masonry Estimator* should have a complete knowledge of every process or operation necessary for the execution of any piece of work.

As individual masons differ in their methods and different systems are employed in different localities, it follows that, in estimating, no two men express themselves in the same way. We therefore find that the systems of estimating and costing differ in various parts of the country; but with the advancement of technical education there is certain to be a tendency towards the standardisation of method.

There are so many kinds of stone, each varying to a greater or lesser degree in texture, hardness, and constituents, that a wide practical experience is required for the estimator to become efficient in this branch of masonry.

ABBREVIATIONS

1. Add.	Additional.	13. Lab.	Labour.
2. B.	Bed.	14. Lin.	Lineal.
3. B. and Jt.	Bed and joint.	15. Mld.	Mould.
4. Cir.	Circular.	16. No.	Number.
5. Cir. F.	Circular face.	17. P.F.	Plane face.
6. Cir. Cir.	Circular circular.	18. P.	Polish.
7. Cir. Sk.	Circular sunk.	19. Prep.	Preparatory.
8. Ddt.	Deduct.	20. Rebd.	Rebated.
9. E.O.	Extra over.	21. S.	Sawing.
10. E.O. ½ S.	Extra over half sawing.	22. Sk. F.	Sunk face.
11. F.	Face.	23. Sk. Jt.	Sunk joint.
12. ½ S.	Half sawing.	24. Sup. ft.	Superficial foot.

MODERN PRACTICAL MASONRY

SECTION I

CONSTRUCTIONAL AND PRACTICAL MASONRY

CHAPTER I

DESCRIPTION OF TOOLS

The Axe—The Patent Axe—Batting Tool—Bevel or Shiftstock—Wood-handled Chisels — Gouge — Mallet-headed Chisels — Drafting Chisel — Boasters — Claw Chisels—Gouges—Points—Waster—Fillet and Moulding Chisels—Hammer-headed Tools—Hammer-headed Punch—Pitching Tools—Cup-headed Chisel—Marble Tools—Fauld's Patent Tools—Steel Wing-compasses—Centre Bob—Cock's Combs—Dummy—Drags—Braces and Archimedean Drills—Flat Bits—Iron Hammers—Spall Hammer—Jumpers—The Diamond Jumper—Flat Jumper—Lewising Tools—Masons' Spirit Level—Boxwood Fourfold Rules—Saws—Cross-cut Saws—Frig-bob Saws—Hand Saws—Whip Saws—Fillet Saws—Grub Saw—French Scraper—Steel Set-squares—Sinking-square—Double Stock Square—Straight-edge—Trammel and Scriber—Box Trammel—Fixers' Bedding Trowel.

TOOLS used by masons vary according to the stone which is to be operated upon. They should be made from best *cast* or *tool* steel.

Tools are sometimes made from *mild* steel, but they lack the edge necessary for the clean cutting of the stone.

Each craftsman accumulates various implements, to facilitate the execution of difficult intersections and undercut mouldings.

For a clear understanding of the various methods of working stone, a brief description of the principal hand tools used by the mason is given.

The Axe (Fig. 1) is used by granite masons for reducing the faces of the stone to the correct level, after the punch and before the patent axe.

The Patent Axe (Fig. 2) is a double-headed tool. A set of chisel plates is bolted to each end of the hammer head. It is used for forming the final surfaces on the faces of granite, the finished surface being known as "*fine axed*" or "*medium axed*," according to the grading of the cutting plates used.

Batting Tool (Fig. 3).—This tool is used chiefly for finishing the surfaces of sandstones. The faces of padstones or template stones are usually specified as "*batted, so many bats to the inch*."

There is a tendency to introduce "*batted*" or "*tooled*" surfaces, as the finished dressing for "*string*" and "*plinth*" courses in Portland stone.

Bevel or Shiftstock (Fig. 4).—This tool consists of a double stock, between which is inserted a thin blade and connected to the stock by means of a running *thumb-screw*, inserted in the centre slot, thus allowing the blade to be adjusted to any angle. It is usually made of brass or gun-metal, and used for testing *bevels*, *chamfers*, *sinkings*, etc.

The principal degrees are often marked on the blade, so that correct angles may be readily obtained by adjusting these degree lines to the edge of the stock.

Wood-handled Chisels.—These chisels are used for cutting soft stones, such as Bath stone. They are similar to carpenters' chisels, but stouter at the neck. They are of various forms and sizes, and are used according to the work required of them. These tools are ground when required to be sharpened. Fig. 5 is a sketch of a *wood-handled drafting chisel* about $\frac{3}{4}$ in. wide, used for working the marginal or other drafts on the stone.

Fig. 6 shows a **Driver**. This tool is similar to the *drafter*, but slightly heavier and about 2 in. wide. It is used for dressing the surfaces to a true plane.

Gouges (Fig. 7).—These are of various curves and sizes, ranging from $\frac{1}{4}$ to $1\frac{1}{2}$ in. wide, and are used for obtaining the contours of mouldings. Wood-handled chisels from $\frac{1}{8}$ to $\frac{3}{4}$ in. wide are also used for fillets and intersections.

Mallet-headed Chisels.—These tools are in general use for Portland and similar stones. They are made of *cast steel*, and used in conjunction with the mallet. When they are required to be sharpened, they are forged, drawn out, and tempered, and brought to a knife edge by rubbing backwards and forwards on a grit-stone.

If a hammer is used on these tools, the head will snap off, the steel being very brittle at the head of the tool.

Fig. 8 shows a **Drafting Chisel**, which is used chiefly for working the drafts on the various surfaces. The cutting edge is about $\frac{3}{4}$ to 1 in. wide.

There are also mallet-headed chisels from 1 to 2 in. wide, these being named according to their width.

When they become 2 in. wide they are called **Boasters** (Fig. 9). They are then chiefly used for boasting over the surfaces of the stone, in a series of drafts, until the surface is in a true plane. Often the surfaces are finished to show the boasted chisel marks, making a very desirable finish to the surface of the stone.

Claw Chisels (Fig. 10).—These tools are from $1\frac{1}{2}$ to 2 in. wide, and are used in sequence to the "*Point*" or "*Punch*." The teeth are provided to prevent the stone from *plucking* or *lifting* in holes over the surface.

After the superfluous stone between the marginal drafts has been pointed in furrows, a series of drafts is worked across the surface, parallel to the furrows, with the mallet and claw chisel.

Gouges (Fig. 11).—These are of various sizes and curves, and are used in working mouldings to their required shapes. Occasionally mallet-headed

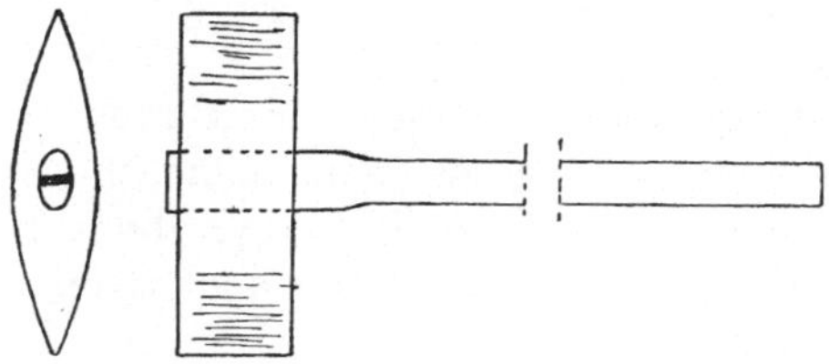

FIG. 1.

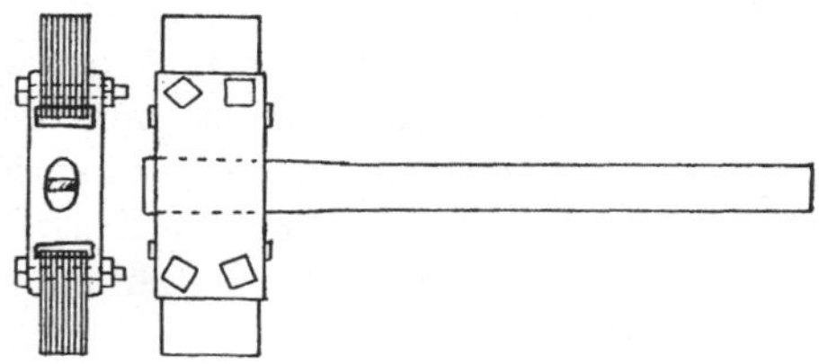

FIG. 2.

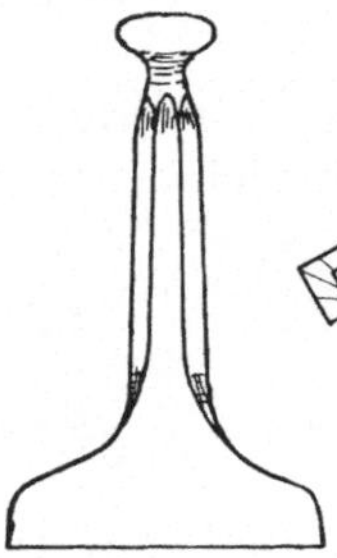

FIG. 3.

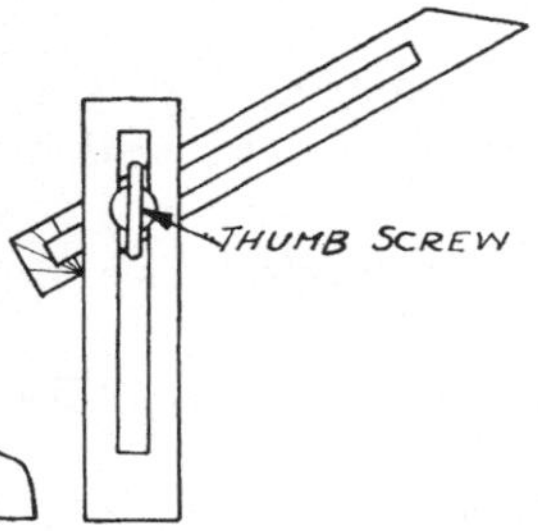

FIG. 4.

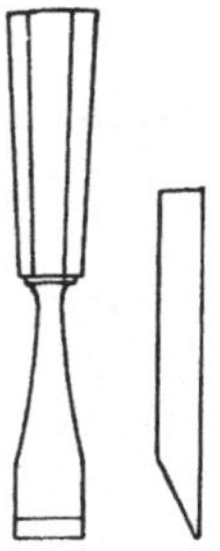

FIG. 5.

FIG. 6.

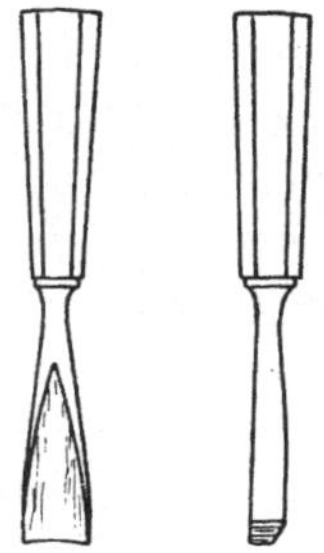

FIG. 7.

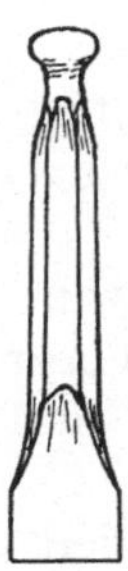

FIG. 8.

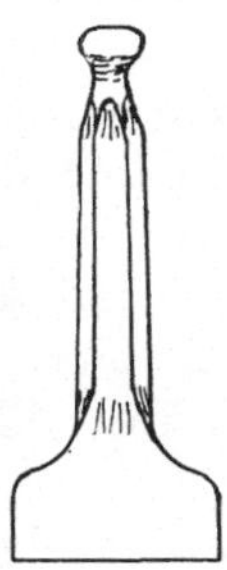

FIG. 9.

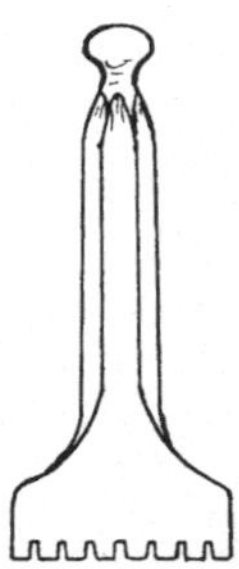

FIG. 10.

FIG. 11.

FIG. 12.

FIG. 13.

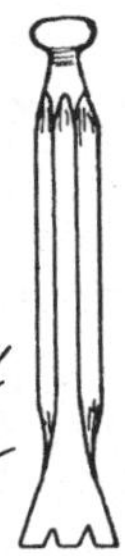

FIG. 14.

FIG. 15.

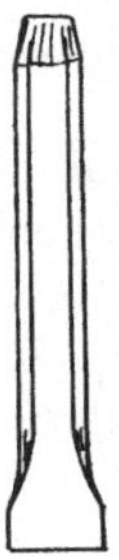

FIG. 16.

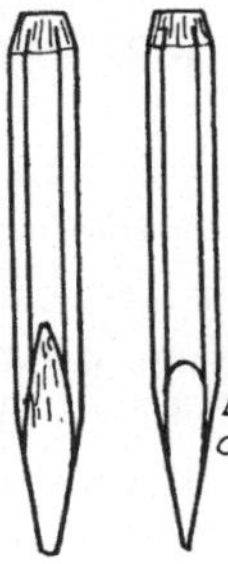

FIG. 17.

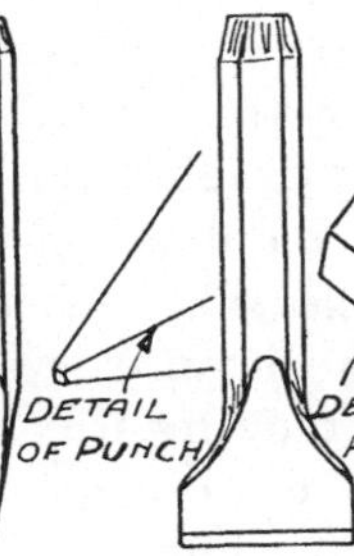

FIG. 18.

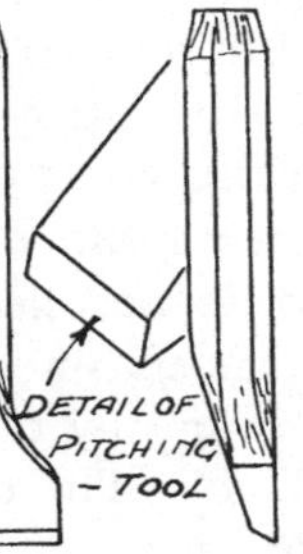

FIG. 19.

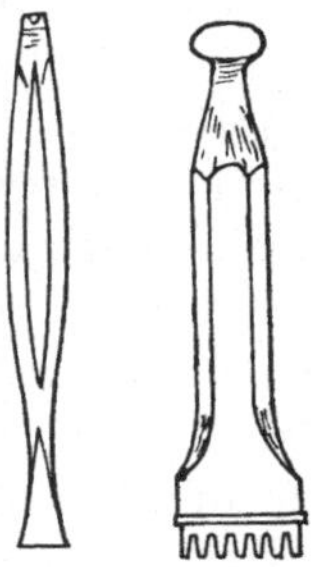

FIG. 20.

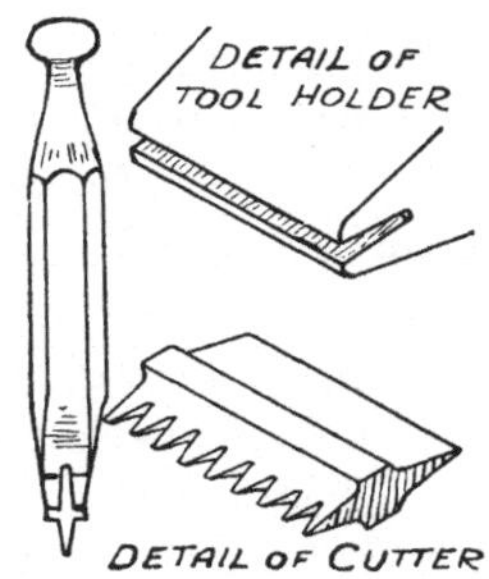

FIG. 21.

MASONS' TOOLS.

chisels of various widths are rubbed up to a circular cutting edge and used for the same purposes in preference to the gouges.

Points (Fig. 12).—These are used for reducing the superfluous stone almost to the level required in a series of furrows across the surface between the marginal drafts. They are drawn out to a chisel-cutting edge about $\frac{1}{4}$ in. wide, and are used with the cutting edge parallel to the surface being cut. This cutting edge is shown in Fig. 13.

Waster (Fig. 14).—This tool is similar to a drafting chisel, the only difference being in the cutting edge, which is formed in a series of teeth. It is used on Bath and similar soft stones for the removal of the superfluous stone in preference to using a point.

Fillet and Moulding Chisels.—Varying from $\frac{1}{8}$ to $\frac{1}{2}$ in. wide, delicately shaped, as Fig. 15, they are used for working details and enrichments, intersections, etc.

Hammer-headed Tools are used on hard sandstones and granite or similar stone.

Fig. 16 is a typical form of drafting chisel, and is used for purposes similar to the mallet-headed drafting chisel.

Fig. 17 shows a **Hammer-headed Punch**. This tool is used considerably on Portland and similar stones, as well as on granite, for the removal of the rough stone. When sharpened for granite and marble, it is drawn out to a point, similar to that shown in detail, but when it is sharpened for working on Portland stone, a cutting edge is formed as for the point.

Pitching Tools (Fig. 18).—These are used for *spalling* or bursting the superfluous stone. They are provided with an edge bevelled to the back of the tool, which is straight in section, as in Fig. 19.

Fig. 20 shows one type of **Cup-headed Chisel**. The head of this tool is formed especially for use in conjunction with an iron hammer. The tools are of numerous shapes, *points*, *chisels*, and *gouges*, $\frac{1}{16}$ to 1 in. wide, and are used chiefly for carving, lettering, etc.

Marble Tools are of similar shape to the above, but hammer-headed, and used in conjunction with a small steel hammer from $1\frac{1}{2}$ to 2 lbs. in weight.

Fauld's Patent Tools (Fig. 21).—The handles of these tools are formed with a socket to receive a steel cutting piece, having a double cutting edge. When one edge is worn, it can be removed and reversed. They are eminently suitable for grit-stones, but not suitable for working on Portland and similar stone because of the shells, which quickly destroy the cutter.

Fig. 22 shows a pair of **Steel Wing-compasses**. These tools are used by masons for measuring distances from the moulds and transferring them to the stone, also for describing circles on the stone where required. The wing is provided to fix the correct radius or distance, when obtained, the thumb-screw is tightened, thus preventing any movement of the compass legs.

Fig. 23 is a sketch of a **Centre Bob**. These are made of gun-metal provided with a steel point. A screw-cap, through which the line is passed and knotted, keeps the steel point perpendicular when in position.

They are used by fixers for determining various points down from a *datum line* or any other fixed position.

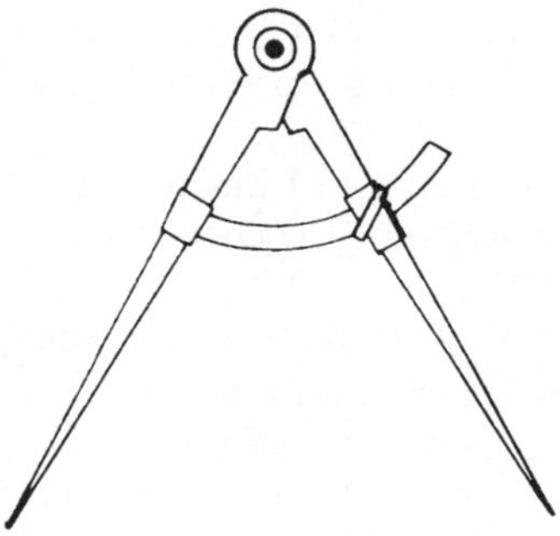

FIG. 22.

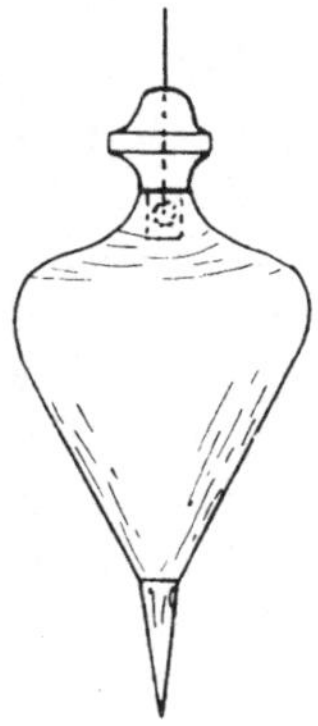

FIG. 23.

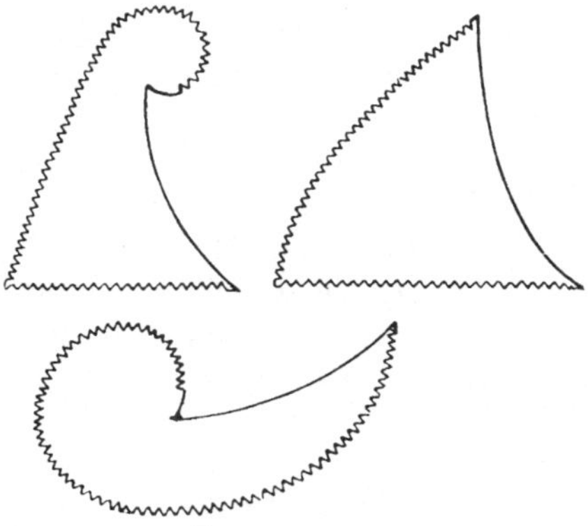

FIG. 24.

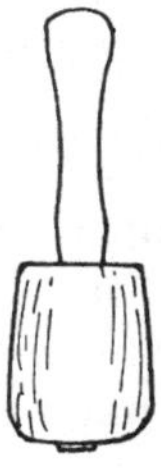

FIG. 25.

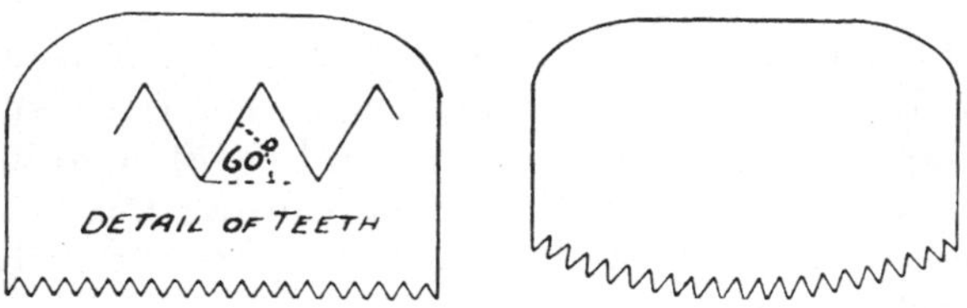

FIG. 26.

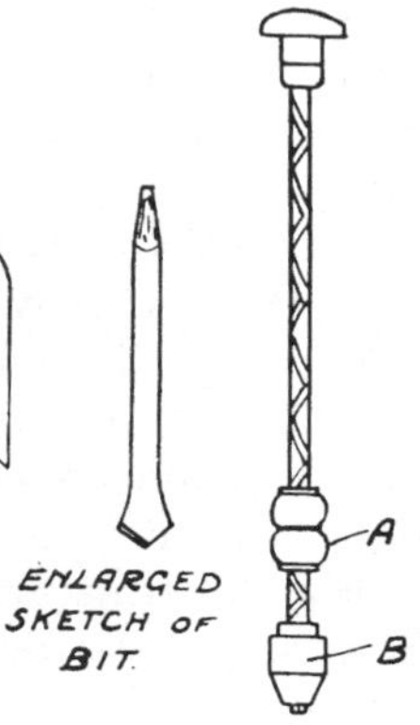

FIG. 27.

FIG. 28.

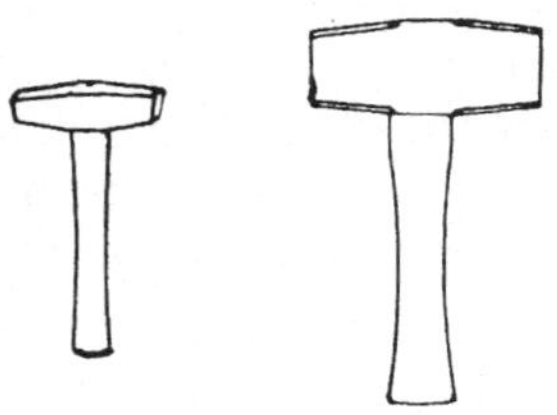

FIG. 29.

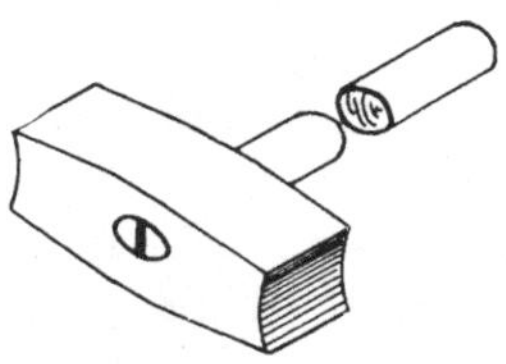

FIG. 30.

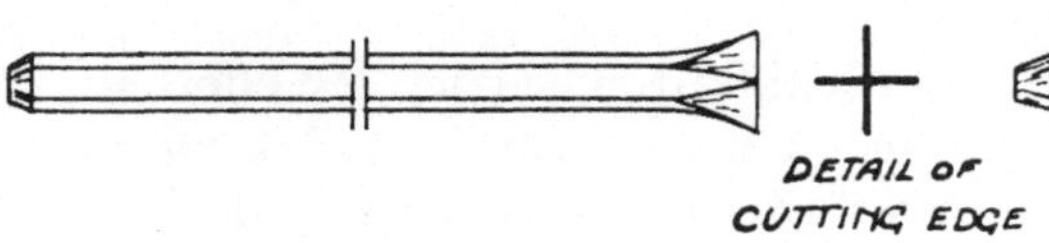

FIG. 31.

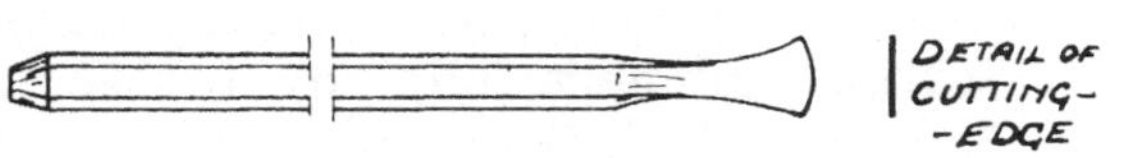

FIG. 32.

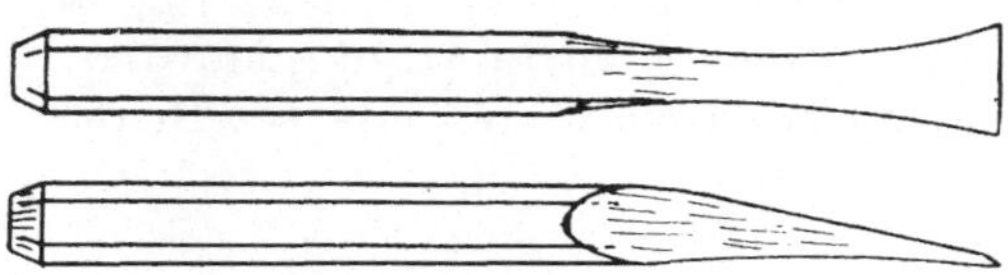

FIG. 33.

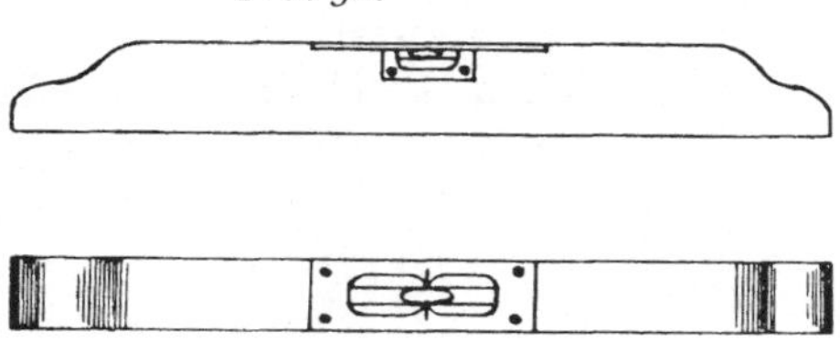

FIG. 34.

MASONS' TOOLS.

Fig. 24 shows two sketches of **Cock's Combs**. They are made of thin steel plate, and are provided with fine teeth round the edges. They are made in the form of *French curves* and used for *combing* and cleaning up mouldings and intersections in Bath and similar stones.

Fig. 25 is a sketch of a **Dummy**. It is in the form of a round mallet, but made of a mixture of zinc and lead, or of zinc, and is used for striking wood-handled chisels. The weight of the dummy is about 3 lbs.

Drags (Fig. 26) are used for finishing the surfaces of Bath and similar stones. They are made of steel plate and provided with teeth along one edge. There are *straight* and *circular* drags, graded according to the size of the teeth, and known as "*coarse*," "*second*," and "*fine*." When the surface has been reduced with the dummy and chisel almost to a true plane, or to the level required, the *coarse drag* is drawn across the surface in a backward and forward movement, or from side to side, until the chisel marks have disappeared.

The *second drag* is next used in a similar manner, thus creating a finer surface. The *fine drag* is then used to remove all scratches, and to bring the surface to the required finish. All drag marks should be at right angles to each other.

Braces and **Archimedean Drills** (Fig. 27) are used in combination with **Flat Bits**, chiefly for drilling holes in marble. These are very convenient tools for drilling holes for securing the lead filling of lettering in marble. The handle A is drawn up and down the stem causing the bit to revolve in one direction only. An adjustable chuck B is fitted to the drill, so that bits of various sizes may be used.

Fig. 28 illustrates the principal **Hammers** used by masons, and although these are of various sizes and weights, according to the material, the shape in nearly all instances is similar.

The Portland mason uses a hammer from 4 to 6 lbs. in weight, and from 6 to 7 in. in length, whilst the marble mason usually works with a hammer about 2 lbs. in weight and from 3 to 4 in. long.

It is natural to imagine that the harder the material to be cut, the heavier should be the hammer for striking the tools; but this is not the case, for usually a granite mason works with a much lighter hammer than a Portland stone mason.

Hard material, such as marble and granite, requires coaxing. If force is applied to these materials by the use of heavy hammers, a great amount of energy is spent, with very poor results.

The heads of the hammers are of cast steel, and the striking faces incline towards, instead of parallel to, the handle.

Iron Hammers (Fig. 29) are made of malleable iron. They are short in the head, and provided with a larger face-surface than the *mason's steel hammer*. They are used in conjunction with *cup-headed chisels* for carving, lettering, and delicate enrichments. The faces of the hammer, being soft, are soon cut away by the cup head of the cast-steel tools.

Fig. 30 is a sketch of a **Spall Hammer**. This tool is from 10 to 15 lbs. in weight, and is used for removing superfluous stone when in large quantities. The faces are *concave*, thus forming a cutting edge when in contact with the stone.

Figs. 31 and 32 show two forms of **Jumpers**. These tools are made so that the cutting edge clears the stem of the tools, and are used for *jumping* or *boring* holes in stone. They are lifted and turned with one hand, and struck with a hammer after each turn.

The Diamond Jumper (Fig. 31) is the most useful for drilling Portland or similar stone, whilst the **Flat Jumper** (Fig. 32) is most convenient for the harder stones, such as granite.

Lewising Tools (Fig. 33).—These are made in the form of mortising chisels, the cutting edge being made wider than the stem of the tool, so that the clearance necessary for the cutting of the lewis mortise is obtained. The back of the tool is straight, whilst the front tapers to the cutting edge, thus enabling the mortise to be undercut or dovetailed.

Fig. 34 shows a typical form of **Masons' Spirit Level**. These tools are used for testing horizontals, and should be applied to the stone, both parallel and at right angles to the face of the wall. The ends of the level should not be decorated with metalwork, as this might cause damage to the arrises of the stone when the level is applied to the surface.

Masons' Mallet (Fig. 35).—These tools are very important to the mason. They are usually made of *beech* or *hickory*. *Pearwood*, if well selected, is also excellent for this purpose. The *beat* of the mallet, or that portion coming into contact with the chisel head, is on the end grain of the wood. It is a speculation, when buying a mallet, as to whether it will prove a success or not. When a craftsman secures a good mallet he prizes it beyond all the other items in his tool kit. They are used in sizes according to the material or the detail to be cut, and in conjunction with the mallet-headed tools.

Mitre Tools (Fig. 36).—These are used for cleaning or straightening mitres. They are made from octagonal steel and formed into various cutting edges to suit curves of different radii.

Mitre Square (Fig. 37).—This is made of thin steel plate about 12 in. long and 4½ in. wide, and provided with one edge making angles of 45° and 135° with the long edges of the tool. It is used as a means of determining the position for holding the straight-edge when *striking* or drawing a mitre line.

Pinch Bars (Fig. 38).—These are used by the fixer for *prising* or lifting the stones, and to assist in adjusting them to their correct position in the building. They are made of steel, and are provided with a *heel*, which forms a *fulcrum* when pressure is brought to bear on the handle end of the bar.

Rifflers (Fig. 39) are a form of rasps of various shapes, and used for cleaning up intersections in difficult positions, and delicate carving details, etc. They are used considerably in France, where the soft stones are worked to the required details *in situ*.

Boxwood Fourfold Rules (Fig. 40) are now in general use by craftsmen, either 2 or 3 ft. long. Their uses are so well known that description is unnecessary.

Saws.—Tooth saws are chiefly used on soft limestones. It is cheaper to cut up these stones by hand than by machinery.

Cross-cut Saws (Fig. 41).—To cut up the quarry blocks into sizes suitable for the stones required, *hand sawyers* are employed. Cross-cut saws are used for this purpose; they are provided with teeth set at an angle of 60° and drawn through the stone in a backward and forward movement. The saws are of various lengths to suit the size of the stone to be cut. In the Bath stone mines, single-handed saws, called *Frig-bob* saws, sometimes measuring 12 ft. long, are used for separating the rock into blocks whilst in the strata.

Hand Saws (Fig. 42).—These are used by the banker-mason for sawing the various surfaces, faces, joints, beds, etc.

Whip Saws (Fig. 43) are convenient for use on the banker for sawing out concave and convex surfaces.

Fillet Saws (Fig. 44).—These are used for cutting direct into the angles of fillets when working groups of mouldings, etc.

Fig. 45 is a sketch of a **Grub Saw**, used in conjunction with sand and water for sawing into the fillets of hard stones.

Fig. 46 is a sketch of a **French Scraper**. These tools are in common use in France for finishing the surfaces of the soft limestones when in position. They are excellent tools for removing the chisel marks in Portland and similar stones. Into a piece of hard wood are fitted several pieces of hardened steel provided with teeth along the two edges. The teeth are arranged so that when the *scraper* is drawn across the surface of the stone, one-half of the number is dragging the surface, whilst the other half is facing the direction of movement, thus preventing a waviness which would result if the teeth were fitted in one direction. A detail of one of the teeth is shown in Fig. 46.

Figs. 47 and 48 show **Steel Set-squares** with angles 45°, 60°, 30° and 90°.

Fig. 49 is a sketch of a **Sinking Square** with an attachment or extra blade forming a *double sinking square*. The stock is usually made of gun-metal, the blades being made of steel. The blade passes through a mortise in the stock, and is held in position by a *thumb-screw*. They are used for *testing sinkings*, *gauging depths*, and *testing intersections*, *breaks*, *ashlar-stops*, etc.

Fig. 50 shows a **Double Stock Square**. This type of square is preferable for use in masonry. They are of various sizes to suit the stone being worked. Their chief use is for testing square surfaces, etc. Upon the accuracy of the square largely depends the accuracy of the finished piece of work.

Fig. 51 shows the usual kind of **Straight-edge**. A bevelled edge is provided for convenience when applying the edge to the surface of a stone.

Fig. 52 shows a **Trammel** and **Scriber**. The trammel is arranged at one end, and is used for drawing lines parallel to the arrises of the stone. It is operated by placing a chisel in the flattened portion and gripping both tools tightly. It is then passed along the arris of the stone, the head of the tool travelling along the surface, which is at right angles to the one upon which the line is required to be trammelled.

The Scriber is used as a means of marking hard lines on the surfaces of the stone. Often a *box trammel* is used for the same purpose. It is not advisable to draw parallel lines with a trammel if more than 4 in. apart. It is then preferable, for the sake of accuracy, to measure numerous points and draw the line by means of a *pliable straight-edge* or a templet of the curve required.

Fig. 53 shows a **Fixer's Bedding Trowel**. These craftsmen prefer a trowel of medium size, slightly rounded at the ends, for spreading the mortar beds.

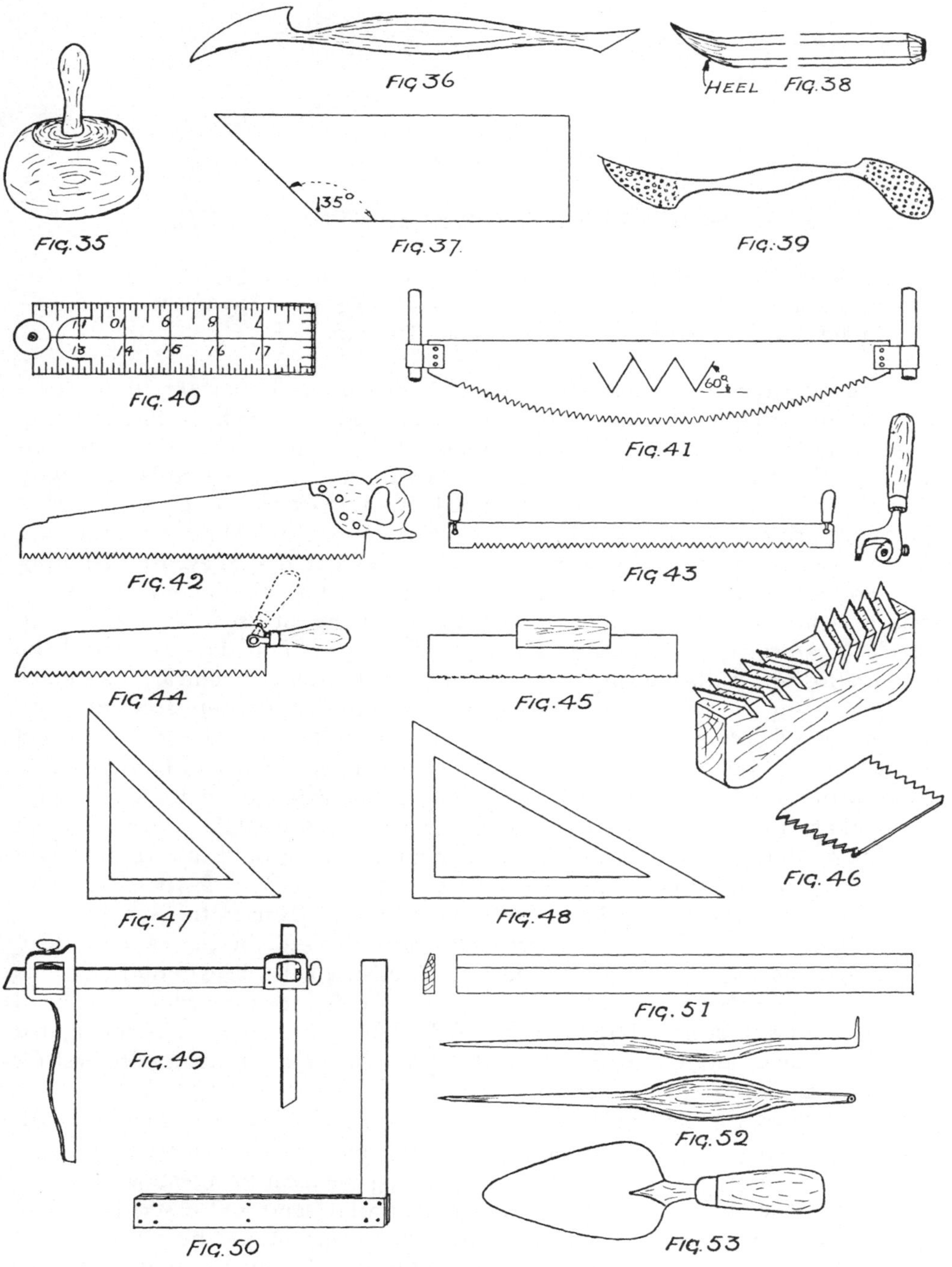

Masons' Tools.

CHAPTER II

DESCRIPTION OF MACHINES IN USE

Frame Saws—Diamond Saws—Duplex Saws—Twin-blade or Duplex Saws—Cross-cut Diamond Saw—Carborundum Saws—Planing and Moulding Machines—Rubbing Beds—Polishing Machines—Stone Turning.

THE installation of efficient modern machinery is as important in the production of wrought stonework as it is in any other branch of the building industry. It has been the custom to adapt old types of machines for the conversion of block stone into wrought stonework, but this practice is gradually dying out. Machines specially designed for the execution of the various processes are now made by engineering firms of repute. These machines are being installed in masonry works to such an extent that they are assuming the aspect of large engineering shops. At one time all the labour on a stone was executed by hand, whereas now machines are utilised for *sawing* the blocks, *surfacing*, *checking*, and *working mouldings*. This does not entail the elimination of craftsmanship, as some suggest, but on the contrary nowadays the craftsman is called upon to execute the more difficult intersections, etc., which necessitates a greater knowledge and aptitude. Buildings are erected at such speed to-day that, without the aid of machinery, it would be impossible to keep the supply of wrought stonework up to the rate required. This would mean that other materials instead of stone would be used for the façades of buildings, thus gradually reducing the demand, which would in turn naturally react upon the craftsman.

A list of the various machines, with their uses, is given in this section.

Frame Saws (Fig. 54).—These machines are used for the conversion of the rough blocks of stone or marble into slabs or to the sizes required. The blocks are placed on trolleys and run under the frame, into which are fitted steel saw blades, either plain or corrugated, about 5 in. wide. The corrugated blades are most suitable for Portland stone, whilst the plain blades are best for sawing marble.

The blades are distanced according to the size of the slabbing required, and are drawn through the stone in a forward and backward motion, making a cut or kerf from $\frac{1}{4}$ to $\frac{3}{8}$ in. wide. The cutting medium for these saws is provided by the addition of sharp *Bridport* or similar sand and a steady flow of water over the top surface of the stone. For some of the hard stones, the addition of a small amount of steel grit facilitates the cutting, but great care is required in the use of this grit in order to prevent rust stains on the stones.

The cutting medium is carried to the bottom of the *kerf* by the water, thus

FIG. 54.—SLABBING UNDER FRAME SAW.

Bath and Portland Stone Firms Ltd.
From "Building."

FIG. 55.—DIAMOND SAW.

FIG. 56.—" ANDERSON " DUPLEX DIAMOND SAW.

South-Western Stone Co.

Fig. 57.—The "Anderson" Simplex Cross-cut Diamond Saw.

South-Western Stone Co.

getting under the bottom edge of the saw-plate and wearing away the stone. When the cutting of the block is finished, the frame is raised and the trolley upon which the stones have been cut is run out clear of the frame and unloaded, or the stones are turned and again placed under the frame for the sawing of other surfaces.

Diamond Saws (Fig. 55).—These machines are specially designed for the rapid sawing of the various stones to shape and to the exact sizes required after they have been slabbed in the frame saw.

Many important improvements and developments in the design of these machines have taken place during recent years. The original type of diamond saw was fitted with a large circular blade, with only a lateral or side movement, whilst the stones, which were placed on a moving table, travelled under the saw blade at a uniform rate, thereby coming into contact with the blade, which revolved at a high velocity. These machines are now superseded to a large extent by the **Twin-blade** or **Duplex Diamond Saw**.

A heavy **Anderson Duplex Saw** is shown in operation in Fig. 56. The stone to be sawn is placed on the table, the motion of which is controlled by the necessary gears, operated by conveniently placed clutches. The cutting of the stone is done by feeding the table up to the revolving blades, and upon the stone being cut through, the table is run back by a quick return motion. A quick approach is fitted to bring the stone up to the blade ready for feeding at the desired cutting speed, thus saving a great deal of time. Diamonds which are fixed in renewable sockets are fitted round the periphery of the saw blades. The blades may be traversed independently or together in the same or opposite direction across the table by means of clutches and cut gear, with precise adjustment. The design of these machines permits of the whole of the motions running while the blades are standing. These motions are controlled through an enclosed gear-box containing machine-cut gears, giving four feeds, which may be instantly changed as desired by means of a lever placed conveniently for the operator. Not only can the rate of feed be adjusted to suit the depth and hardness of the stone, but it can also be manipulated to save the end arrises of the stone as the blades come through. A very high speed is permissible while the blades are in the body of the stone, but if this is maintained right through, damage is likely to occur. It has always been the practice to keep down the table speed to what was considered suitable from both points of view, but with the latest gearing, the rate can be adjusted over a range of 1 to 3 while cutting. This permits of a slow commencement until the blades have got a steadying grip of the stone, followed by a quick increase to obtain a maximum speed until almost through. The rate is then brought down to keep within the ability of the stone to stand against the cutting pressure. This facility makes a very big difference to the rate of output, especially on long pieces of stone.

The saw illustrated in Fig. 56 is fitted with two 6-ft. diameter blades, allowing a cut 30 in. deep. Both blades may operate over one table, as in the photograph, or one over each table. The vertical adjustment of the spindle is 30 in., whilst the tables are 12 ft. long by 6 ft. wide. A 35 H.P. motor is required to drive the saw spindle, and a 6 H.P. variable speed motor, with range 1 to 3, for moving the tables. A separate motor may be used for each table, thus giving greater freedom for handling, and is to be preferred, since the ability to operate both tables in unison is secured by mechanical as well as electrical coupling of the gear units.

In order that higher cutting speeds may be obtained without detriment to the quality of the work turned out, it has been found necessary to evolve an improved arrangement of spindle bearings comprising heavy ball journals. These are carried on separate sleeves which allow the spindle to be drawn endways for replacing the blades.

Safety trips are provided to prevent overloading, and also to stop the table motors in the event of anything going wrong with the main spindle drive. These machines are constructed with blades up to 98 in. diameter. A 90-in. blade has about 200 diamond sockets, and is able to cut 3 ft. 6 in. in Portland stone.

To assist in the production of accurate sawing, both the method of supporting the tables and their construction have received attention. Whereas in the older type of machines the tables were constructed of wood and supported on steel rollers, these are now made of iron cast in one piece and supported on continuous cast-iron slides with inverted vees.

Anderson's Simplex Cross-cut Diamond Saws (Fig. 57).—These machines mark a very great improvement in the mechanical equipment of modern stone and marble works. The cross-cut saw is arranged so that the blade moves across a stationary table, this facility meeting a need for adaptability which effects great economy in working as well as in speeding up some of the vital operations of masonry production.

FIG. 58.—"ANDERSON" DIAMOND BLADE WITH U-TYPE SOCKETS. Diamonds proportioned two "angles" to each "centre," with opposing side diamonds after every second group of face diamonds. U-type sockets are spaced 4 per inch of blade diameter.

The blade, which may be any size up to 5 ft. diameter, would cut a stone 2 ft. thick, whilst the spindle traverse is usually 12 ft., so that a stone 9 ft. by 2 ft. can be cut with ease.

The adaptability of these machines for ending off or jointing increases their value as an item of equipment.

Either diamond or carborundum fitted blades may be used on these machines for cutting Portland and similar stones. The diamond fitted blades are best, especially if fitted with the latest U-type diamond socket, as shown in Fig. 58.

Carborundum blades are most suitable for cutting marble and granite. With these blades the cutting speeds are about half of that obtained with diamond saws.

A photograph of a **Gravity Saw** using a *diamond blade* is shown in Fig. 59. The machine has a rise and fall motion of 15 in. and blade overhang of 29 in. The table surface is 6 ft. 6 in. by 3 ft. 9 in., and requires a 10 H.P. motor for driving.

Fig. 59.—Gravity Saw Fitted with Diamond Blade.

Fig. 60.—A Gravity Saw Fitted with Carborundum Blade.

FIG. 61.—12 FT. × 5 FT. × 5 FT. "ANDERSON" CANTING ARM PLANER, SLIDE BED PATTERN.

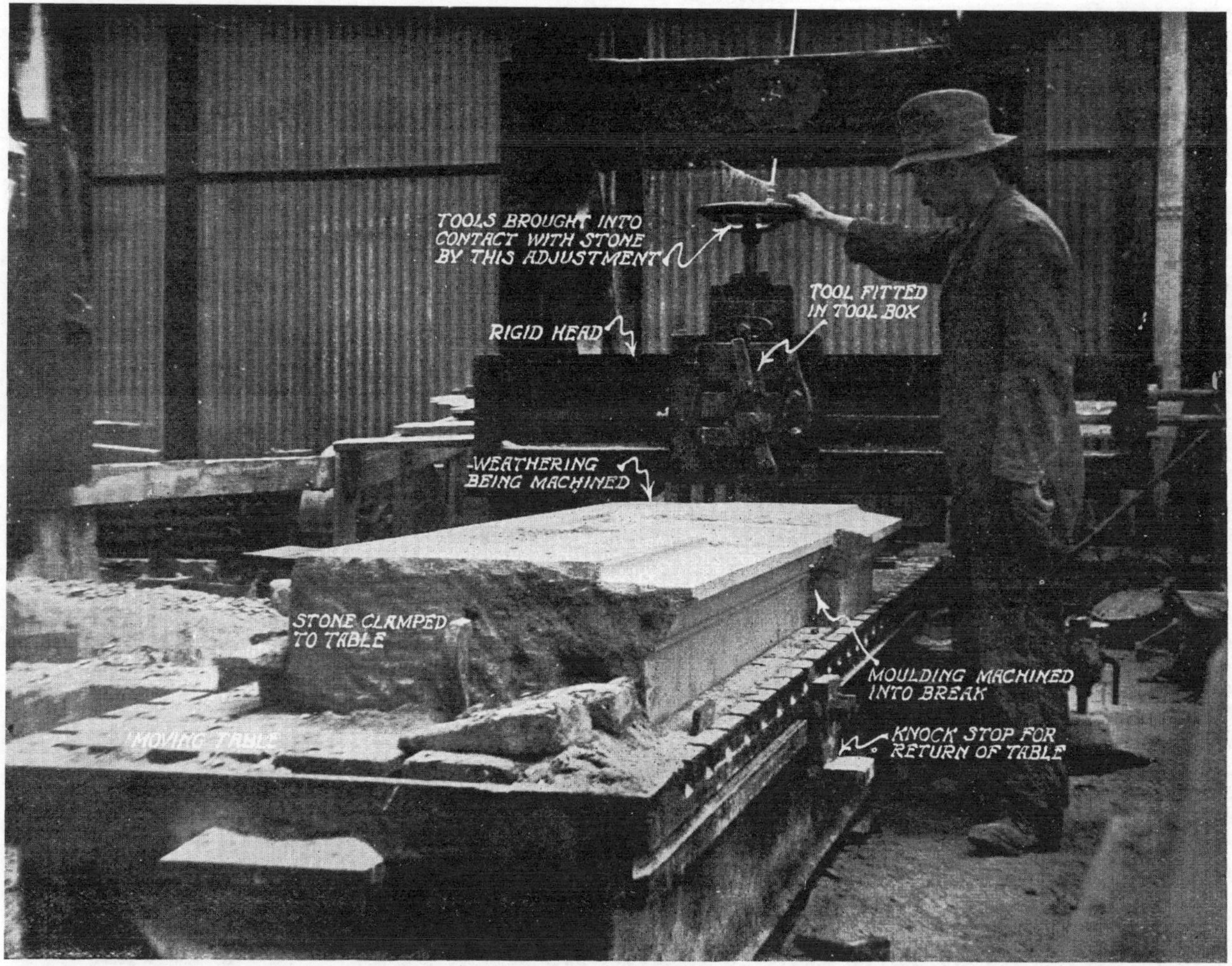

FIG. 62.—RIGID HEAD TYPE OF PLANING MACHINE.

Bath and Portland Stone Firms Ltd.

From "Building."

Carborundum Saws.—The use of these saws is extending as rapidly as is the case with diamond saws, especially in connection with light work such as *jointing*, *squaring*, and cutting smaller stones ready for the banker-mason or for finished work. Fig. 60 shows a **Carborundum Fixed Height Jointer Saw** in operation. These saws are very suitable for cutting marble slabs and tiles, especially when *twin blades* are used. Although the cutting speed is only about half that of diamond saws, the smoothness of the cutting surface left fully justifies the extra time taken. For many years carborundum has been widely used for sawing marble and similar stones, but it is now an established practice to employ it for any class of material, including granite. The style of machine to take carborundum blades does not differ very much from those used as diamond saws.

The carborundum is mixed with shellac, and whilst hot is pressed round the rim of a steel blade.

For really effective production the range of work to be executed by these saws is limited to work up to 10 to 12 in. deep. It is considered that beyond 10 in., machines of the large *cross-cut* or *beam-type* varieties, fitted with *diamond blades*, are more economical on account of their quicker cutting, with less expenditure of power.

The carborundum blade finds its best field both in lighter work and where the quality of finish is of more importance than speed.

The tables of these machines are of cast iron, rigid in design, and machined on the top surface. Squaring lines are marked on the table to enable the stones to be set accurately and easily in the least possible time. A slot is provided for the blade to pass well below the table surface, so that the material being sawn may be cut clean through, and also to enable the full wearing depth of the carborundum rim to be utilised.

The table has a cross movement of 16 in., which is particularly serviceable in combination with a machine provided with a *rise and fall motion* of the saw spindle. The *traverse motion* is obtained by a screw and hand-wheel. One placing of the stone on the table is sufficient for any number of parallel cuts within a width of 16 in.

The cutting speed of these machines is very high in relation to the power consumed.

PLANING AND MOULDING MACHINES

The machines so far dealt with are all used for the sawing of the material. The machines under the above heading are specially designed for the surfacing and finishings for wrought stonework. There are various types of *planing* and *moulding machines* in use at the present time, and much has been done in recent years by the various makers to improve the design and capabilities of such machines.

The Reversible Head Type or Canting Arm Planer, as shown in Fig. 61, cuts the stone during the forward and backward movement of the table. The head in which the tools are fixed swings over in time to meet the stone on the return of the table in either direction.

The fall of the *canting arm* as it turns over at the end of each stroke is broken by *pneumatic cushioning*.

So promptly is the *canting arm* reversed that, with a table speed of about 35 ft. per minute, only about 5 in. over-travel beyond the end of the stone is required. The *canting arm* can be raised or lowered by power as well as by hand control.

The tables of these machines are now supported on inverted Vee slides, as shown in the photograph, instead of rollers. This type of slide is an improvement on the recessed type of the Vee, as supplied with some machines, which are liable to be damaged by the dust which obviously collects in the groove.

The Revolving Table, which is usually fitted to these machines, is a convenient means whereby the various surfaces of the stone can be brought into contact with the tools. This is of great advantage in working a deep cornice, as it minimises the vibration caused by the tools having to project too far from the tool head in order to meet the lower members of the cornice. A convenient locking arrangement is provided to secure the revolving table at any angle, whilst the stone is gripped to the revolving table by means of cast-steel vices.

Rigid Head Planers (Fig. 62).—Though these machines cut the stone in one direction only, they have certain advantages over the *reversible head* or *canting arm type* in that they have a fast return movement of the table, and delicate mouldings may be executed to a fine degree of accuracy. Two or more tool-boxes are often used on these machines, thus permitting the tools to act simultaneously on the stone, thereby reducing the time for the execution of any piece of work to a minimum.

These machines are often converted into Carborundum Abrasive Machines for working mouldings, etc., in marble and granite. Owing to the hard and brittle nature of these stones, the use of steel tools tends to bruise the finished surface, so that carborundum wheels, shaped to the profile of the moulding required, replace the tool-box, and by revolving at a high velocity in a steady flow of water, wear away the stone to the required section by abrasion.

Fig. 63 illustrates an **Anderson Four-head Planing Machine** in operation. All the tool-boxes are independently controlled, functioning as two separate machines. The full capacity of this machine is 12 ft. by 6 ft. by 5 ft.

Fig. 64 shows an **Open-side Planing Machine**. Though these machines are little used in this country, they are invaluable, as the limitation of width which is apparent in the types of machines already described is absent. It is often a great advantage to be able to operate on the ends of a large stone as well as on the beds and sides.

Rubbing Beds (Fig. 65).—These machines were at one time extensively used in masonry works, but, owing to the improved quality and finish of the work produced by the various machines, they are to-day seldom used. It has become more economical to *plane* the surfaces on the *planing machines* rather than to rub them on the *rubbing bed*. They are, however, extensively used in marble works for sanding to size small pieces, being very suitable for such purposes.

Fig. 63.—Rigid Head Planing Machine Using Four Tool-boxes Simultaneously.

FIG. 64.—" ANDERSON'S " OPEN-SIDE PLANING MACHINE.

FIG. 65.—THE "ANDERSON" RUBBING BED.

FIG. 66.—THE "ANDERSON" JENNY LIND POLISHER.

FIG. 67.—TURNING A PORTLAND COLUMN STONE IN CENTRE LATHE.

Bath and Portland Stone Firms Ltd.

MACHINE POLISHING

Jenny Lind Polisher (Fig. 66).—These machines are used for polishing the surfaces of marble and granite. The polish is produced by a gradual "*fining*" of the surface by various processes.

To accomplish this, a cast-iron disc is fitted to the machine and pressed lightly on the surface of the material whilst revolving at the rate of 300 revs. per min. in a steady flow of water, sand or steel grit being added as a cutting medium. This process continues until the required surface is produced. These *discs, or polishing-heads* as they are called, are sometimes fitted with carborundum blocks, the carborundum being graded from coarse to very fine, the latter producing a surface ready for polishing. For the final polishing, *heads* fitted with felt pads are used. These are damped, and a small quantity of *putty powder* (oxide of tin) is applied as a polishing medium.

Surfaces 5 ft. square may be polished by these machines at one setting.

Rotary Polishers are also used for this purpose.

Pneumatic Hammers are in general use for working marble and granite, but their use on stones softer than marble is not attended with success. The tool is attached to a pliable hose placed conveniently to each banker, and connected to the main pipe from the compressor plant. A turn-off cock is fitted to the hose, close to the hammer and within easy reach of the mason, for turning off and on when required.

A steel tool provided with a *shank* is fitted into the hammer and held on the surface of the stone to be cut. The mason guides the tool in the direction required, whilst the piston in the hammer continually strikes the head of the steel tool, thus providing the impact required for the removal of the waste stone.

Stone Turning.—Most modern masonry works are equipped with centre lathes for turning *column stones*, *finials*, *balusters*, *bases*, etc. Fig. 67 shows a column stone being turned in a lathe. If *flutings* are required, they are worked by hand on the banker or on a planing machine.

CHAPTER III

DETAILS AND CONSTRUCTION

Foundations—Footings—Depth of Concrete—Width of Footings—Stone Footings and Walls—Stone-faced Walls—Damp-proof Courses—Construction of Piers—Stone Columns—Rubble Walls—Surface Finishings—Masonry Joints—Masonry Mitres—Ashlar Stops—Moulding Stops—Details for Window Openings—Window Jambs—Treatment of Stoolings—Construction of Lintols and Arches—Heads or Lintols supported by R.S.J.'s—Stone Lintol with Secret Keyed Joints—Construction of Flat Arches—Construction of Pediments—Construction of Fascia Courses—Marble Wall Linings—Details of Marble Chimney-piece—Soffit Courses—Details for Gable—Details for Buttresses—ARCHES, Technical Terms and Descriptions—Construction of Oriel Windows—Tracery—Entablatures—Construction of Cornices—Coverings to Cornices—Details of Masonry Doorway—Construction of Stone Domes and Vaults—Construction of Stone Stairs.

ALTHOUGH the craftsman is not called upon to design, it is important that he should be able to appreciate the various forms which he has to execute in stone. A great amount of knowledge in this respect can be gained by observing the details of various buildings in the vicinity, taking notice of the grouping of the mouldings, the proportion of the members, and the numerous methods of forming delightful intersections. If possible, he should measure up and draw to scale units of a design, such as a *doorway* or *window opening*. This, combined with the study of a good book on the *styles* of architecture, will be of great assistance to the young craftsman in helping him to appreciate how closely his craft is linked up with architectural forms.

A craftsman is able to execute a piece of detail in stone with greater aptitude if he understands the fundamental principles underlying the combination or grouping of the mouldings and intersections.

To enable the craftsman to visualise the required form of the unit, and the position it has to occupy in the building, a general knowledge of building construction is necessary.

In this work the author has combined masonry details with construction, so that the student may the more readily appreciate the importance of their combination.

Building construction is a branch of masonry frequently neglected by the craftsman. If he is to become efficient in the various branches of the craft, it is a matter of the utmost importance for him to acquaint himself with the principles involved, and the methods adopted in modern building.

The introduction of steel into the construction of buildings has revolutionised orthodox methods, and we find that masonry, instead of being the art of building in stone, has become the means whereby the carcass of the building may be pleasingly covered with stone. An outline of construction specially biased towards masonry, starting with the building of walls from their foundation, will be dealt with in this section.

Foundations.—A good foundation for any structure is of primary importance. To ensure this, the subsoil should be strong enough to bear the load to be placed upon it. A good foundation bed should possess the following qualities. It should be :—

(1) Hard and solid.

(2) Equally compressible or incompressible throughout.

(3) Such that the stratification is horizontal.

(4) Protected from atmospheric changes by making the foundation bed sufficiently deep, according to the nature of the soil.

Footings.—As most soils are compressible, it becomes necessary to distribute the weight of the wall and the various loads it has to carry over a larger area than the wall itself. To provide for this increase in area, "*footings*" should be formed by projecting courses on each side of the wall until the required width is obtained. These footings should rest on a bed of concrete as a foundation, which should project about 6 in. beyond the bottom course of footings on either side of the wall, thus distributing the weight.

L.C.C. (General Powers) Act, 1909.—The following safe loads on soils must not be exceeded :—

Natural bed of soft clay or wet or loose sand . .	1 ton per sq. ft.
Natural bed of ordinary clay or confined sand . .	2 tons per sq. ft.
Natural bed of compact gravel, London blue clay, or chalk	4 ,, ,, ,,
The pressure on concrete foundations shall not exceed	12 ,, ,, ,,

The pressure on any brickwork shall not exceed the following :—

Blue brick in cement mortar	12 tons per sq. ft.
Hard brick (including London stock) in cement mortar	8 ,, ,, ,,
Ordinary brick in cement mortar	5 ,, ,, ,,

Loads on the following stones shall not exceed :—

Granite	45 to 60 tons per sq. ft.
York stone	20 tons per sq. ft.
Portland	16 ,, ,, ,,
Bath	8 ,, ,, ,,

Having these figures, it is only necessary to find the total weight to be carried, and ascertain, according to the soil, the superficial area required for the concrete.

Depth of Concrete.—The depth of the concrete must be such that it will not fail or fracture when the weight is received. Such fractures occur in the direction of 45° from the base of the wall.

Fig. 68 is a section through an 18-in. wall, showing the method of determining the depth of the concrete. The bottom course of footings should be twice

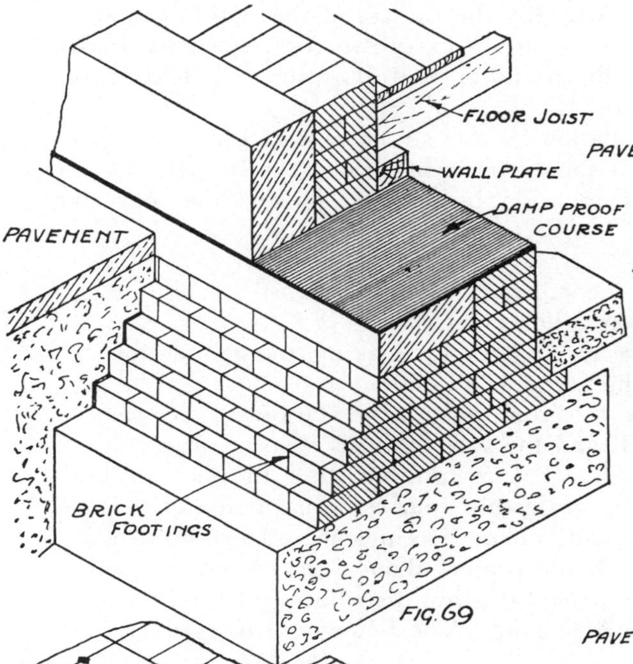

FIG. 69

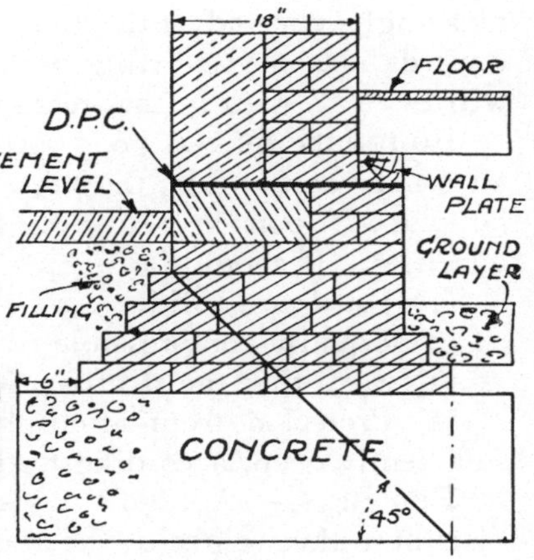

FIG. 68

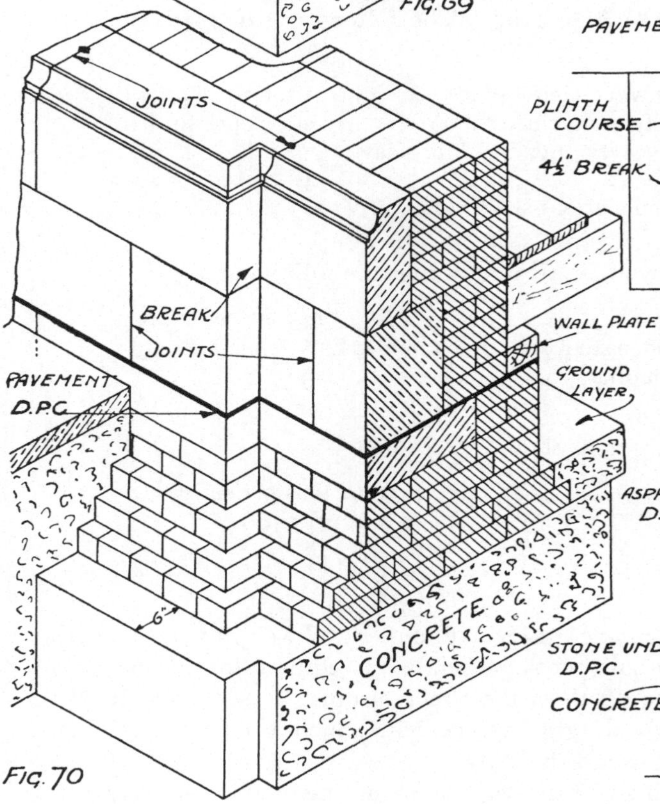

FIG. 70

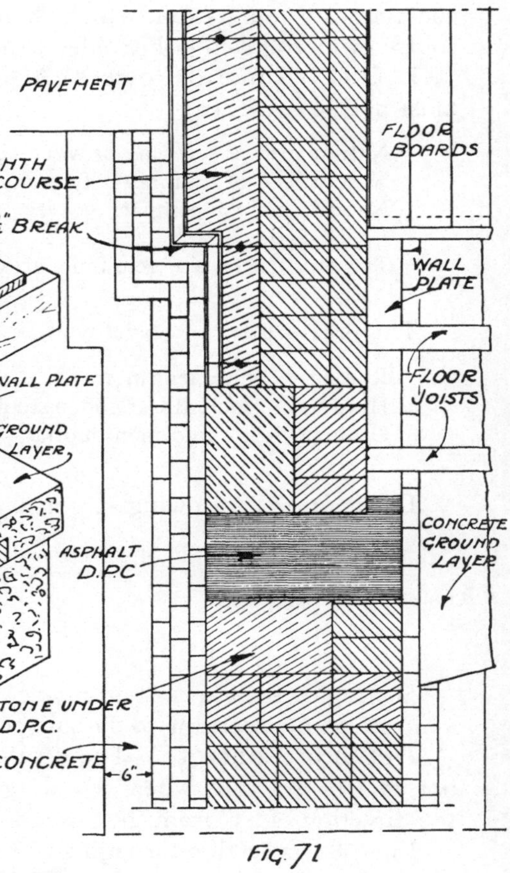

FIG. 71

CONSTRUCTION OF STONE-FACED WALLS.

the thickness of the wall, the width being increased equally from the centre line.

Width of Footings.—Brickwork footings are constructed by projecting each course of footings $2\frac{1}{4}$ in. on each side of the wall. Hence an 18-in. wall should be four bricks, or 36 in. wide at the lowest course.

The whole construction is shown in Fig. 69, including the position of damp-proof course in relation to ground and pavement level, concrete ground layer over the site, and the method of supporting the floor. Means for ventilating the space under the floor should be provided.

Stone Footings and Walls.—In districts where stone is plentiful, the foundation and walls may be built with stone throughout.

When stone is used for foundations, projections of 6 to 12 in. may be formed on either side of the wall. The stones should be as large as practicable to reduce the number of joints and beds. Where walls are built of rubble masonry instead of wrought masonry, the thickness of the walls should be one-third thicker than the schedule thickness for brickwork. A proportional increase of thickness beyond that required for rubble walls should be also allowed for walls built of flint.

Stone-faced Walls.—Wrought stonework is chiefly used in combination with brick walls. The walls are in reality built of bricks and faced with stonework, as shown in Fig. 68. It is important that the sizes of the stones, when in combination with brickwork, should be arranged to suit the brick dimensions. The standard measurements of bricks, including joints, are 9 in. by $4\frac{1}{2}$ in. by 3 in., arranged in heights of four courses to 1 ft. Six of the northern counties use larger bricks, making four courses of bricks = 13 in. in height.

All the stones should be properly and effectively bonded with the brickwork. The depth of the stone on the wall should be $4\frac{1}{2}$, 9, $13\frac{1}{2}$, 18 in., and so on. It is important where possible to arrange the heights of the stones or courses so that they conform to the heights of the several brick courses.

Damp-proof Courses.—All walls should be provided with a layer of non-absorbent material throughout their entire thickness to prevent the damp rising from the soil through the wall by capillary attraction. The omission of an effective damp-proof course not only allows the dampness to penetrate into the inside of the building, but it is often the cause of decay in the lower courses of masonry.

This course should be placed 6 in. above ground level. In the examples this distance is assumed from the underside of the pavement stones.

Asphalt is an excellent material for this purpose. There are other suitable materials to be obtained, such as *bituminous felt*, *slates*, etc.

When **Plinth Courses** are formed at the bases of walls, the projection of the plinth is added to the thickness of the wall. The footings and bonding are arranged as shown in Fig. 70.

Breaks.—Fig. 70 also shows the construction of the wall where the "breaks" occur. Great care must be given to the position of joints and the bonding with the brickwork.

Fig. 71 is a plan of the wall, showing the courses raked back to show the bonding.

Quoins.—Fig. 72 shows the construction for *plinth*, *ashlars*, and *cornice* at the quoin of a building, also the treatment where an opening occurs for a doorway. All the stones should be properly bonded, both on the face and through the wall. The *bonding* or *vertical joints* of the cornice should be arranged to line perpendicularly with the bonding of the ashlar courses below. If desired, the *quoin piece* of cornice may be further secured by inserting *metal cramps* across the joints, as shown in the figure.

Joggles should be cut in the stones to ensure that the cement grout entirely fills the space between the joint surfaces of the stones.

When stonework is backed by brickwork in cement mortar, it is advisable to coat the back of the stones with a layer of lime mortar to prevent stains working through the wall to the face of the stonework.

Piers (Fig. 73).—Independent piers are constructed in a manner similar to that described for walls. The *footings* and *foundations* should extend to cover an area sufficient to carry the calculated load to be borne by the pier and the weight of the pier itself. Great care should be exercised in the bonding of the stones and also in the craftsmanship. The material selected should be of such a quality as to resist effectively the tendency to crushing. All the stones should be bedded solid, with the joints well filled with *cement grout*. The brick footings should project in each direction in offsets of $2\frac{1}{4}$ in. until the correct area is obtained. It is often necessary to allow a greater area for the concrete owing to the condition of the soil. In such cases steel may be embedded in the concrete, or introduced in the form of grillages, which consists of several parallel steel joists, used to assist in spreading the concentrated load from a pillar or stanchion over the necessary area of the foundation. They may be composed of one, two, or three tiers of joists.

Reinforced Concrete Raft.—For the foundation of towers, etc., it is usual to design a raft of combined steel and concrete over the whole area, either in the form of steel grillages, composed of rolled steel sections, or reinforced with bars or expanded metal of the various forms obtainable. The raft should be of sufficient area to allow the more concentrated load of the tower to exercise the same pressure on the soil as the surrounding lower portions of the building.

To calculate the necessary area of the foundation, the total load is divided by the safe pressure of the soil, given on p. 17. The area obtained can be either square, round, or oblong.

$$\frac{\text{Total load}}{\text{Safe bearing value of soil}} = \text{Area required.}$$

Steel Stanchions in Piers (Fig. 74).—In modern construction, the stonework of piers forms a casing for a core of steel, called a stanchion, which continues vertically up through the piers from the foundations. The steelwork carries the loads, including the weight of the stonework, which is notched for the flanges of the steel section. The size of the piers may be reduced under these conditions if required.

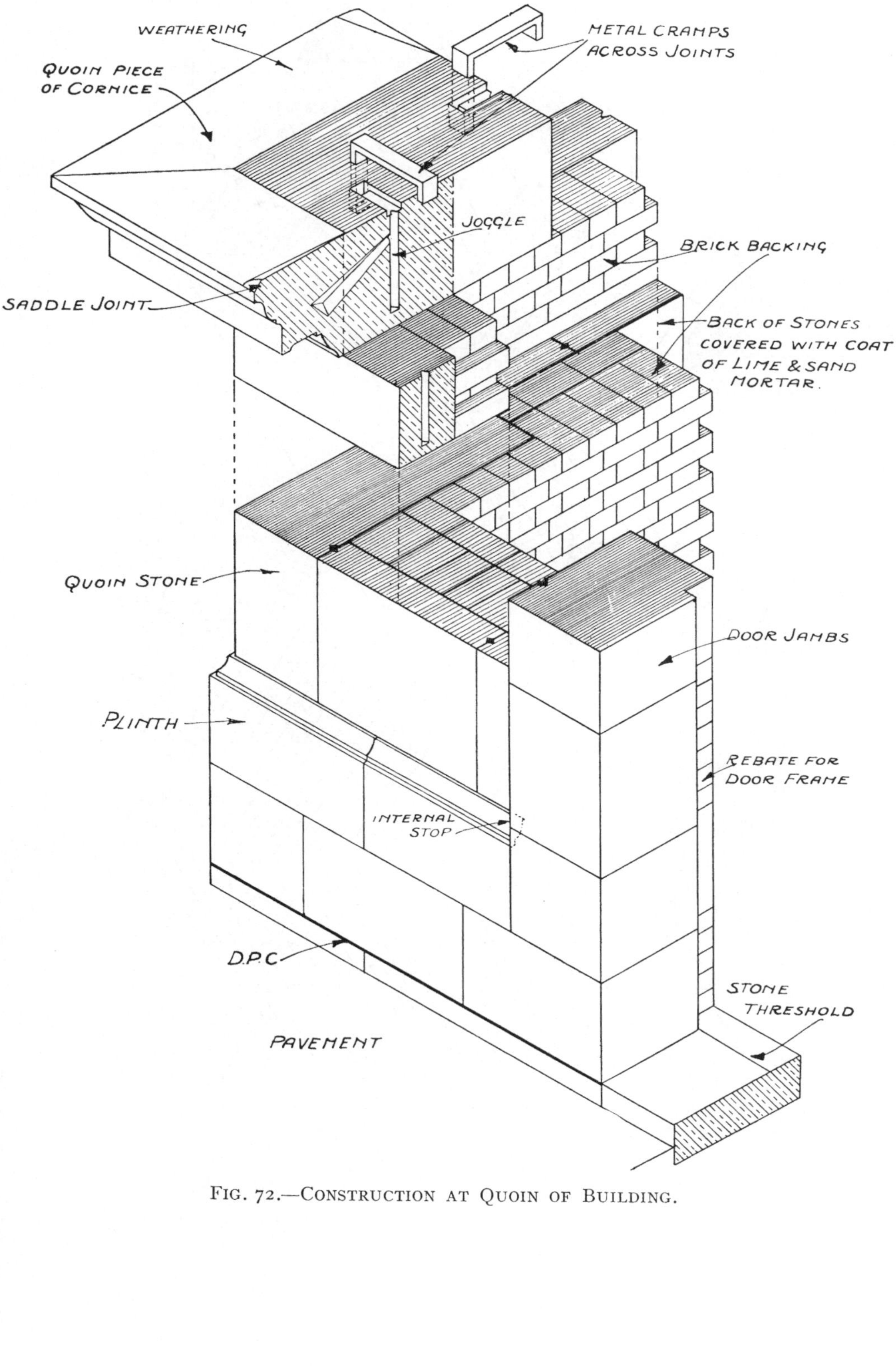

FIG. 72.—CONSTRUCTION AT QUOIN OF BUILDING.

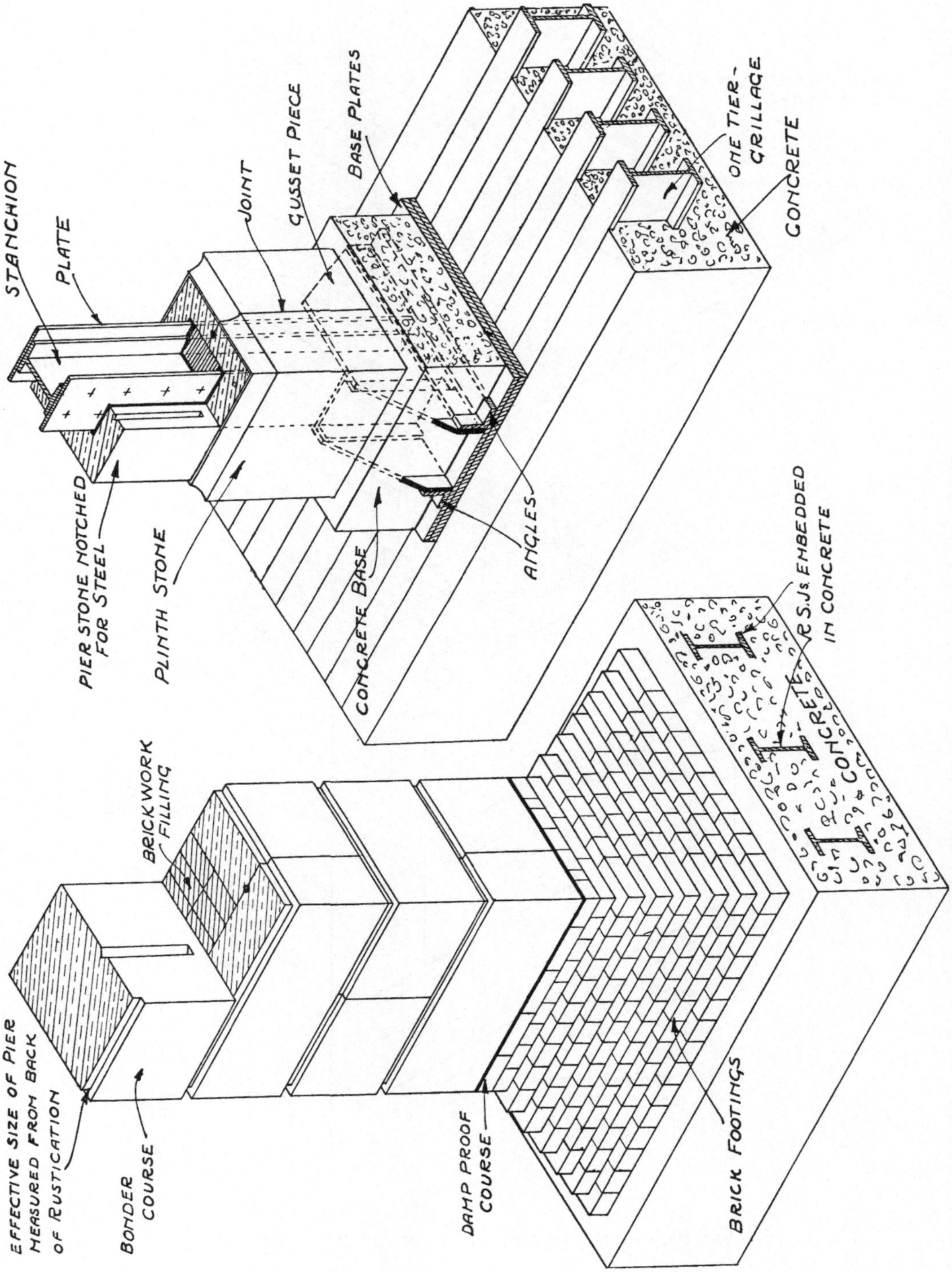

FIGS. 73 AND 74.—CONSTRUCTION OF PIERS.

Fig. 75.—Columns in Conjunction with Steelwork.

Fig. 76.—Method of Bedding Column Stones.

FIG. 77.—COURSED RUBBLE WALL.

FIG. 78.—RANDOM RUBBLE WALL BUILT TO COURSES.

Stone Columns in modern construction are not intended to support the loads of the building as they were previously, although they are designed to convey that impression. Usually they are an architectural detail, forming a covering to steel supports or stanchions. A photograph illustrating this is given in Fig. 75.

When the column is required to carry the superimposed loads, great care should be exercised in the *bedding* of the individual stones. The beds should be worked slightly round, and the mortar, or *masons' putty*, spread in the centre of the bed surface, so that when the next stone is placed in position, the putty squeezes towards the outer rim of the shaft. The bedding material should be raked out clear of the arris, and the joint pointed when cleaning down, thus preventing spalling or fracturing the stones when the load is concentrated on the column. Fig. 76 shows a column stone being fixed.

Granite columns are often bedded on a disc of sheet lead of less diameter than the bed surface, but this practice is not to be recommended. It is preferable to bed these stones as described above.

Rubble Walls.—There is not sufficient rubble walling executed at the present time to justify a lengthy description in this work. It would not, however, be permissible to disregard its association with masonry in general. Whereas in walls built of wrought stonework the strength depends almost entirely upon the manner the stones are bonded together, scarcely any strength being derived from the bedding material, in rubble walls the strength depends to a very large extent upon the mortar. The appearance or finish of wrought masonry walls is predetermined by the shape and the dressing of the stones, whereas in rubble walls the resulting appearance depends upon the skill and taste of the waller. These walls are generally arranged to fill in between wrought stone quoins and dressings; thus the plumbing of the wall is ensured by the dressed stones.

The bonding of the stones is left to the craftsman, but he should arrange a *bonder* or *through stone* at intervals, in order to tie the work together.

Coursed Rubble Walls (Fig. 77).—The stones in this class of wall are selected for size, and roughly worked to course heights.

Random Rubble Walls.—In this class of walling the stones are set in the condition they come from the quarry, or roughly squared with a hammer, the horizontal coursing not being studied.

Random Rubble Built to Courses (Fig. 78).—Stones of irregular height are placed so that they level up to a horizontal bed at intervals.

Snecked Rubble Walls.—Small rectangular filling stones are inserted at intervals to equalise the varying heights of adjoining stones. Sometimes the stones of these walls have roughly chiselled faces.

Range Walling is a term given to Bath stone walling which forms the wall surface for interior church work. The stones are sawn to heights varying from 4 to 9 in., the whole of the wall surface being *dragged* during the cleaning down of the work. The vertical joints are placed at random.

Polygonal Walling (Fig. 79).—The stones are set in the wall any shape, or roughly pitched to fit the adjacent stones. This type of walling is often called *rag-walling*.

Flint Walling (Fig. 80).—This type of walling is usually arranged between wrought stone dressings, more as ornamental face-work than for strength. The flints are split or knapped, the broken surface of the flint forming the face of the wall. Fig. 81 shows a portion of a wall faced with *knapped* and *squared* flints between stone dressings.

Flint work is fairly general in Norfolk and Suffolk, where a peculiar and distinctive treatment called *flush-work* may be seen.

Diaper Flint Walling.—A photograph illustrating this type of walling is given in Fig. 82. Flint panels alternating with square wrought stones form an interesting piece of walling. The split flints vary in colour from blue-black to rich yellow. The blue-black variety are chiefly arranged for panelling.

SURFACE FINISHINGS

Face is the term applied to the exposed surface of the stone, usually the vertical face in elevation. It is finished according to the detail or form required.

Return Face is the vertical face exposed to the side elevation.

Beds are the lower surfaces upon which the stone rests and the upper surfaces which support the stone above. They are either inclined or horizontal.

Joints are the surfaces prepared to receive other surfaces butting against them.

Back is the surface entirely hidden in the thickness of the wall.

Back Joint is the jointed surface of a *return* or *quoin stone*, prepared to receive other stones butting against it on the return elevation.

Clean Back.—When stones bond through the wall and a clean surface is required, as in stones for parapets, etc., the inside vertical face is called a *clean back*.

Quoin Stones are the stones placed at the external angle of a building, usually of a larger dimension, and arranged to bond with the other stones in the wall in each direction, as Fig. 72.

The Arris is the sharp edge made by the intersection of two surfaces forming an external angle.

Ashlar is the term used for finely dressed stone which is worked to fit in the general face of the wall. It may be *plain*, *rusticated*, *rock-faced*, or *chisel-drafted*, according to the detail required. It is then described according to the finish.

Plain Ashlars are stones with *rubbed*, *dragged*, or *polished* surfaces.

Boasted Surfaces (Fig. 83) are those left from the *mason's boaster*. The regularity and form of the tool marks depend upon the style of the craftsman.

Furrowed Surfaces (Fig. 84).—Small *flutings*, from $\frac{1}{4}$ to $\frac{3}{8}$ in. wide, are worked vertically or horizontally across the surface. This finishing is often adopted for *plain surfaces*, *string-courses*, *pier-stones*, *stall-boards*, etc.

Chisel-drafted Margin Ashlar (Fig. 85).—The marginal drafts are worked along the four edges of the stone, the centre being left rough or worked into various forms. A photograph illustrating this type of walling with *tooled margins* and *rusticated ashlar* is given in Fig. 92.

Punched or Broached Ashlar (Fig. 86).—The surface is left from the punch after the marginal drafts have been worked.

FIG. 79.—POLYGONAL OR RAG WALLING.

FIG. 80.—FLINT WALLING.

Fig. 81.—Flint Walling. Knapped and Squared Flints between Stone Dressings.

FIG. 82.—DIAPER FLINT WALLING.

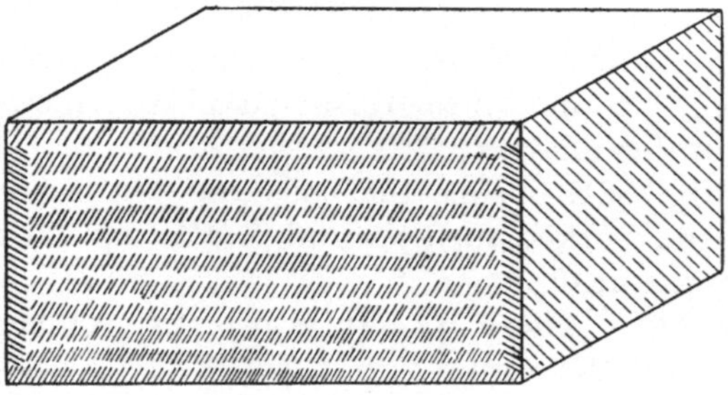

FIG. 83.—Boasted Surface.

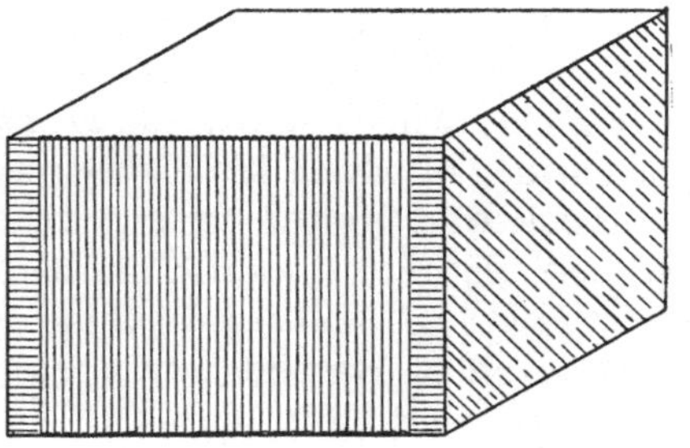

FIG. 84.—Furrowed Surface.

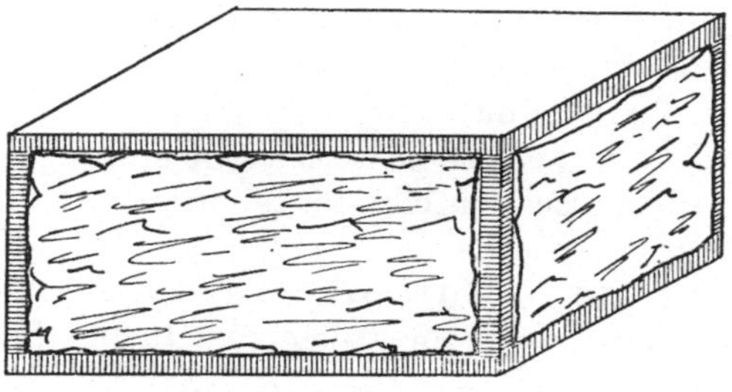

FIG. 85.—Chisel-drafted Margins.

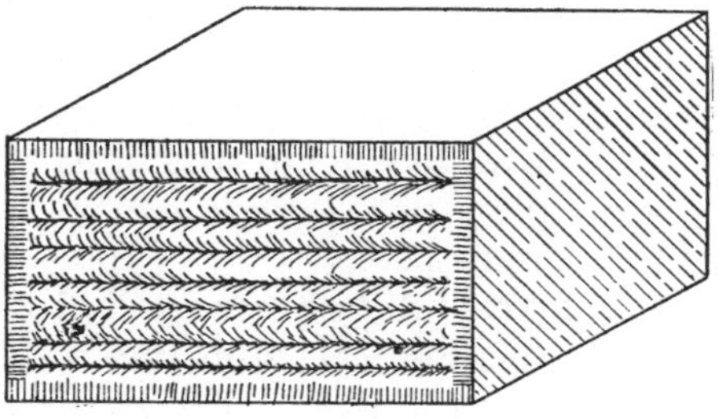

FIG. 86.—Punched or Broached.

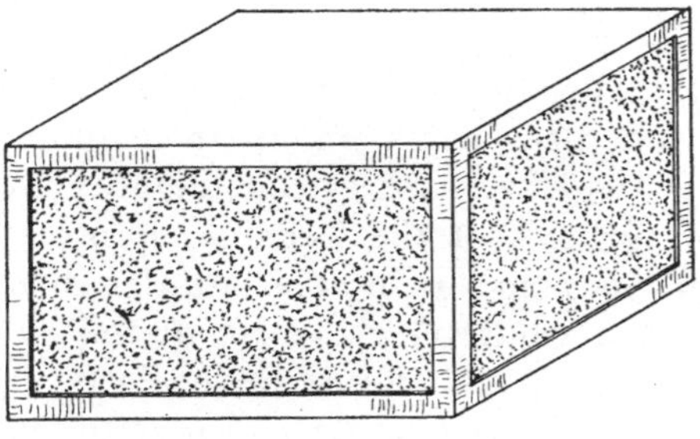

FIG. 87.—Picked Panel Quoin.

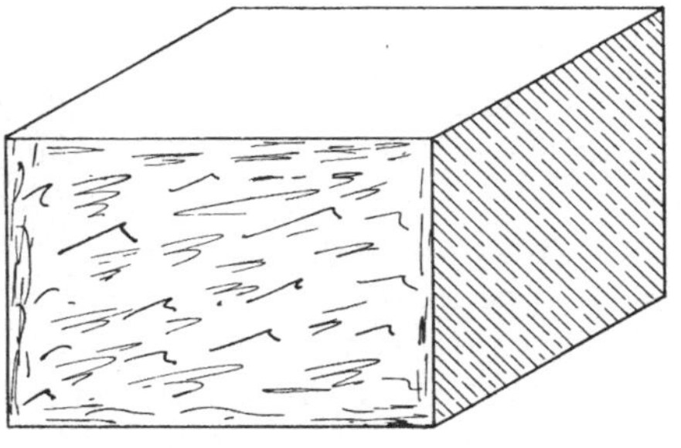

FIG. 88.—Pitched Face.

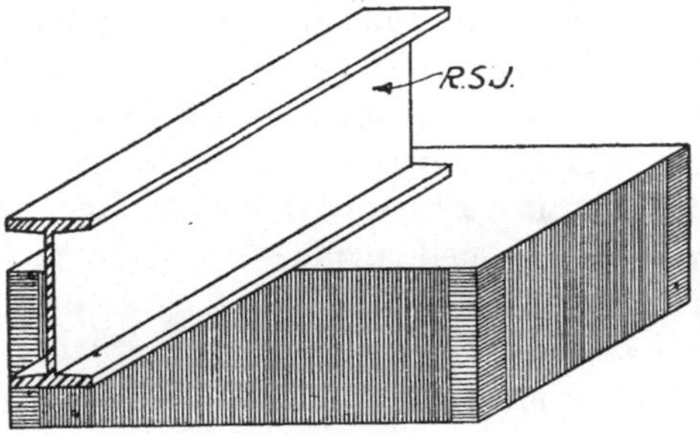

FIG. 89.—Tooled or Batted.

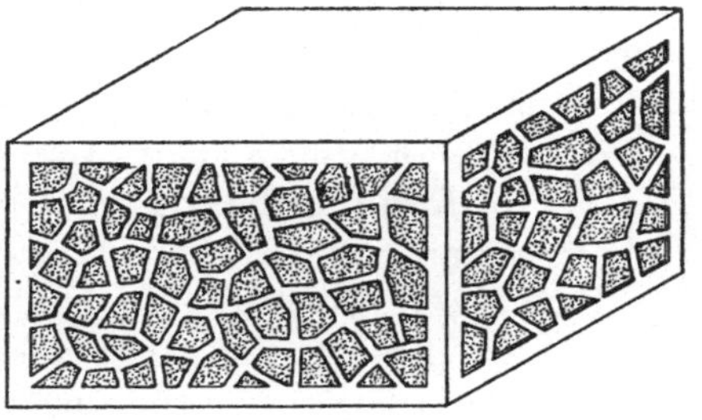

FIG. 90.—Reticulated Quoin.

SURFACE FINISHINGS.

Picked Panel Quoins (Fig. 87).—Panels are sunk in the faces of the stone to a depth of $\frac{1}{4}$ in. The surface of the panel is chiselled true and afterwards *picked* over with a *mallet* and *sharp point*. Sometimes a raised panel is formed, the surface of which is finished in a similar manner.

Rock or Pitched-face Ashlar (Fig. 88).—When the beds and joints are worked, the face lines are marked on the worked surfaces and *pitched up* with a *pitching tool*, and left in this rough condition.

Tooled or Batted Surfaces (Fig. 89).—These are left with regular chisel marks cut vertically across the surface of the stone with a *batting tool* after the face has been brought to a fine surface by rubbing. These chisel marks are usually specified "*so many to the inch*."

Reticulated Quoins (Fig. 90).—The surface is worked true with a series of sinkings cut into the stone about $\frac{3}{8}$ in. deep. The sinkings are separated by bands of regular width. The depth of the sinkings should be worked to a gauge, being picked afterwards with a sharp point.

Vermiculated Quoins (Fig. 91).—The surface is in a series of sinkings, giving the stone a worm-eaten appearance, used chiefly in the position of quoin stones.

Rusticated Ashlars and Piers (Figs. 91, 92, 93, 94).—The surface of these stones projects from the wall face, the back of the *rustication* being the face line of the wall. The *rustications* are of various forms. *Rustic* finish in granite infers that the faces of the stones are left rough, beds and joints only being dressed.

MASONRY JOINTS

The joints in masonry show little variety, and are not numerous or complicated in form. The chief difficulty arises in their arrangement, especially in those masonry details which require considerable setting out. In constructional masonry, the surfaces of the beds or joints should be perpendicular to the pressure. The same principle applies whatever the direction of the pressure may be, and whatever the form of the external surfaces of the stone, whether plane or curved, the joints in the latter should be normal to the curve, thus avoiding acute or sharp angles. Joints in masonry are usually in compression, but occasions arise when they are subjected to a lateral pressure, or there may be a tendency for the stones to pull apart.

Butt Joints (Fig. 95).—These are formed in wrought masonry by two plane surfaces being butted or placed together. These joints are usually combined with *cement joggled joints*. A V-shaped groove is cut into the surface of each stone, forming a cavity into which *cement grout* is poured. An excellent method of forming the joggle is by cutting a deep furrow in the surfaces with a hammer and punch, thereby providing a roughened surface for the cement.

Dowelled Joints (Fig. 96).—Stones subjected to lateral pressure are strengthened by the insertion of either a *metal* or *slate dowel*, half of which is let into a mortise cut in the surface of each stone.

Metal Dowels are needed in the bed joints of *finials*, etc., owing to the length of the dowel required and the small surface area of the stone, but for the bed joints of *mullions*, *columns*, etc., *slate dowels* are chiefly used. They

Fig. 91.—Vermiculated Quoins and Rusticated Ashlar Walling.

Fig. 92.—Rock-faced, with Tooled Margins and Rusticated Ashlar.

Fig. 93.—Bold Form of Rusticated Piers.

Fig. 94.—Rusticated Piers.

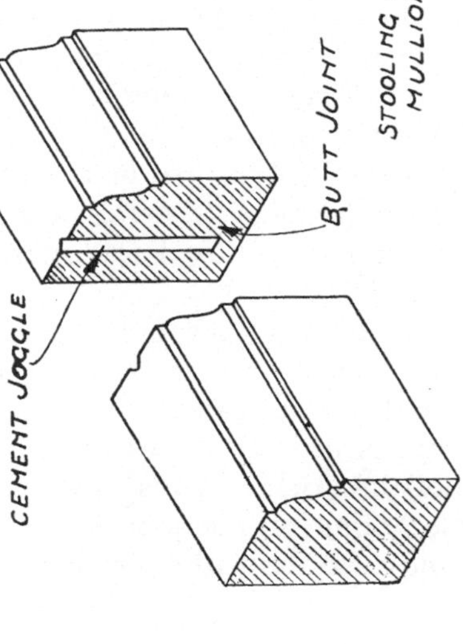

FIG. 95.—BUTT JOINT.

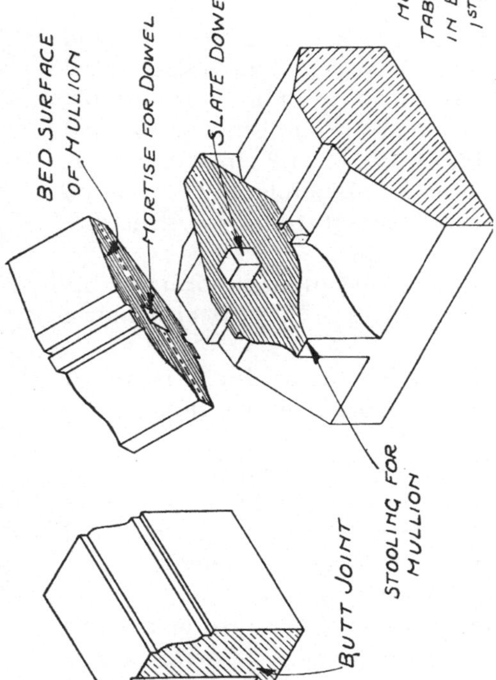

FIG. 96.—DOWELLED JOINT.

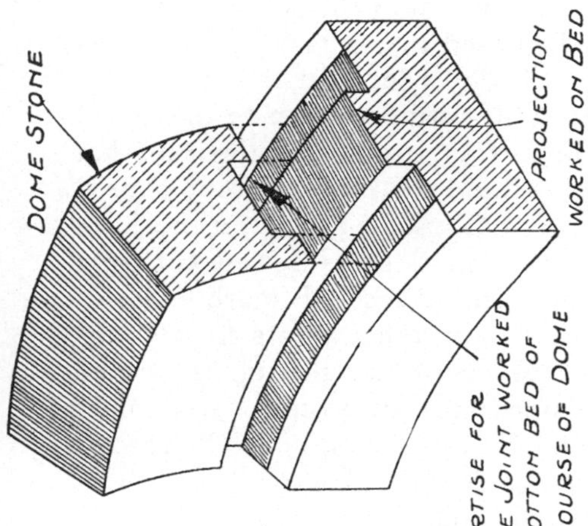

FIG. 99.—TABLE JOINTS.

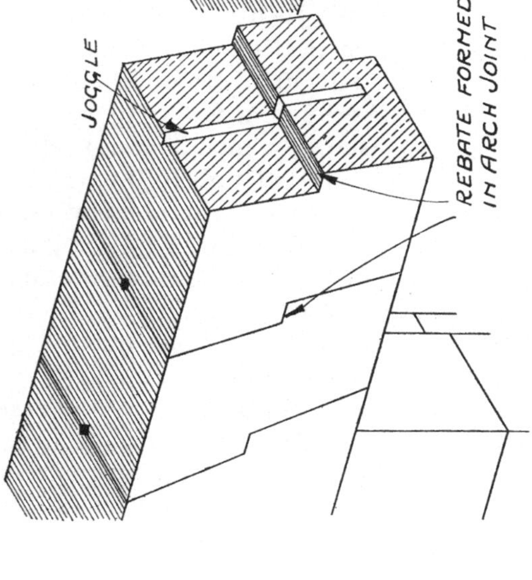

FIG. 98.—REBATED JOINTS IN FLAT ARCH.

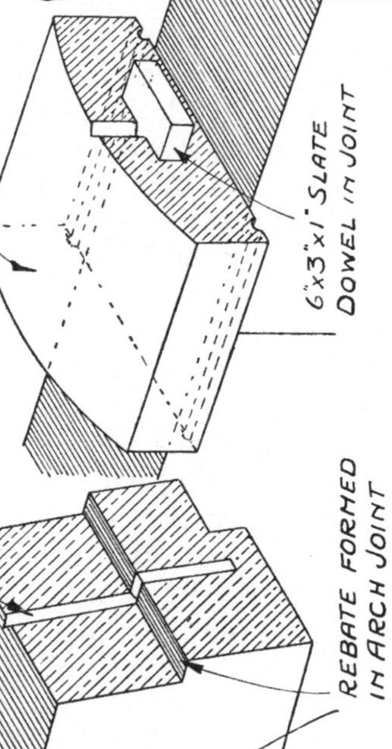

FIG. 97.—SLATE DOWEL IN JOINT OF CAPPING.

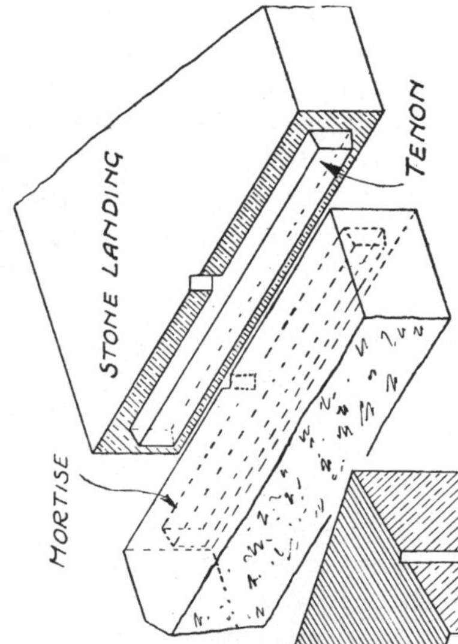

FIG. 100.—MORTISE AND TENON JOINTS.

should be rectangular, the length being twice the width of the section, as 2 in. by 1 in. by 1 in. or 4 in. by 2 in. by 2 in. When used in joints of *coping stones* for dwarf parapet walls they should be about 6 in. by 3 in. by 1 in., as in Fig. 97. The cement joggle should be cut in the joint surfaces down to the dowel mortise, and sufficient space allowed in the cutting of the mortise, so that when the stones are placed together and pointed and the cement grout poured into the joggle, the whole is entirely filled. A small *pebble* placed in the joint is the most practical form of dowel for *tracery*. Small brass dowels cut from $\frac{1}{8}$-in. diameter wire are used to strengthen the joint surfaces in marble work.

Rebated Joints (Fig. 98).—These are used as weather joints for copings to gable walls and for joints in flat arches. A horizontal seating is formed in the normal joints of the arch to prevent distortion in the event of settlement taking place. They are expensive in material and labour, and not to be recommended. As the voussoirs of the arch are wedge shaped, they tend to tighten at the joints when the weight is concentrated on the arch. The seatings naturally hinder this, causing the stones to fracture.

Table Joints (Fig. 99).—These joints are also expensive in material and labour, and are only used where great strength is required to resist lateral pressure. The sketch shows an instance where this type of joint is suitably used in building. The lower courses of a stone dome are subjected to a lateral pressure, which is often counteracted by the insertion of metal rings or chains. This is not necessary for small domes, but the use of table joints, in the position shown, is of undoubted advantage as a means of strengthening the structure, especially when used in conjunction with metal cramps.

Mortise and Tenon Joints (Fig. 100).—These are similar in form to table joints, and are used chiefly in the joints of landing stones. They are also known as *joggle joints*.

Secret Key Joints (Fig. 101).—This form of joint is used as a means of forming an arch joint for a stone lintol comprising more than one stone, when radiating arch joints on the surface are not desired.

Cramped Joints.—Joints that are subjected to sectional stress or are liable to be pulled apart need assistance in counteracting the stress. This is obtained by the introduction of metal ties. These should be formed and fixed across the joint of the stones, so that they tend to draw the stones together. Fig. 102 shows a metal cramp which is used for the purpose stated above. It should be fitted tightly into a mortise cut into the top surface of the two stones forming the joint, and covered completely with cement. Metal cramps should be made of gun-metal or galvanised iron, or brass if used for marble work. Fig. 103 is a detail of the cramp in position.

Slate Cramps (Fig. 104).—These are used for the same purpose as metal cramps, but are not so effective. They are placed loosely in a mortise cut in the top surface of the stone, and completely covered with cement.

Saddled Joints (Fig. 72).—These are sometimes used as a weather joint for cornices. Their object is to prevent the moisture, which runs down the weathering of the cornice, percolating into the joints of the stones. This type of joint is not in common use to-day, as most exposed top surfaces are covered with sheet lead or asphalt.

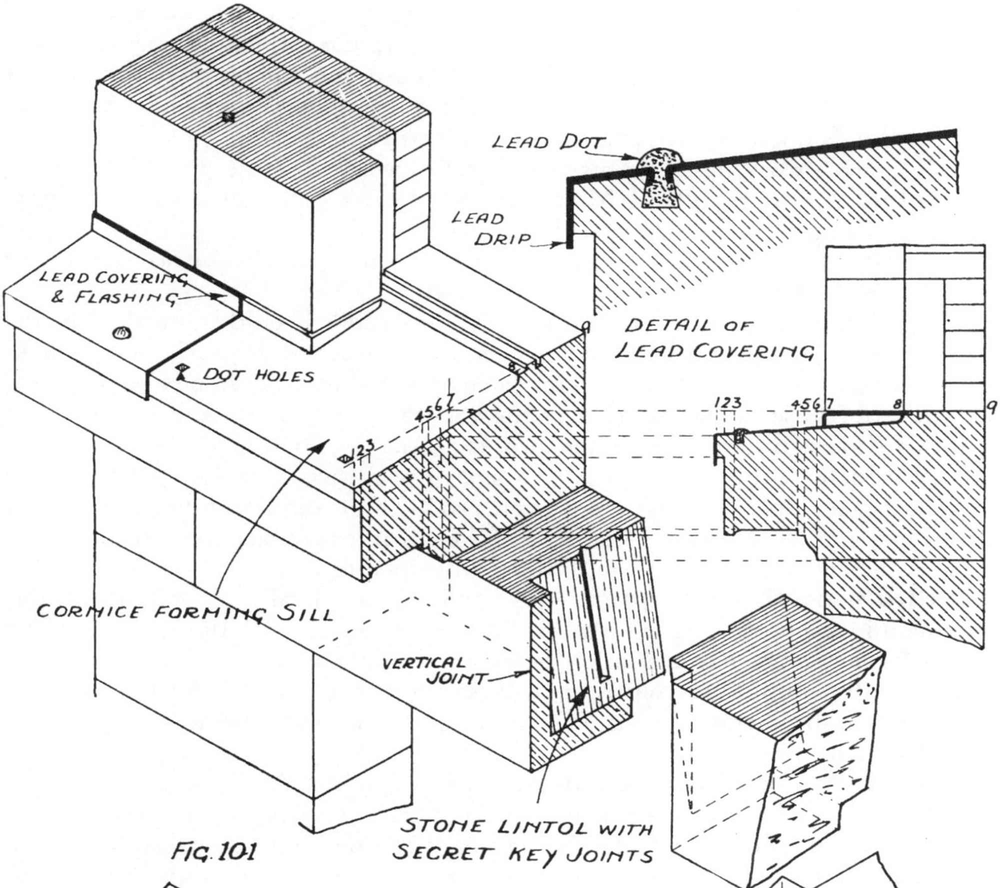

FIG. 101

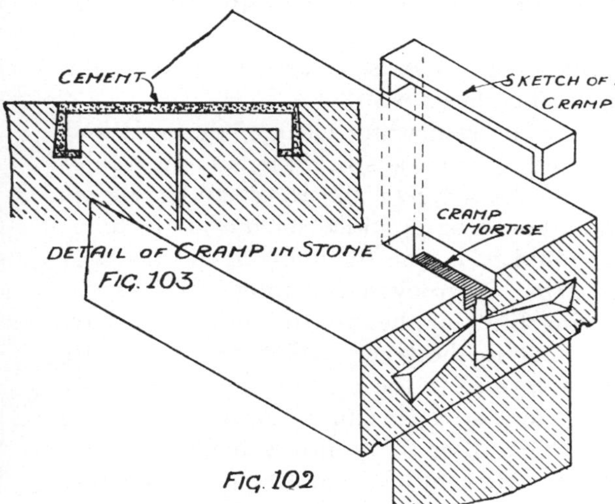

DETAIL OF CRAMP IN STONE
FIG. 103

FIG. 102

METAL CRAMP.

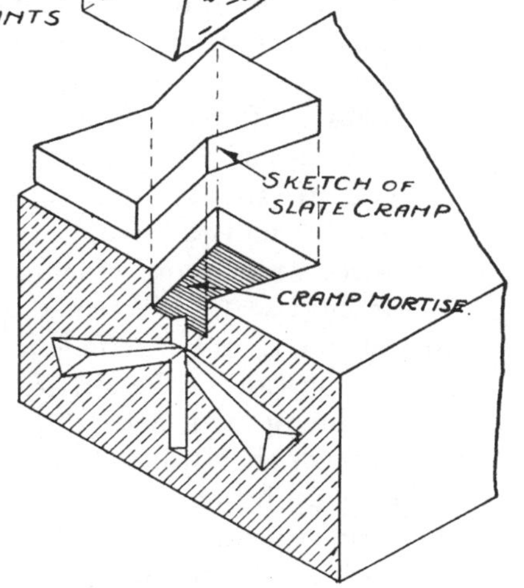

FIG. 104

SLATE CRAMP.

Masons' Mitres.—In constructional masonry, the mitre, or the line of intersection of two planes or mouldings, is cut in the solid stone, being known as either an *internal or external mitre*. When the intersection produces a straight line, viewed in the direction of the mitre, it is known as a *true mitre*. If the line is distorted, it is called a *broken mitre*.

In marble work, mitre joints are often formed as in joinery.

Internal and External Mitre (Fig. 105) is a detail of a *moulded break stone* showing an *external* and *internal mitre*.

Ashlar Stop.—Fig. 106 shows a similar moulding returned on to an ashlar face. This is known as an *ashlar stop*.

The intersection of the moulding with the ashlar should be worked in the solid stone. The ashlar portion is sometimes a separate stone, being back-jointed against the moulding. This manner of forming the stop is unavoidable at times because of the position of the vertical joints above and below. The construction is not to be recommended as a usual procedure.

Moulding Stops.—Fig. 107 is a sketch of a moulding stopped against a splayed face. The line of intersection produces an outline different from the true shape of the section. Numerous forms of intersections occur in masonry, details of each varying in form, but not in principle.

Bird's-mouth Mitre Joint.—Fig. 108 shows a bird's-mouth mitre joint, commonly adopted in marble work for the angles of wall linings, etc. The fillet formed at the angle eliminates the acute angle occasioned by the position of the mitre joints. In construction these slabs forming wall linings are held in position by metal ties fastened into the brick wall, metal cramps being fixed across the joints.

Mitre in Architrave Moulding.—Fig. 109 is a detail of an architrave moulding round an opening, showing the manner of forming the intersections of the jamb moulding with the stone head. The bed-joint is placed level with the soffit line of the head, the intersection being cut in the solid stone, thus forming an *internal mitre*.

Details for a Window Opening.—Fig. 110 is a section showing the construction of a window opening in a stone-faced wall.

The sill should be designed to carry the rain water clear of the wall and to keep the moisture from penetrating under the wood sill of the window frame. To meet these requirements it should be *weathered*, *throated*, and *grooved*. Horizontal seatings, termed *stoolings*, should be provided for the jamb-stones to rest upon, and an efficient *throat* should be cut on the underside of the projecting portion of the sill, forming a *drip*. A groove for a water bar is shown cut in the top bed, behind the weathering line, for the insertion of a metal bar, half of which is placed in the groove cut in the stone, whilst the other half projects up into a corresponding groove formed in the bottom of the wood sill. The bar is bedded in *red lead*. Where stone sills are employed in brick walls, it is advisable, if time permits, to fix them after the building has progressed to a higher level, thus allowing for any settlement of the wall. In stone walls they should be fixed during the erection of the wall, and bedded at the ends under the jambs, the remaining part of the bed being left free from mortar. To ensure this freedom after fixing, a saw blade should be run along the bed

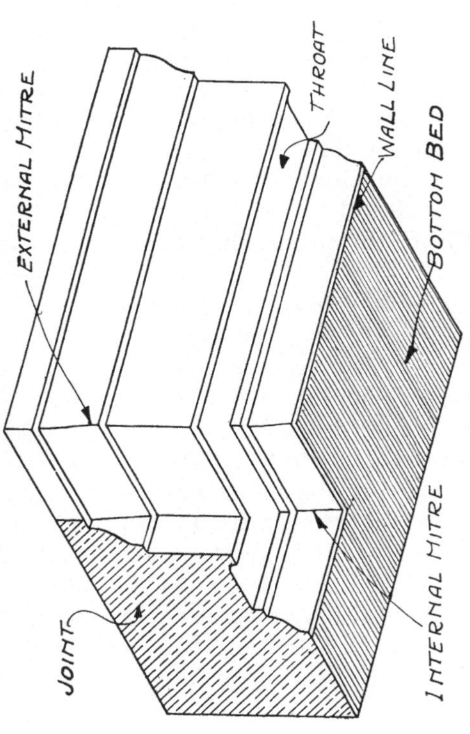

FIG. 105.—INTERNAL AND EXTERNAL MITRES.

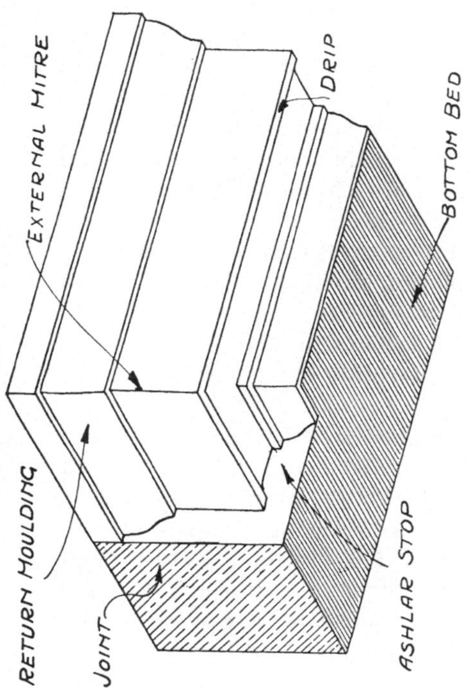

FIG. 106.—ASHLAR STOP.

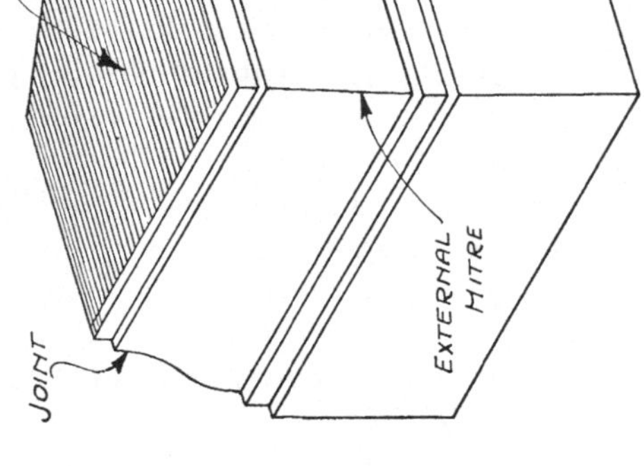

FIG. 107.—MOULDING STOPS.

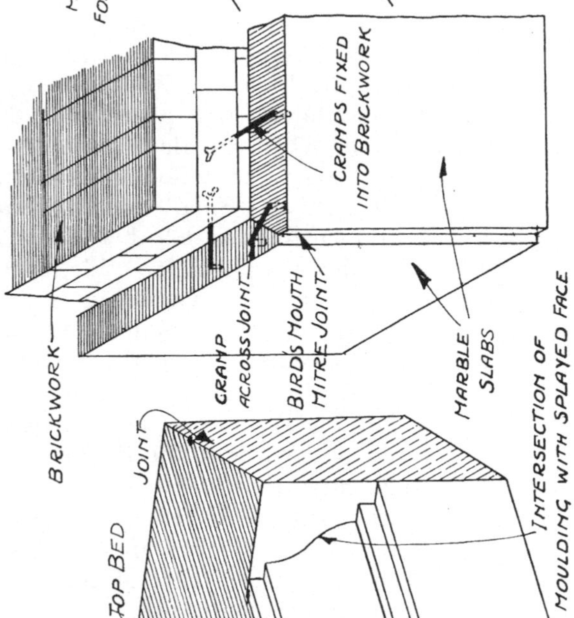

FIG. 108.—BIRD'S-MOUTH MITRE JOINT IN MARBLE.

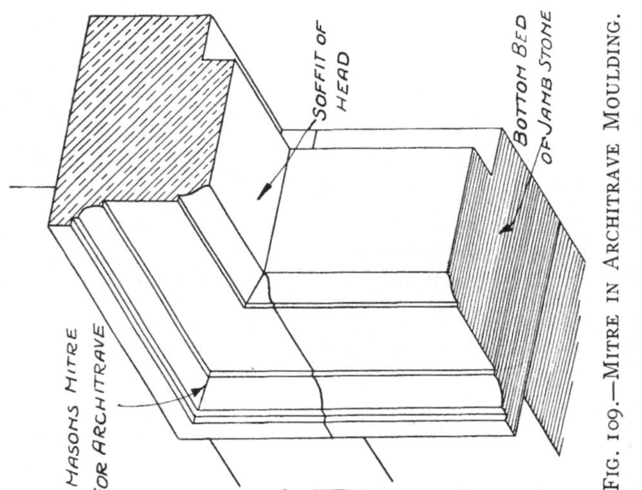

FIG. 109.—MITRE IN ARCHITRAVE MOULDING.

of the stone, clearing away any mortar which may have accumulated under the centre portion of the stone, so that any unequal settlement which might take place in the wall will not cause the sill to break. The open bed-joint should be pointed when cleaning down the stonework.

Window Jambs.—Fig. 111 shows the necessary bonding for the ashlars and jambs at sill level. These should be effectively bonded on the face and through the wall, and provided with a rebate for the frame if required, and bedded direct on to the horizontal stoolings of the sill.

Details at Ends of Sills.—Figs. 112 and 113 are sketch details showing two different methods of treatment for the ends of sills, and the method of stopping the throating.

Treatment of Stoolings.—There are various ways of finishing the weathering of sills against the stooling. Three different treatments are shown in Figs. 114, 115, 116.

Jamb-stops.—The mouldings round an opening are often stopped above sill level, and various forms of stops are introduced to suit the character of the work and the taste of the designer. A suggestion is given in Fig. 115.

Lintols and Arches.—These are employed as a means of spanning an opening in a wall, and should be designed to carry the load placed upon them. An equilateral triangle, having the width of the opening as a base, is a general guide when calculating the amount of material to be carried by the lintol. There may also be loads from joists, etc., resting on a lintol.

Head or Lintol.—Small openings are often spanned by a straight head or lintol. If a head is desired, this may be spanned by one stone, as in Fig. 117, though when the opening is too wide to be spanned by one stone, a number of stones may be employed, as in Fig. 118.

The construction here is similar to that of an arch. The joints are made to converge to a centre, the centre stone forming a keystone.

An invisible arch line is included within the outline of the lintol, from the points of the supports. The portion of the lintol below this arch line could be removed without affecting the stability of the lintol, thus forming a shallow segmental arch.

Concrete Lintol.—It is usual to place behind the stone head a concrete lintol, designed with steel reinforcements, sufficient to carry the weight of the wall and in some cases part of the floor above. The course immediately above the head, which in Fig. 117 is a cornice, should be bonded on the wall, so that it may be bedded partly on the concrete lintol.

Heads or Lintols Supported by Rolled Steel Joists (R.S.J.).—In steel-framed structures horizontal R.S.J.'s connected to the stanchions usually occur at head heights. They are so arranged that they may assist in carrying the floors. These joists are utilised to assist in supporting the stone facings, the method adopted being shown in Fig. 119. The head is notched for the lower flange of the R.S.J., and when placed in position, together with the other stones forming the course, the space between the stone and the steel is filled with cement grout, so that the stone has a solid bearing on the bottom flange of the joist.

The cornice above is also notched for the top flange, and bedded partly

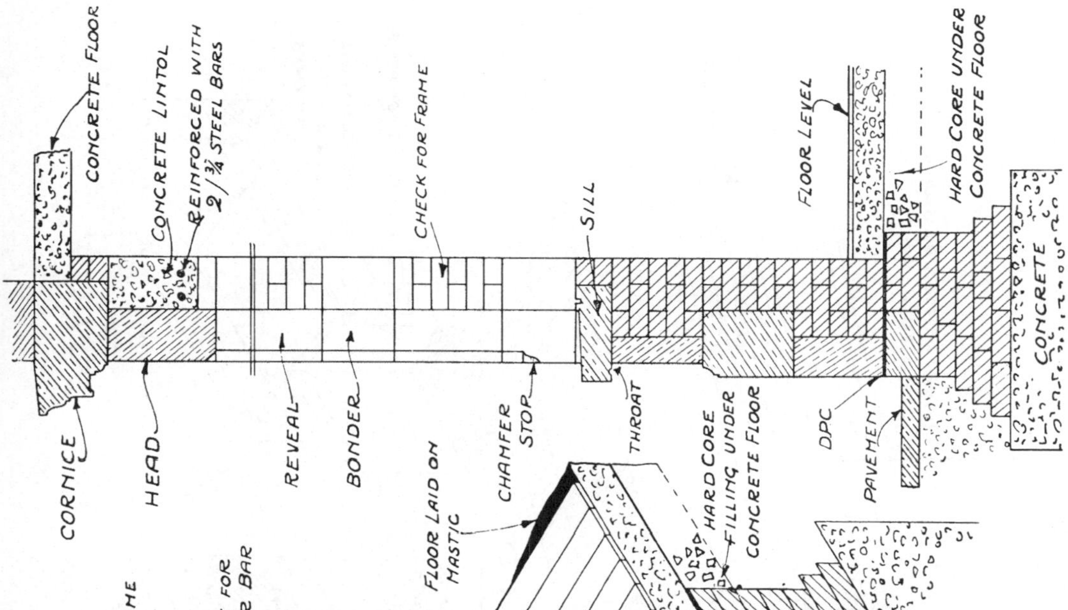

FIG. 110.—SECTION THROUGH WINDOW OPENING.

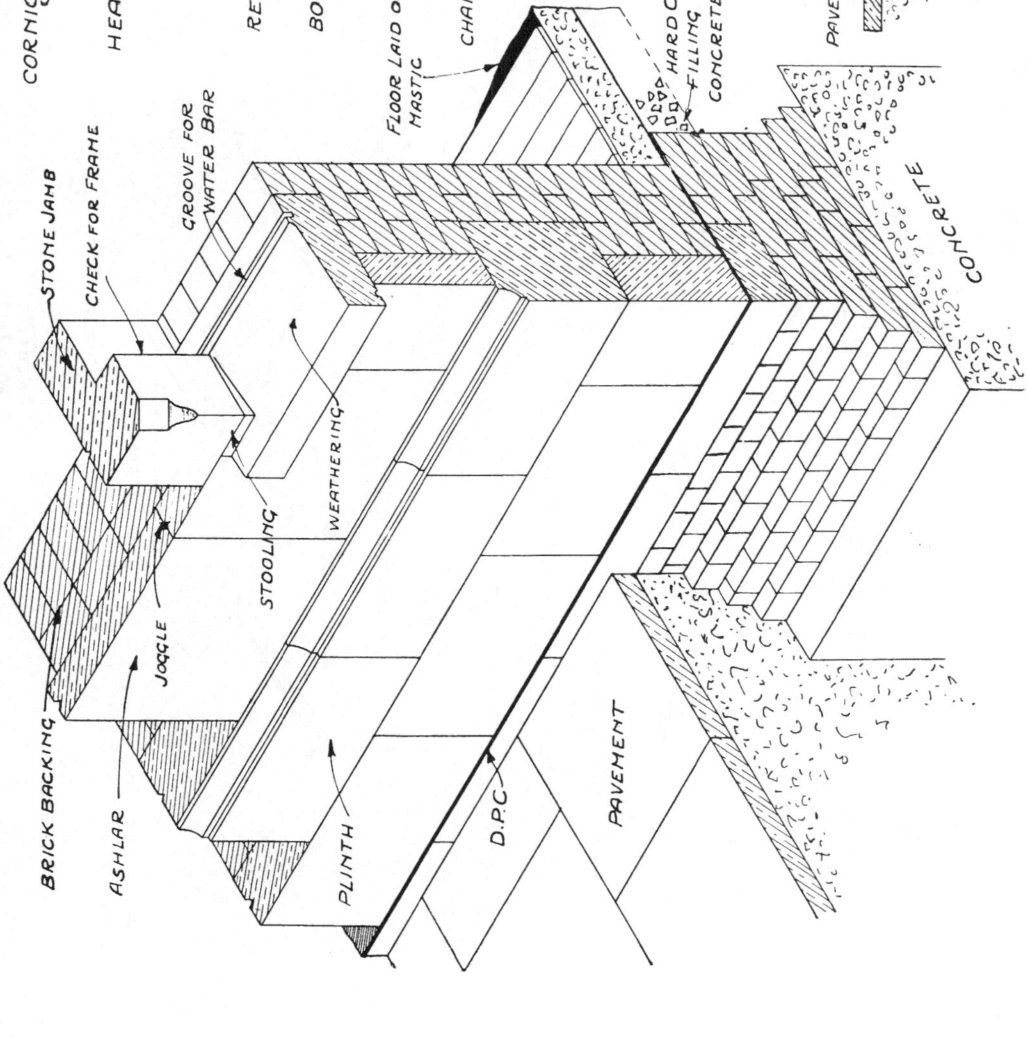

FIG. 111.—SECTION SHOWING BONDING OF ASHLAR, SILL, AND JAMB-STONE.

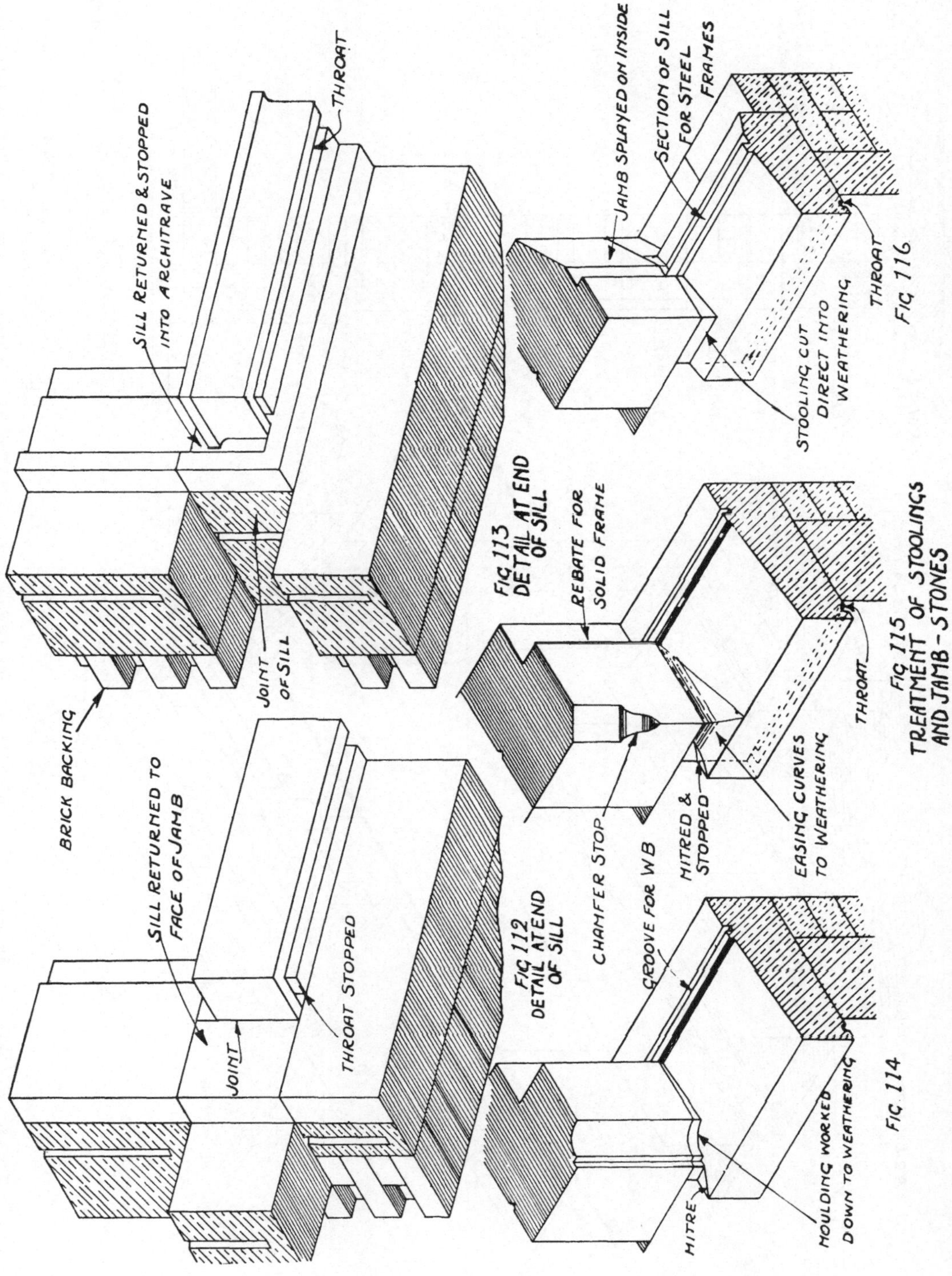

FIG 112 DETAIL AT END OF SILL

FIG 113 DETAIL AT END OF SILL

FIG 114

FIG 115 TREATMENT OF STOOLINGS AND JAMB-STONES

FIG 116

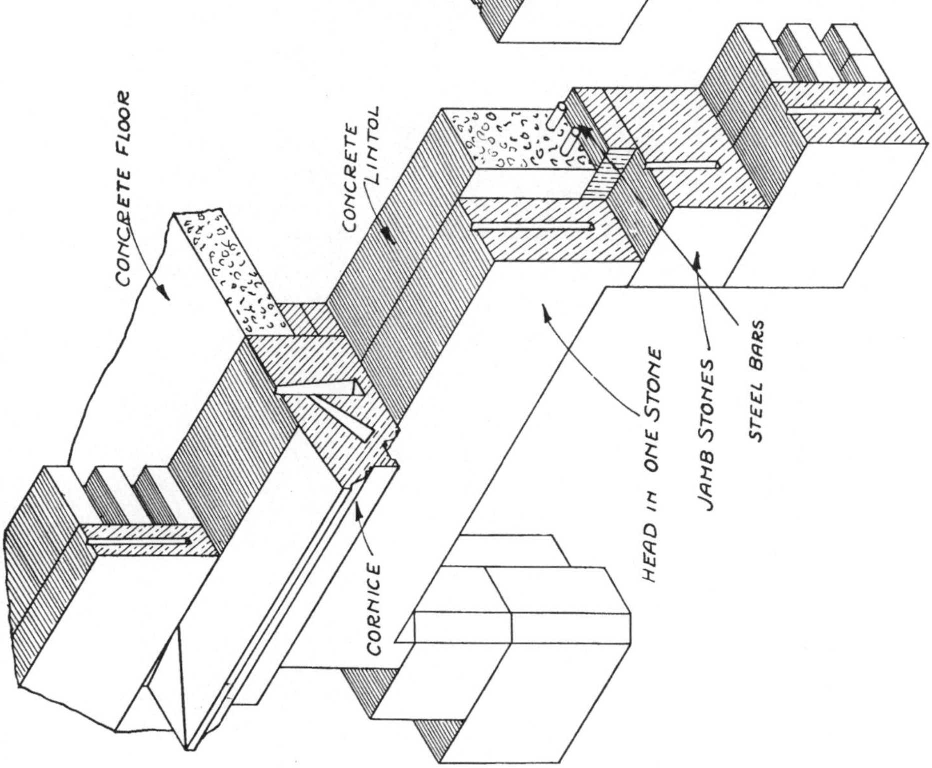

FIG. 117.—STONE HEAD SHOWING CONSTRUCTION.

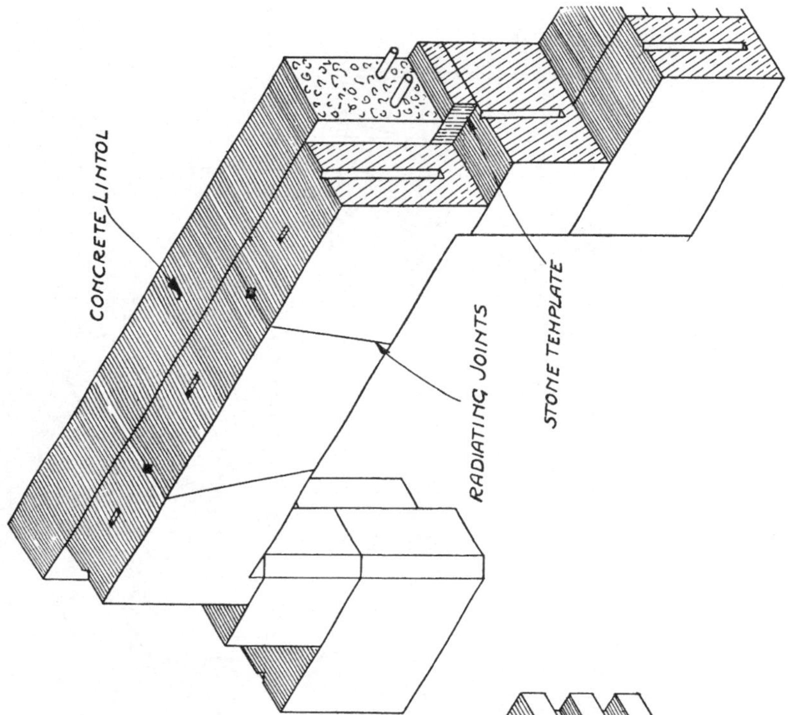

FIG. 118.—FLAT ARCH WITH RADIATING JOINTS.

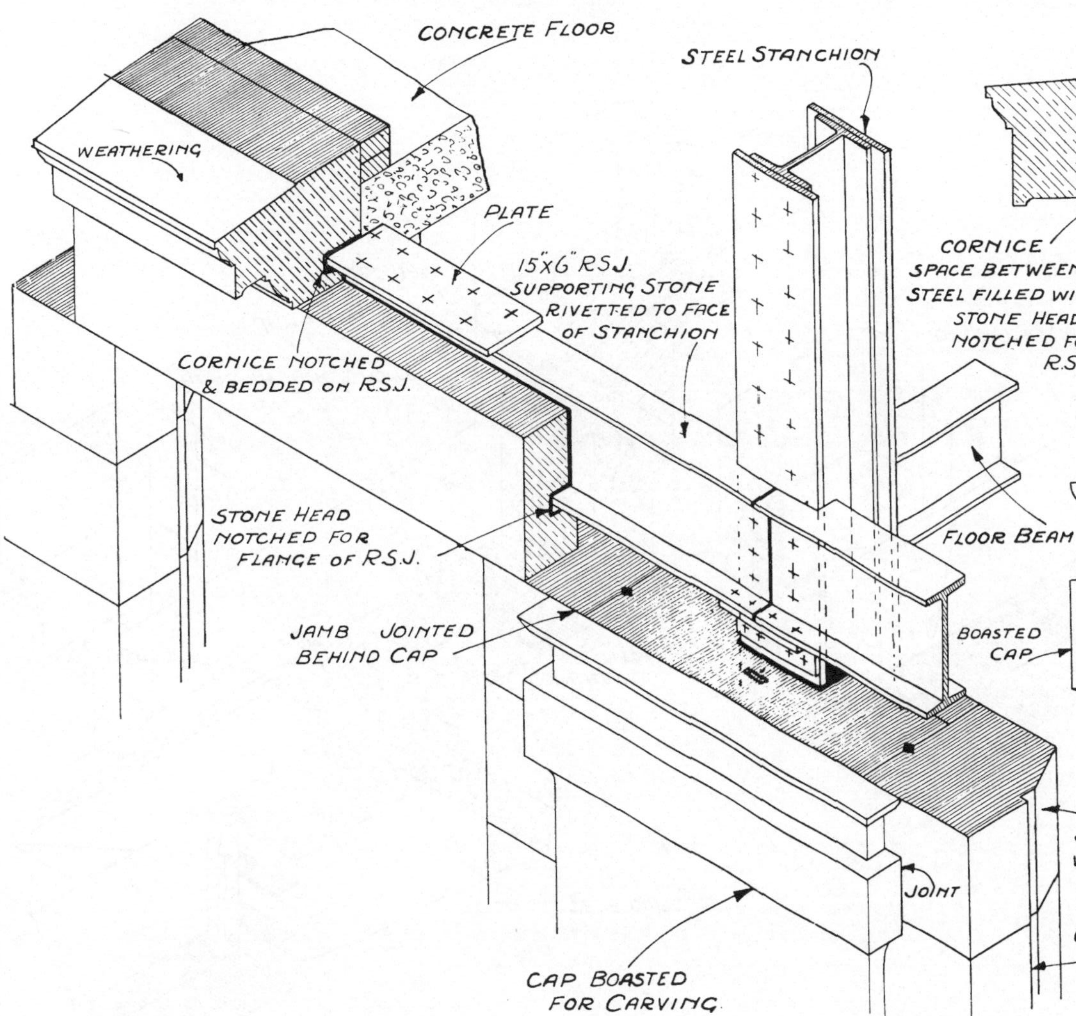

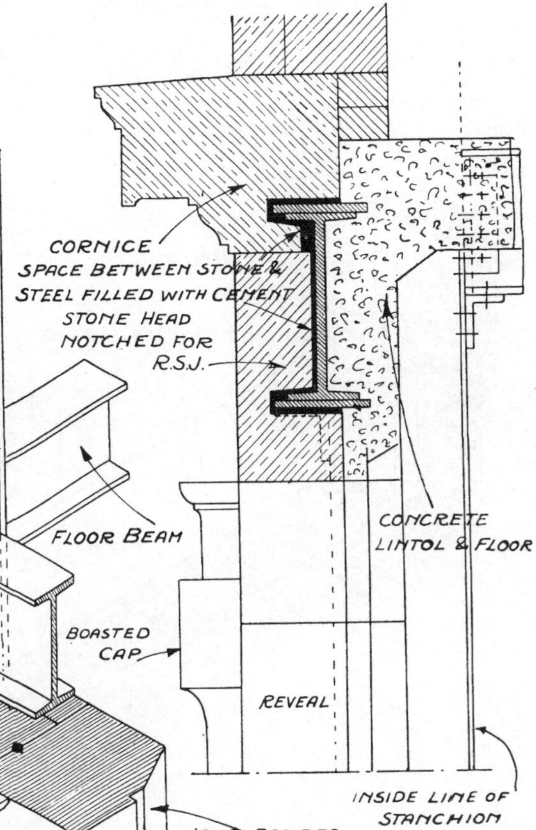

FIG. 119.—STONE AND STEEL CONSTRUCTION AT HEAD OF WINDOW OPENING. FIG. 120.—SECTION THROUGH CENTRE OF OPENING.

FIG. 121.—CONSTRUCTION OF STONE HEAD AND STEELWORK.

FIG. 122.—FIXING THE CORNICE ABOVE THE HEAD.

on the top surface of the joist, and when in position is well grouted, thus ensuring that all the cavities round the top flange are filled with cement. A section through the centre of the opening is given in Fig. 120 and photographs taken on the site in Figs. 121 and 122.

Stone Lintol with Secret Key Joints.—Fig. 101 shows a method of spanning an opening with stone, when steelwork is not utilised. The stonework in this case is required to be sufficiently strong to carry the calculated load coming upon it. The lintol is composed of three stones, and though the joints are shown vertical on the face, they are made to converge to a centre, behind the face of the stone. These joints are called *secret key joints*. A cornice forming the sill for an opening above is shown in the drawing.

Construction and Details of Flat Arch.—Fig. 123 is a part elevation and Fig. 124 the section through the head of a window opening. The stone head is shown, comprising three stones, forming a flat arch. The joints converge to a centre for effect. In construction these stones are shown notched and resting on the flange of a R.S.J., and further strengthened by a *metal plate* and *bolt*, fitted into each joint and secured to the steelwork.

Pediments.—An architectural feature in the form of a moulded pediment is shown over the opening. The individual stones comprising this feature are clearly shown in Fig. 123.

In construction the pediment should bond as far as possible into the wall, so that the weight is thrown back on the R.S.J. Pediments are of various forms according to the taste of the designer. They may be straight or segmental in elevation, and comprised of several stones, termed *springers*, *apex-stone*, *closers*, and *core* or *tympanum*. When pediments occur in brick walls, the top sloping surface is cut through to the back of the stone, so that the brickwork intersects the sloping surface. When they occur in an ashlar wall, they should be bonded with the vertical joints of the ashlars, the sloping surface being stopped against the wall face and a horizontal top bed worked. A sketch showing the construction in detail is given in Fig. 125. The joints of the raking cornice should be normal to the rake or pitch of the pediment.

Fascia Courses.—Strictly speaking, a fascia is a subdivision of an architrave, which is a part of an entablature, but the term fascia is often applied to the horizontal course of stones over a wide opening, such as a shop front. These courses are usually held in position by steelwork, and though assuming the aspect of solid stones carrying the weight of the wall above, they are simply a casing to steel girders, designed to carry the loads. The stones are notched and secured to the steelwork by some device. Fig. 126 shows a *fascia course* in position. Each stone is notched for the flanges of the girder, the bottom flange forming a seating for the stone. It is held in a vertical position by the insertion of a metal plate and bolt in each joint. The bolt passes through the web of the girder, whilst the plate is fitted in mortises cut in the joint surfaces.

A large joggle should be cut in the joint surfaces of the stone, thus ensuring a free access for the cement grout. When the stones are in their correct position and the joints pointed, the whole of the space between the steel and the stone, including the joints, should be filled with cement grout. Great

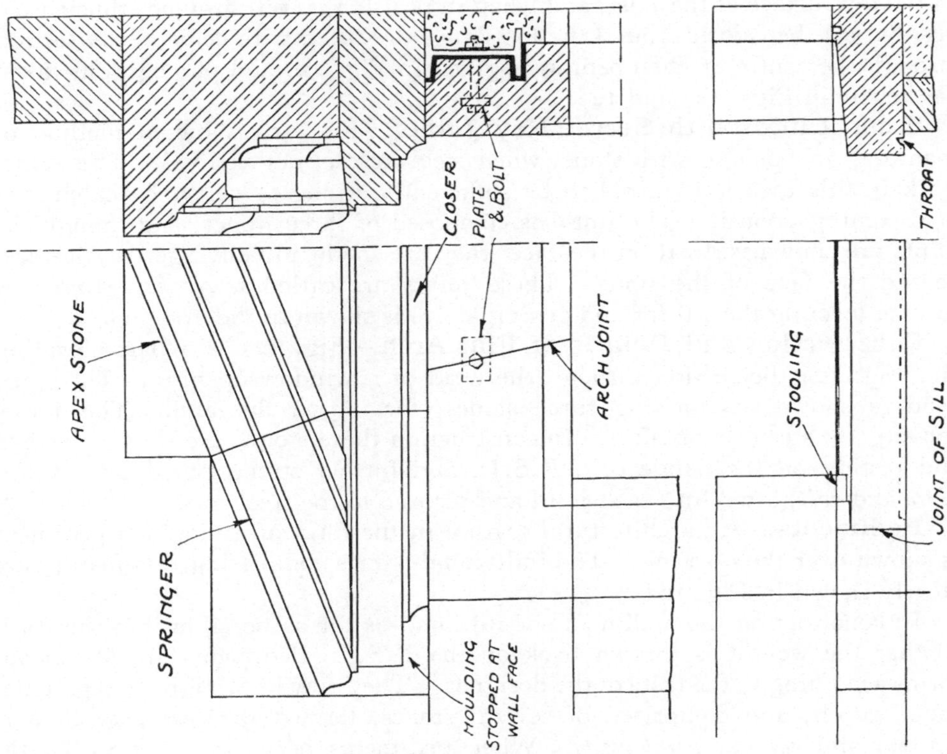

FIG. 123. FIG. 124.
DETAILS OF FLAT STONE ARCH AND STEELWORK.

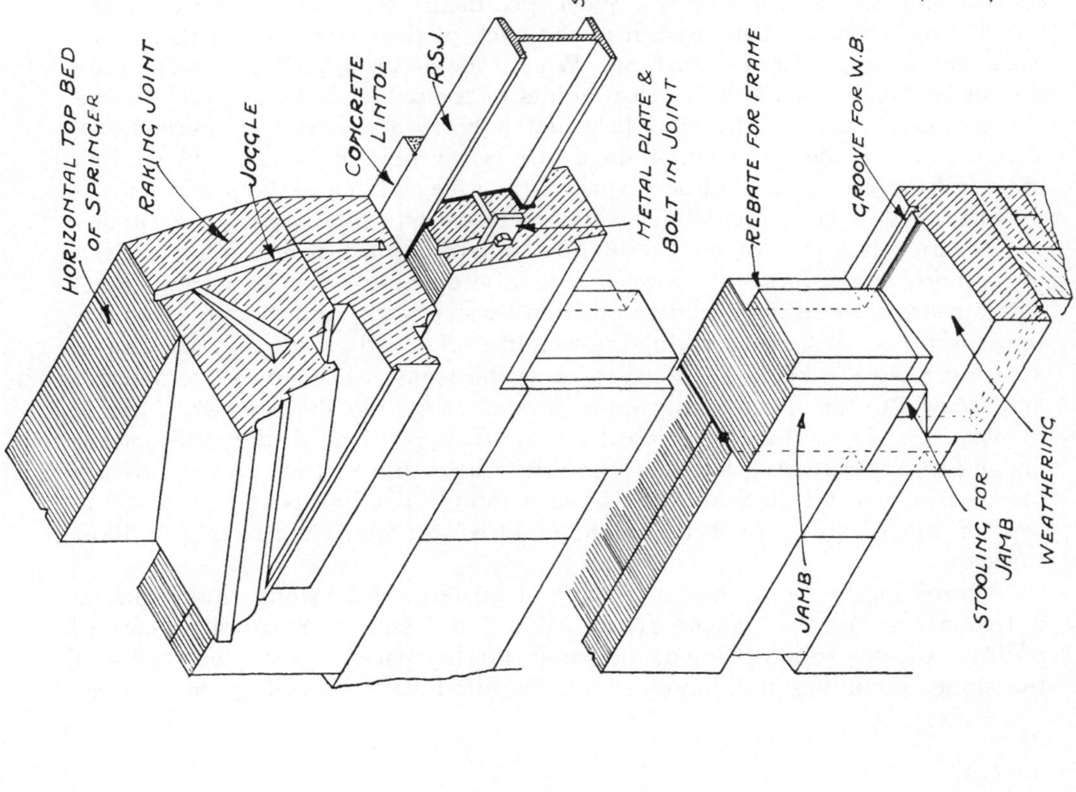

FIG. 125.
SKETCH SHOWING CONSTRUCTION OF ARCH AND PEDIMENT.

care must be exercised to see that the cement fills all the cavities around the plate and bolt in the joints.

Fig. 127 is a section through the fascia course.

Rag Bolts fixed into the back of the stones are sometimes employed for this purpose, but their use is not desirable. The plates and bolts should be made of galvanised iron or gun-metal; painted iron should not be used.

Marble Fascia over Shop Front.—Owing to its high cost, marble is generally used in the form of a *veneer* or thin decorative covering for walls, etc. The construction necessary for this class of work calls for a different treatment from stonework. Fig. 128 illustrates a method of securing a marble fascia, covering a reinforced concrete lintol, over a large opening. The marble slabs are held in position by metal hanging cramps, which are fastened to the concrete. In the sketch these cramps are shown extending to the bottom edge of the fascia, but if this edge is to be clean, the cramps should be made shorter and a mortise cut in the joint surface of the marble for the insertion of the cramp. The slabs are secured vertically by metal pin-cramps, which are built into the wall. A detail of these cramps is given in Fig. 129. The top portion of the pin is allowed to project into the bottom bed of the course above, similar to a dowel, thereby forming a means for securing the moulded course, which is also held in position by metal wall-cramps built into the brickwork.

Some coloured marbles are defective from a constructional point of view, because of the natural flaws and joints which occur throughout the blocks when quarried. Whether these marbles should be used is a matter to be decided by the architect, but often a marble is selected entirely from a decorative point of view to carry out a colour scheme.

When these marbles are used, it is usual to strengthen them by securing them to a "liner" of stronger and cheaper material. Belgian granit is often used for this purpose. Fig. 130 shows the construction of a lined marble fascia over a shop front, the weight of the wall above being carried on two rolled steel joists. The lining is shaped to clip on to a horizontal *metal strip*, which is screwed to the *framework*, close to the bottom edge of the fascia. Holes are drilled through the marble near the top edge for the insertion of brass screws, which hold the fascia slabs in a vertical position, whilst the top course is arranged to cover the heads of the screws. This top course is supported in a similar manner to the fascia, and covered with sheet-lead or copper, as shown in the sketch.

Marble Wall Linings and Flooring for Bathroom.—The plan and sectional elevation as a suggestion is given in Fig. 131, showing the construction and arrangement of wall linings and floor for a bathroom. The concrete floor should be covered with a cement layer about ½ in. thick, or some impervious material, the marble floor slabs being bedded on to this. The wall linings, architrave, etc., should be securely cramped to the brick walls with brass wall-cramps, and pressed into position against *plaster dabs*, placed at intervals as shown, thus allowing air space between the wall surface and the back of the marble.

Details showing the construction and fixing are given in Figs. 132, 133, and 134.

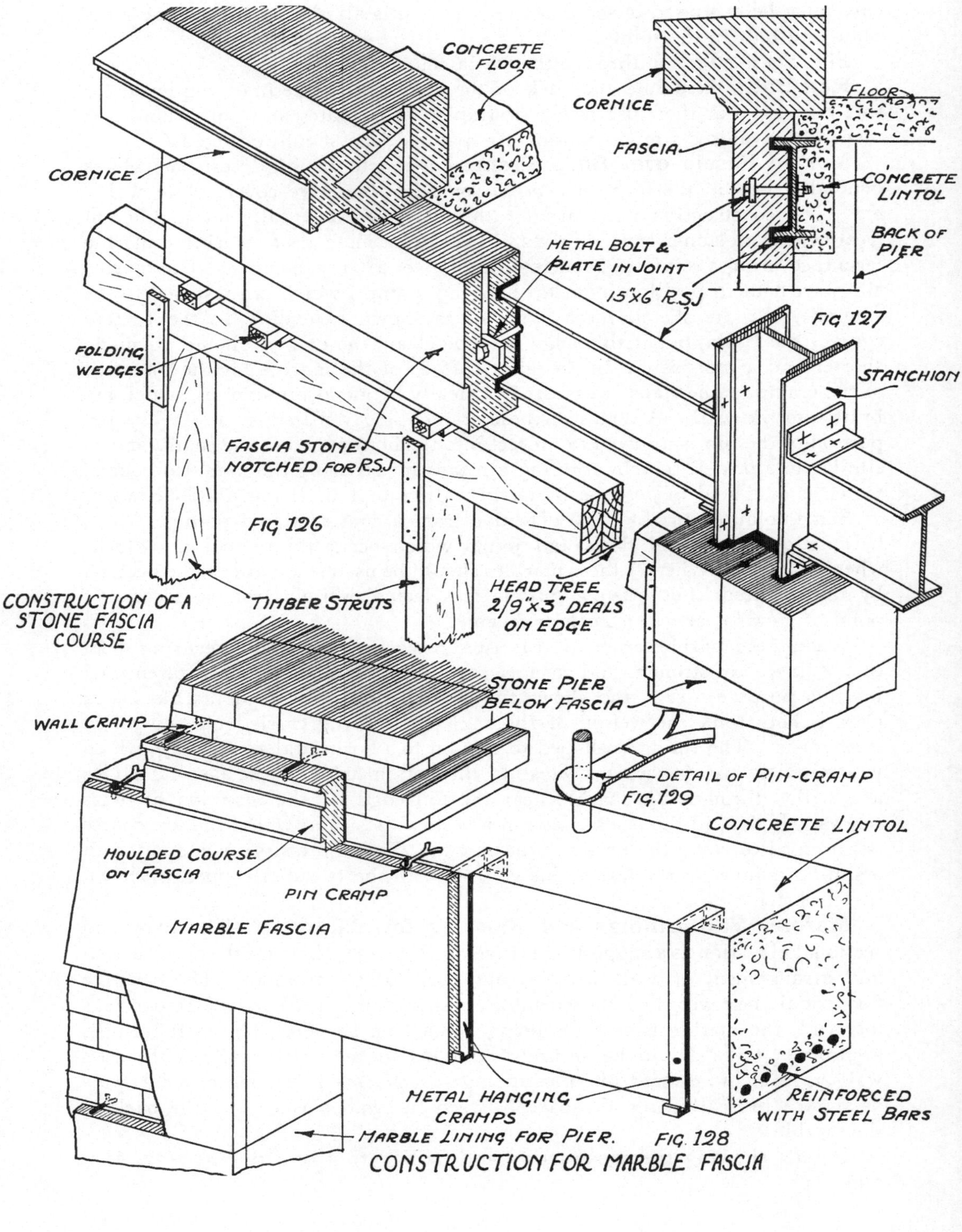

CONSTRUCTION OF A STONE FASCIA COURSE

CONSTRUCTION FOR MARBLE FASCIA

Marble Chimney-piece.—A suggestion for boxing up and securing a marble chimney-piece is given in Fig. 135. The panelled hearth and curb, which is secured to a slate liner, is placed in position, then the bases and boxed-up jambs built up from this and secured to the wall by metal wall-cramps. The cornice forming the mantel is built up in thin courses, thereby economising in marble. The construction through the frieze and cornice is given in Figs. 136 and 137, whilst details showing the jointing and cramping for the bed mould of the cornice and the angles of the frieze above the pilasters are given in Figs. 138 and 139.

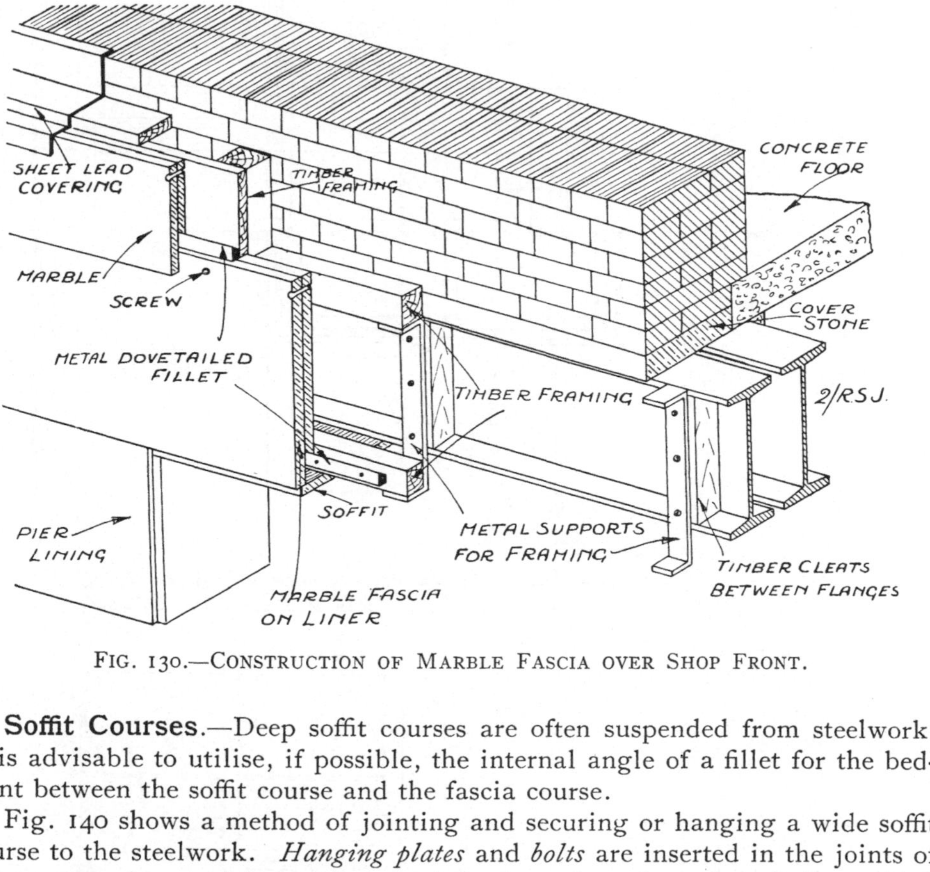

FIG. 130.—CONSTRUCTION OF MARBLE FASCIA OVER SHOP FRONT.

Soffit Courses.—Deep soffit courses are often suspended from steelwork. It is advisable to utilise, if possible, the internal angle of a fillet for the bed-joint between the soffit course and the fascia course.

Fig. 140 shows a method of jointing and securing or hanging a wide soffit course to the steelwork. *Hanging plates* and *bolts* are inserted in the joints of the stones and secured by passing the bolt through a piece of steel plate which is allowed to rest on the bottom flanges of the R.S.J.'s, as shown in sketch. A section is given in Fig. 141.

An alternative method is shown in Fig. 142. In this instance *lewis bolts* are fixed into the top surface of the soffit stones at intervals, and bolted direct to the bottom flanges of the steelwork.

Details of a lewis bolt and plate and bolt are given in Figs. 143 and 144.

Fig. 145 shows a soffit course suspended from a reinforced concrete lintol upon which is bedded the cornice. This is shown as an alternative method

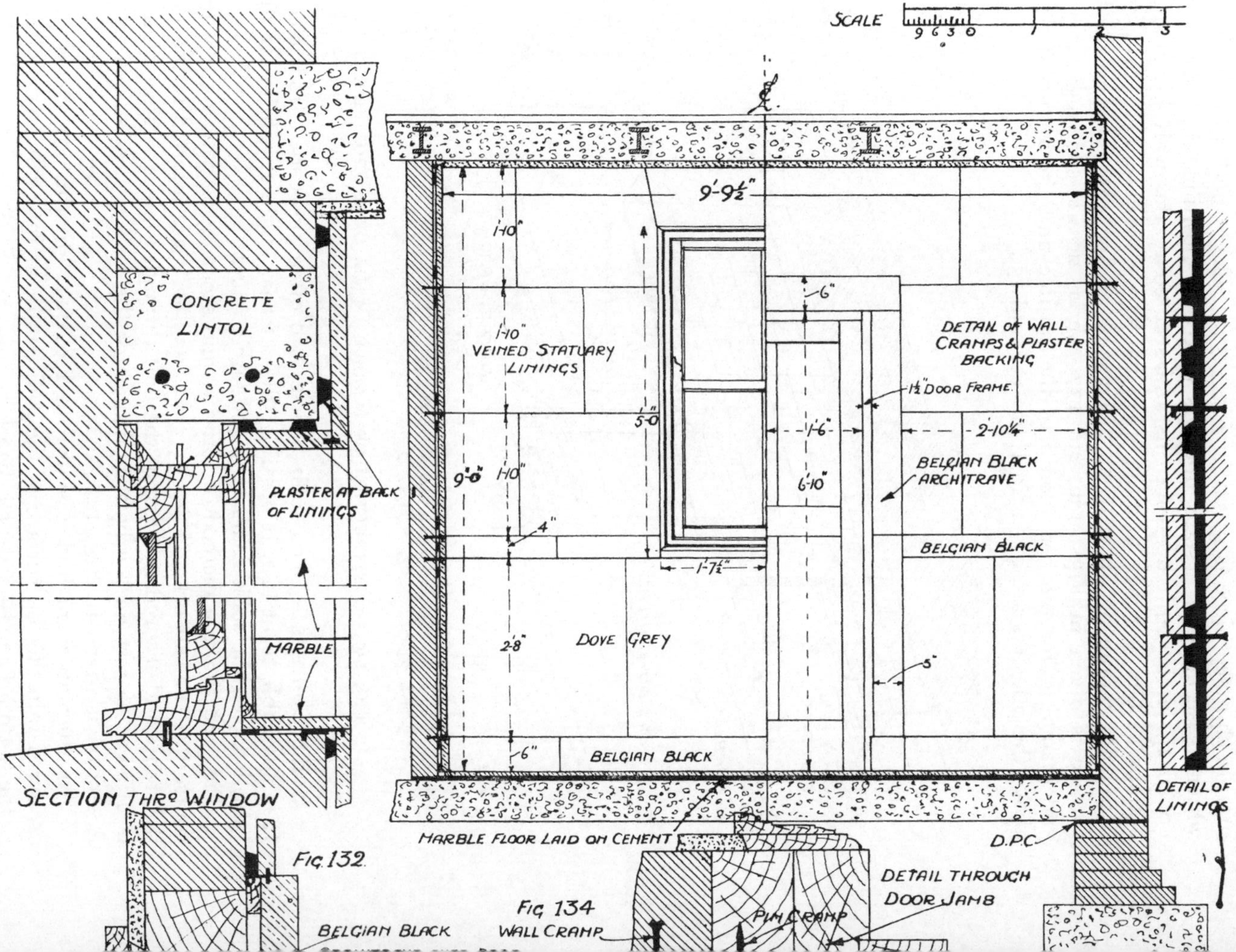

Fig. 132.

Fig. 134

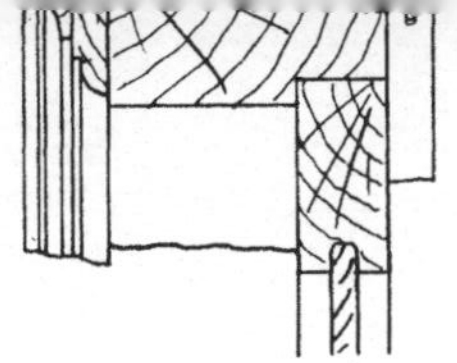

DETAIL THROUGH
HEAD OF DOORWAY

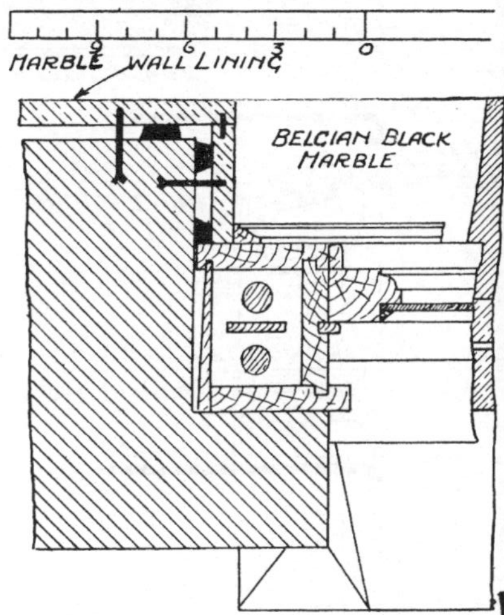

DETAIL THROUGH WINDOW
OPENING SHOWING MARBLE
LININGS FIG 133

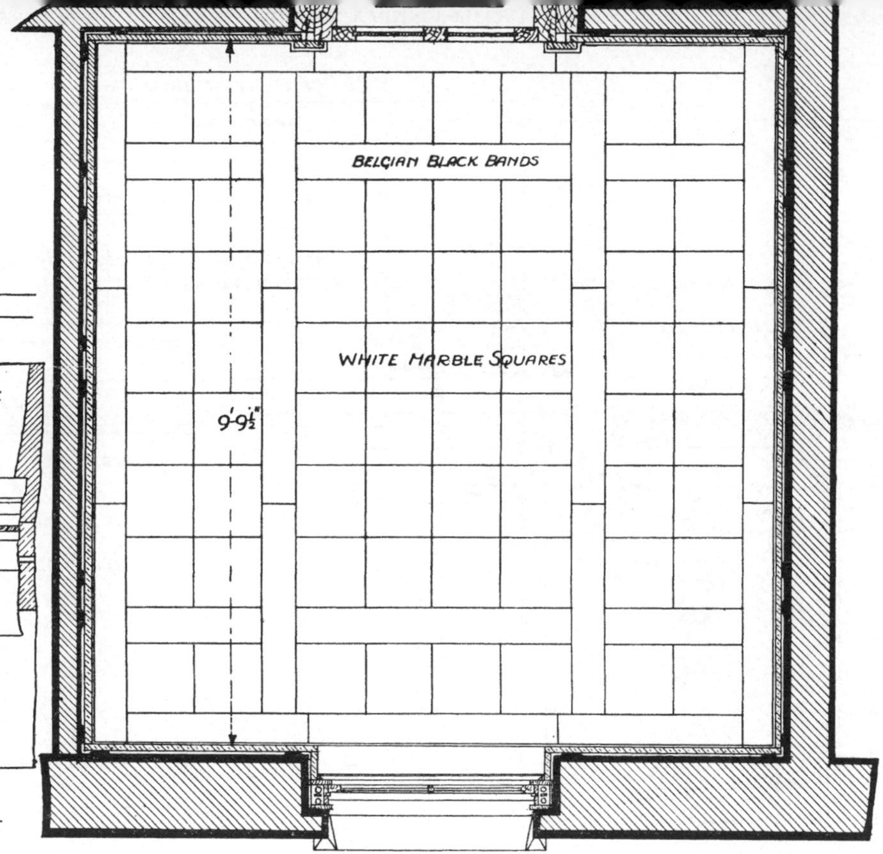

FIG 131

MARBLE FLOOR AND WALL LININGS TO BATHROOM.

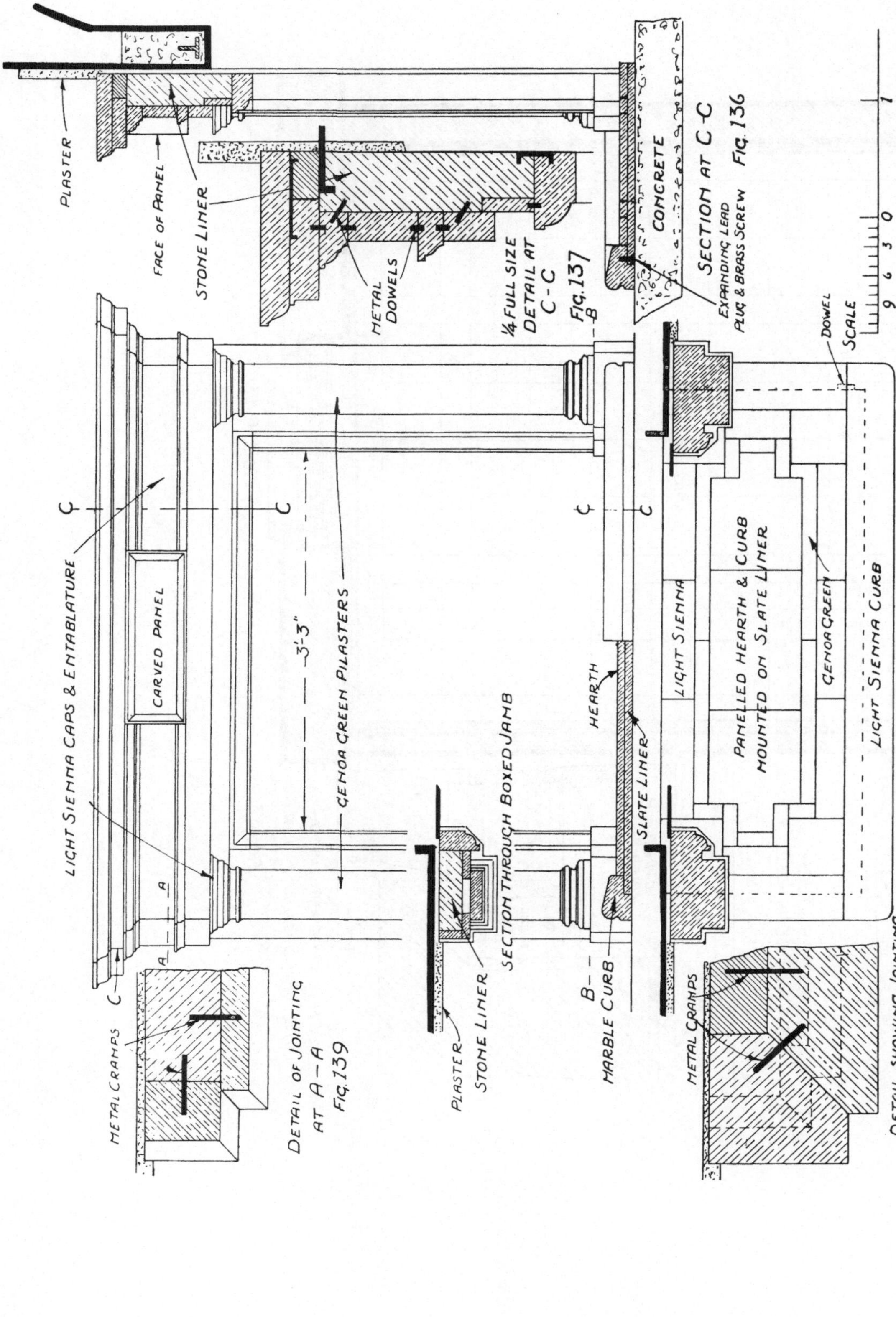

METHOD OF BOXING UP MARBLE CHIMNEY-PIECE.

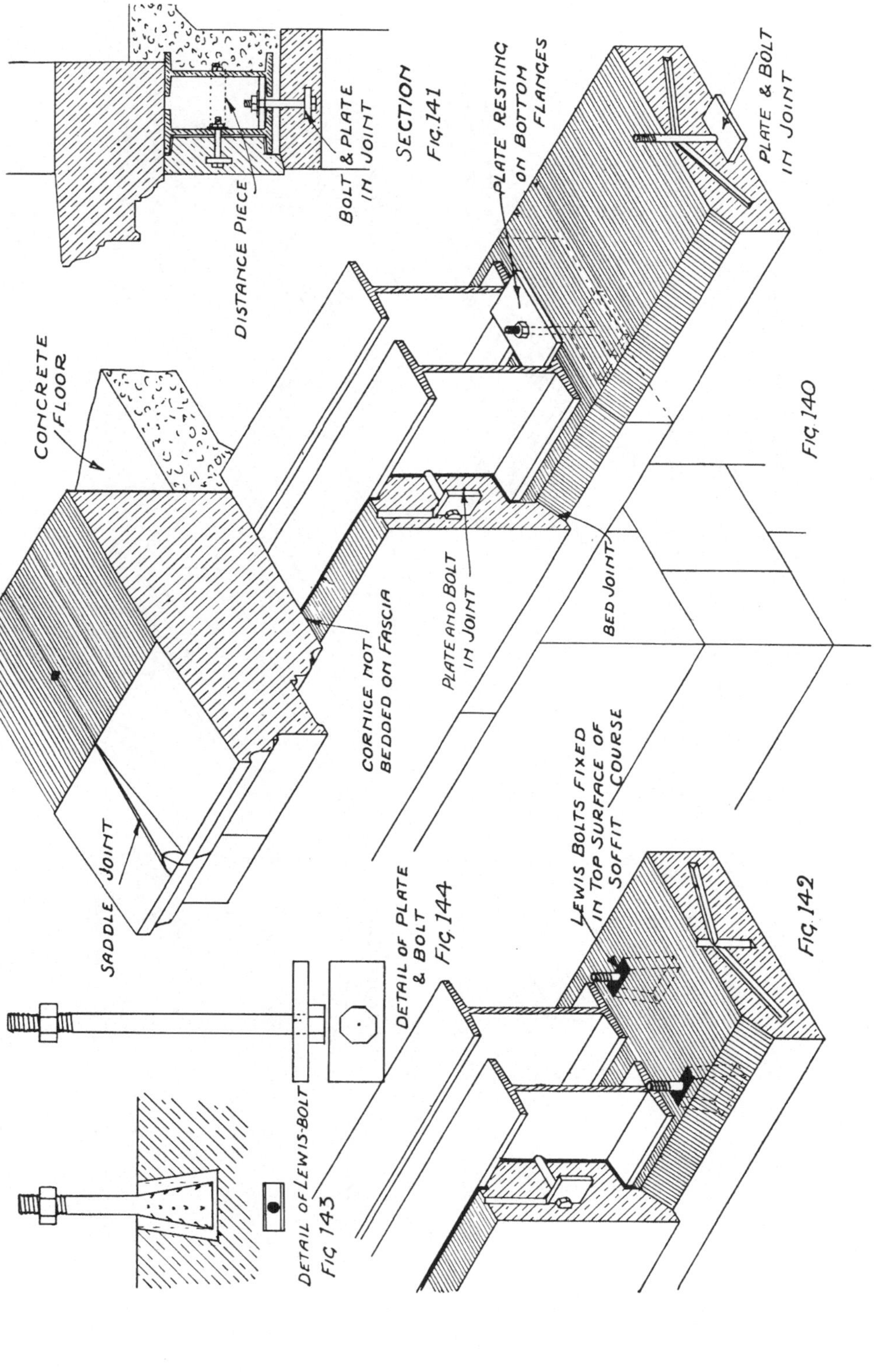

METHOD OF SECURING SOFFIT COURSE.

AN ALTERNATIVE METHOD OF CONSTRUCTION.

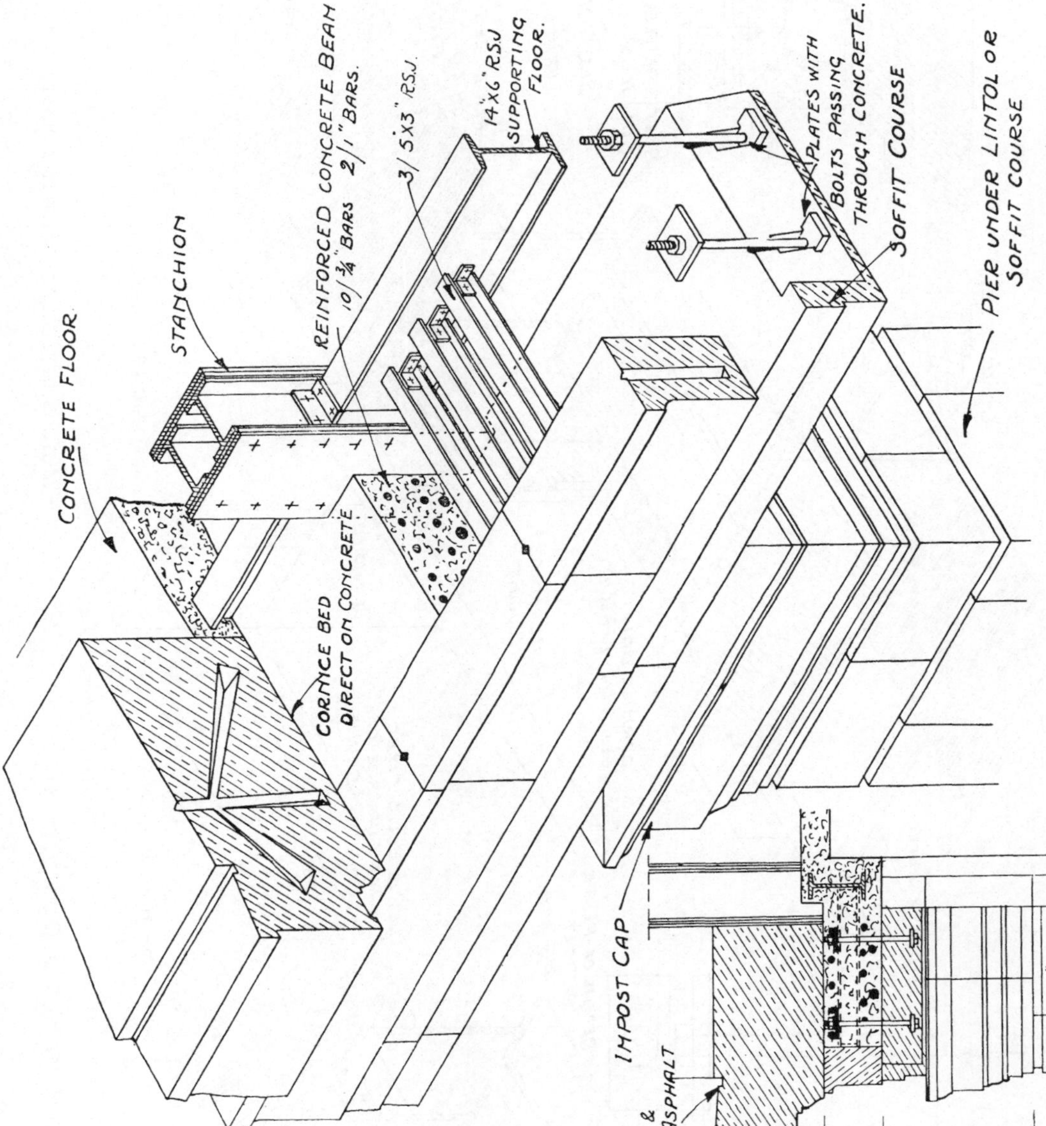

FIG. 145.—DEEP SOFFIT COURSE SUSPENDED FROM REINFORCED CONCRETE LINTOL.

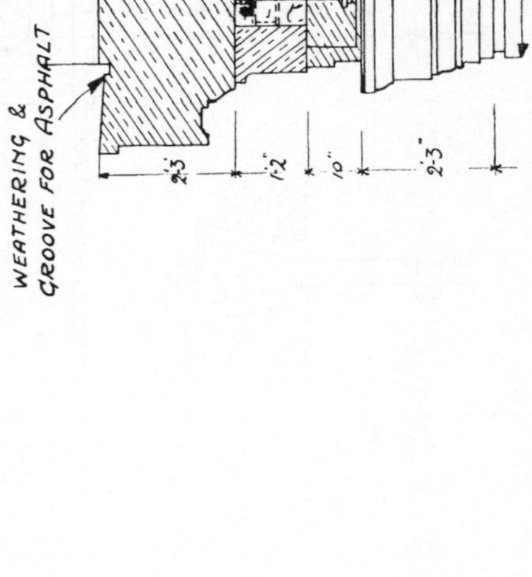

FIG. 146.—SECTION THROUGH SOFFIT AND CONCRETE LINTOL.

FIGS. 147 AND 148.—SOFFIT COURSE SUPPORTED FROM REINFORCED CONCRETE LINTOL.

FIGS. 149 AND 150.—CONSTRUCTION OF PIER AND FIXING IMPOST CAP, VICTORIA HOUSE.

Chas. W. Long, Esq., F.R.I.B.A., Architect.

to bolting to steelwork. The disadvantage of this method of construction is that it does not permit of speed in the erection of the building. Sufficient time is required for the cement to set before the work above can proceed. The bolts in this case pass through the concrete lintol and are fastened by means of a nut and a large steel washer on the top surface of the concrete. The small R.S.J.'s shown in the drawing over the pier are embedded in the concrete lintol to form a tie for the masonry to the steelwork. The fascia course is supported on the soffit stones, whilst the cornice is bedded direct on the top surface of the concrete and clear of the fascia course, thereby transmitting all the weight of the work above on to the concrete. Fig. 146 is a section through the soffit course, and photographs are given in Figs. 147 and 148 showing the actual work in progress.

Impost Cap is a capital of a pier or pilaster which receives an arch, varying in design according to the Order. In Fig. 146 it is shown as the top moulded course of the pier immediately under the soffit course. Photographs showing the construction of the pier and the fixing of the *impost cap* are given in Figs. 149 and 150.

Details for Gable.—A gable is the vertical triangular piece of wall at the end of a roof from the eaves to the summit. The stone units are similar in almost all cases, although the detail and form may vary considerably.

Fig. 151 is an elevation, and Fig. 152 is a side elevation of a gable, showing *springer*, *knee stone*, *apex stone*, *coping*, and *lacing course*. The joints of the *springers* and *knee stones* should be arranged in construction so that they effectively resist the thrust of the coping stones.

Apart from forming an architectural feature, the object of the coping stones is to keep the wall dry.

The joints of the coping stones are often *rebated* or *weather jointed* to assist in this function, as shown.

Fig. 153 is a sketch of one of the *springers*, showing the detail and construction, whilst Fig. 154 shows the apex stone.

Lacing courses are often provided in brick and rubble walls. They are ashlar courses, and usually placed to coincide with the *springers* and *kneeler stones*, as shown.

Buttresses.—The construction of a buttress is shown in Fig. 151. These are designed to support the side of a wall, or in Gothic architecture, to resist the thrust of the vaulting. The buttress in the figure is placed at the angle, the treatment of which shows the top *weathering* or *tabling stones* intersecting the wall face. A sketch of this stone is given in Fig. 155.

Fig. 156 is the plan of the buttress, showing the footings and bonding for the stones.

A five-light window is also given. The units comprising this feature are the *sills*, *jamb-stones*, *mullions*, and *tracery head*. It is a common practice in this style of window for the openings to be glazed with leaded lights, in which case a glazing groove is worked around the openings for the fixing of the flanges of the lead cames which contain the glazing.

A rebate instead of a groove should be cut in the sill, and weep holes provided for the discharge of the moisture resulting from the condensation which takes place on the inside glass surface.

A detail of the section through the sill is given in Fig. 157.

A horizontal section of the mullion is given, showing the glazing groove, in Fig. 158.

A *mullion* is the vertical post or stone dividing the window into several lights. Sometimes the windows are divided into two lights in height. The dividing stone is called a *transome*. The top surface of the transome is worked in a manner similar to the sill, with weatherings for the discharge of the rain water.

A moulded string-course is shown stopped against the side of the buttress.

A sketch of the stone forming the stop is given in Fig. 159.

A *hood moulding* or *drip course* is usually placed over the head to throw the rain water clear of the window. It is usually returned vertically and stopped at the side of the opening, and formed to suit the taste of the designer. The *string-course* in this example forms the *hood moulding* over the window opening.

ARCHES

An arch is a mechanical arrangement of blocks of any material supporting one another by their mutual pressure. The stability of an arch depends largely upon its having sufficient opposition to the thrust, the supports being strong enough to resist effectively any outward movement. All the *bed-joints* should be normal to the curve of the arch.

It is necessary that the student should appreciate the various parts and terms connected with arches.

Technical Terms.

Abutments are the arch supports or the solid part of a pier from which the arch immediately springs. They should be of sufficient solidity and strength to resist the thrust of the arch.

Skewback is the surface of the abutment on which the arch rests, or that part of it which recedes on the springing from the vertical line of the opening.

Voussoirs or **Arch-stones** are the wedge-shaped stones forming the arch.

Keystone.—This is the highest central stone of the arch, so named because when placed in position it "*keys*" or *locks* the voussoirs of the arch.

Springers are the lowest voussoirs of an arch, having a horizontal bottom bed surface.

The **Span** is the clear horizontal distance between the supports, or the width of the opening.

The **Rise** is the perpendicular distance from the springing to the crown.

The **Springing Line** is an imaginary line between the points from which the curve of the arch springs on the intrados.

The **Intrados** is the inner curve of the arch ring.

The **Extrados** is the outer curve of the arch ring.

The **Soffit** is the lower or concave surface of the arch.

The **Depth of the Arch** is the distance between the inner and outer curves of the arch on the face.

The **Haunch of the Arch** is the name given to the lower part of the arch, between the springing and the crown.

Bed-joints are the normal joint surfaces of the voussoirs or arch-stones.

The **Spandrel** is the triangular space between the extrados of the arch, and a horizontal line from the top point of the arch, and a perpendicular line from the springing of the outer curve.

A **Respond** is a corbel or half-pier projecting from the wall face to receive the springing stone of an arch at the end of an arcade. The aisle arches of a church usually terminate upon a respond or half a capital projecting from the wall face.

An **Archivolt** is the ornamental band of mouldings round the voussoirs, terminating horizontally upon the impost.

Arches are classified according to their shape and the manner of treatment.

Semicircular Arch.—Fig. 160 is a semicircular arch, showing the bonding necessary to connect with the ashlar courses of the surrounding wall. An alternate method of bonding is shown.

Semi-elliptical Arch.—Fig. 161 is a detail of a semi-elliptical stone arch.

Approximate Semi-elliptical Arches are similar in form to the semi-elliptical arch, but for convenience in setting out the curve is often struck from centres, and is therefore compounded of circular arcs instead of following the true elliptical curve.

Segmental Arch.—Fig. 162 shows a segmental stone arch bonded to suit the ashlar courses of the wall and illustrating the position of the skewbacks.

Gothic or Pointed Arches.—Fig. 163 shows an equilateral arch.

Fig. 164 shows a lancet arch.

Fig. 165 shows a drop arch.

Fig. 166 shows a four-centred or Tudor arch.

A rampant arch is shown in Fig. 167. This description is given to an arch whose springings are not at the same level, and which is used chiefly under a flight of steps.

Fig. 168 shows a shouldered arch.

The geometry of arches is fully explained in the "Geometry of Masonry" section.

Corbel.—Corbels are stones projecting from the wall face for the purpose of receiving a load. They are designed usually in keeping with the architectural forms connected with the building.

Oriel Window.—An oriel window is a large bay projecting from the outer face of a wall, commencing from an upper floor, the projection being supported by a series of corbellings from the wall face. In Gothic and Tudor work elaborate details were introduced, making it a special feature for ornamentation.

The projecting stones were arranged in a series of beds, each bed projecting beyond the course below, the projecting portion being enriched by mouldings and carvings. This corbelling continued until the projection required was obtained. In modern construction oriel windows usually call for a different method of construction owing to their angular form and the necessity for supporting the stones by the steelwork of the structure.

A sketch for an oriel corbelling is given in Fig. 169.

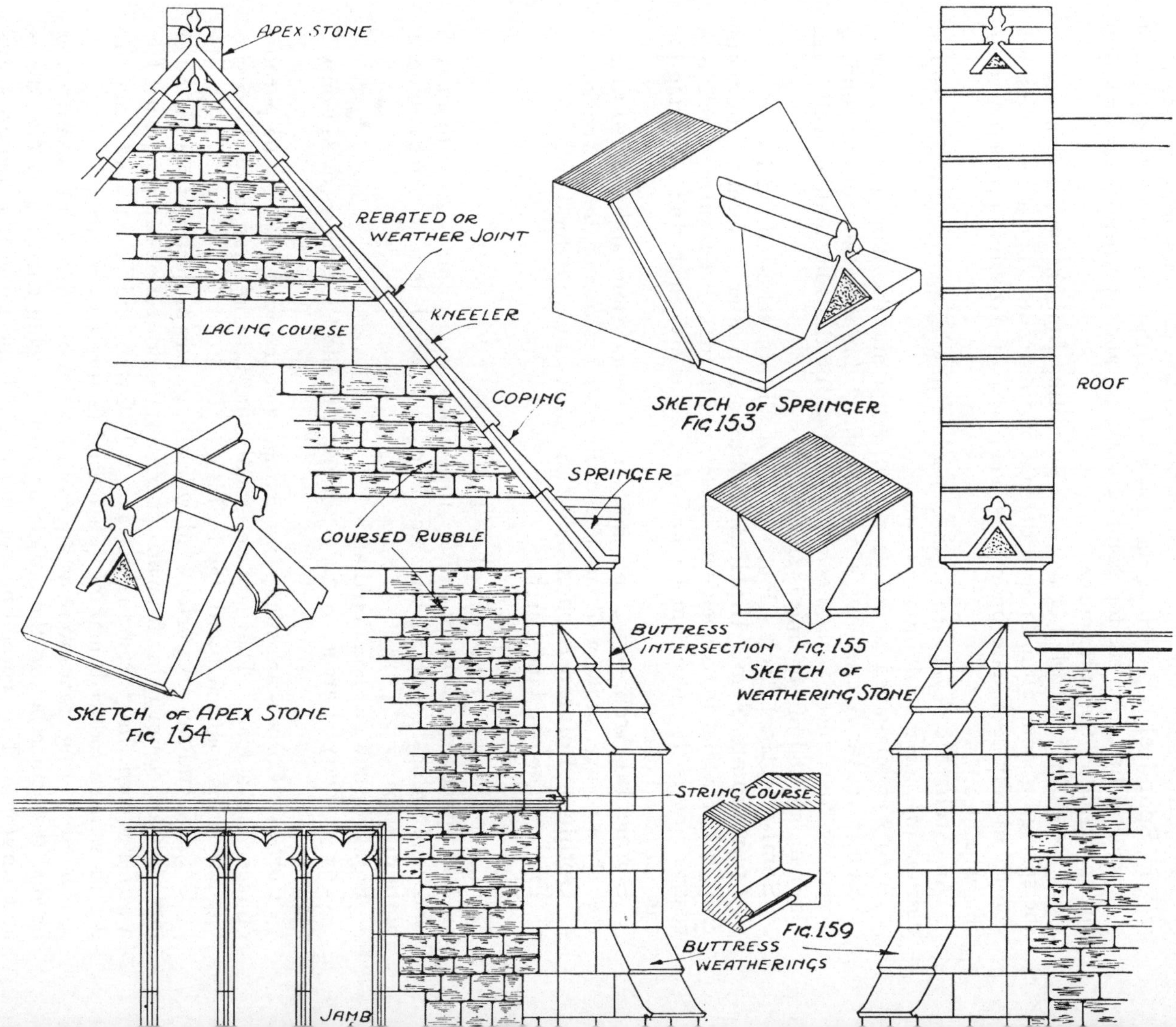
APEX STONE
REBATED OR WEATHER JOINT
KNEELER
LACING COURSE
COPING
SPRINGER
COURSED RUBBLE
SKETCH OF APEX STONE
FIG 154
SKETCH OF SPRINGER
FIG 153
ROOF
BUTTRESS INTERSECTION
FIG. 155
SKETCH OF WEATHERING STONE
STRING COURSE
FIG. 159
BUTTRESS WEATHERINGS
JAMB

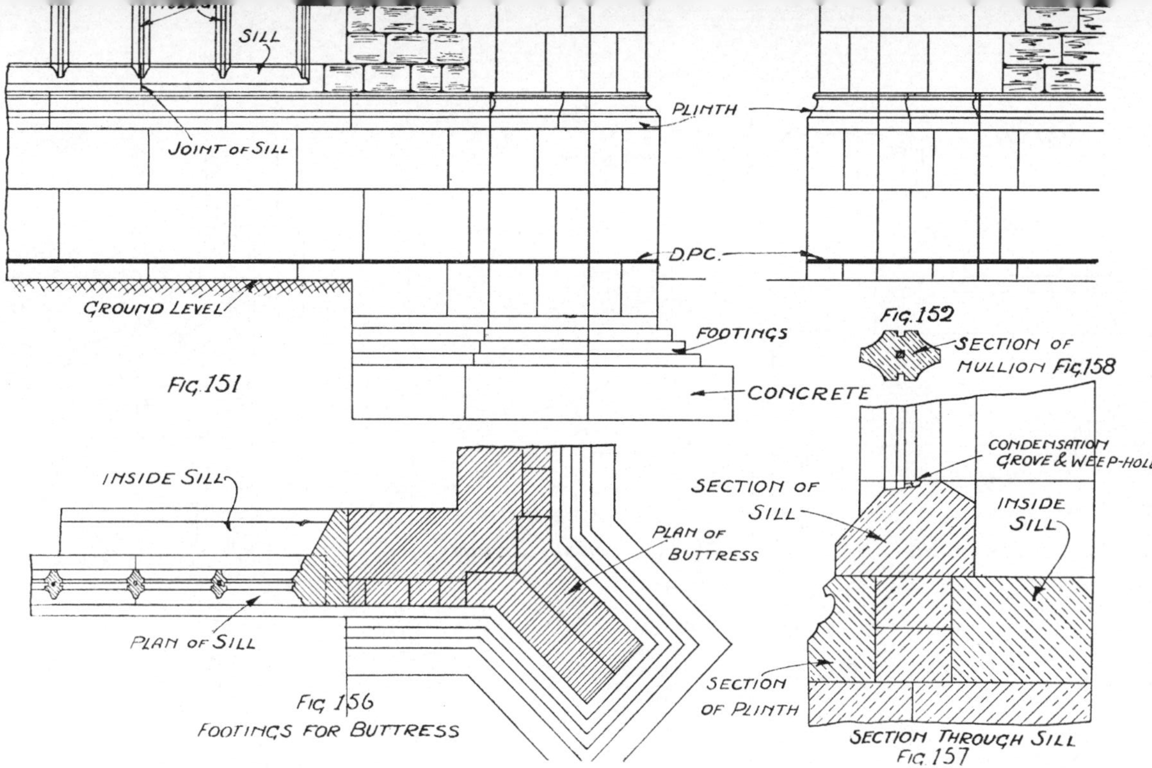

Construction of Gable and Angle Buttress.

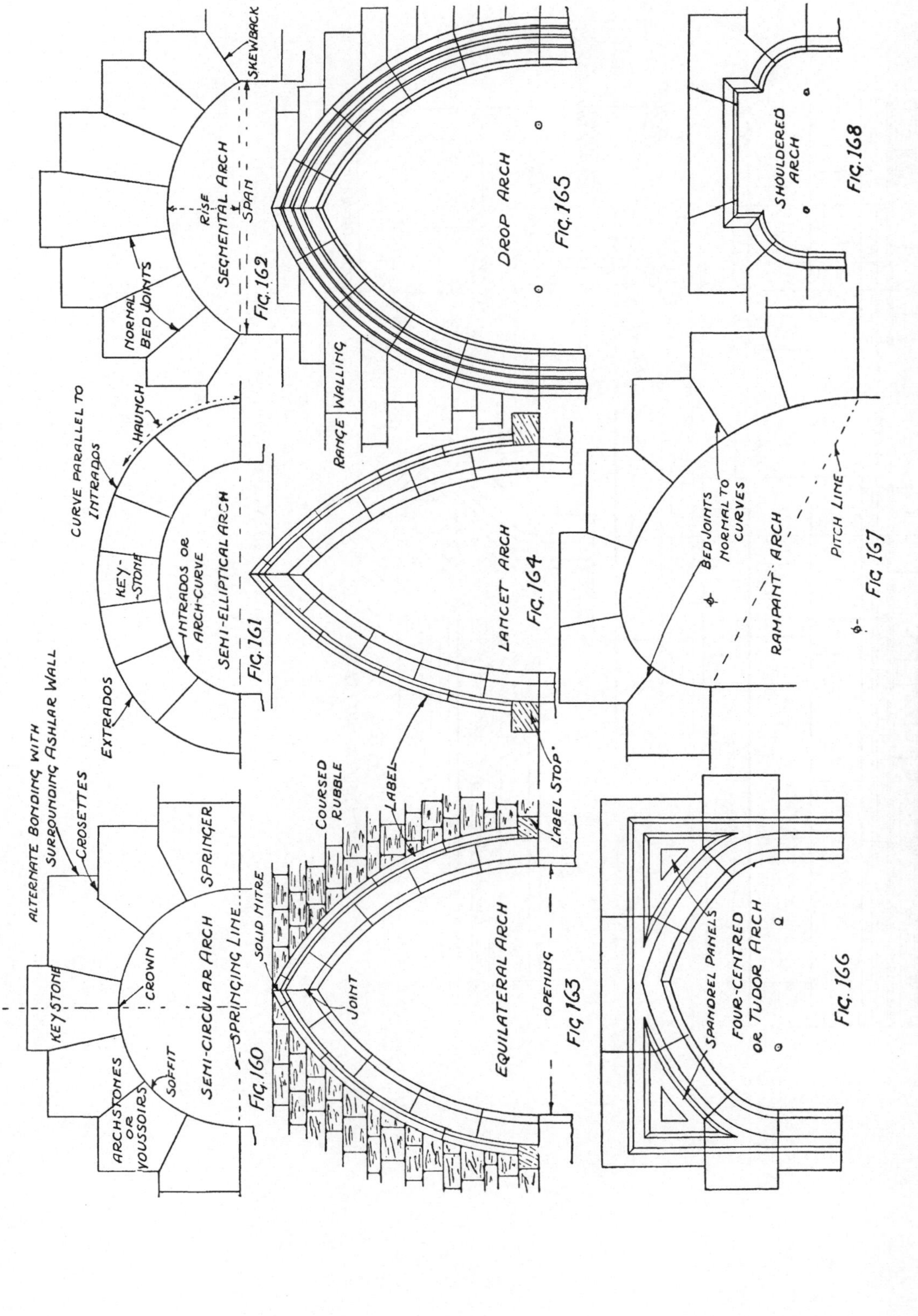
ALTERNATE BONDING WITH SURROUNDING ASHLAR WALL
CROSETTES
KEYSTONE
CROWN
SPRINGER
ARCHSTONES OR VOUSSOIRS
SOFFIT
SEMI-CIRCULAR ARCH
SPRINGING LINE
FIG. 160
EXTRADOS
KEY-STONE
CURVE PARALLEL TO INTRADOS
HAUNCH
INTRADOS OR ARCH-CURVE
SEMI-ELLIPTICAL ARCH
FIG. 161
NORMAL BED JOINTS
RISE
SEGMENTAL ARCH
SPAN
SKEWBACK
FIG. 162
COURSED RUBBLE
SOLID MITRE
JOINT
EQUILATERAL ARCH
OPENING
FIG 163
LABEL
LABEL STOP
LANCET ARCH
FIG. 164
RANGE WALLING
DROP ARCH
FIG. 165
SPANDREL PANELS
FOUR-CENTRED OR TUDOR ARCH
FIG. 166
BEDJOINTS NORMAL TO CURVES
RAMPANT ARCH
PITCH LINE
FIG 167
SHOULDERED ARCH
FIG. 168

The form gives the impression of support to the work above, but in reality it is simply a decoration assuming this function.

The construction of the oriel is given in Fig. 170. Each stone is supported and held in position by the steelwork. The stone head is notched for the flanges of the R.S.J. over the opening, and also notched for the stone brackets, in the joints of which are inserted *metal plates* and *bolts* connected to the R.S.J. The overhanging portions of these brackets are supported from *steel cantilevers*, placed as shown. The brackets are connected to these by *metal lewis bolts*, which are fixed into the top bed of the stone and bolted to the cantilevers.

A *lewis mortise* is cut into the stone to receive the dovetailed portion of the bolt, and when in position it is cemented or caulked with lead. The

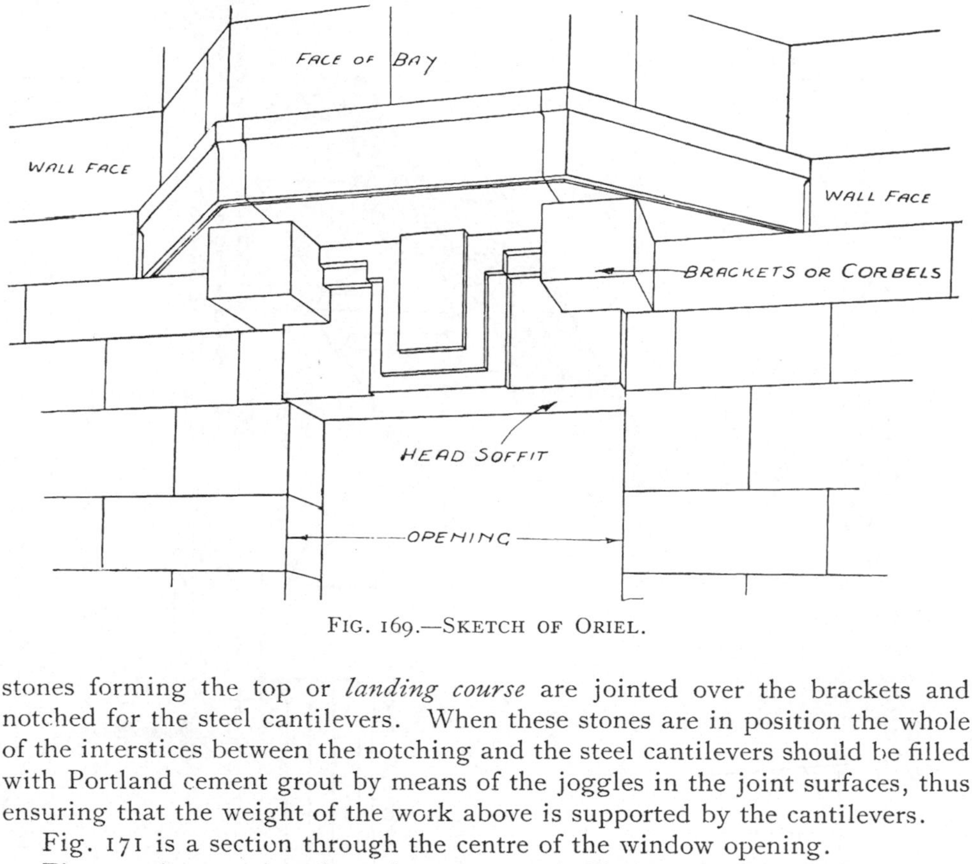

FIG. 169.—SKETCH OF ORIEL.

stones forming the top or *landing course* are jointed over the brackets and notched for the steel cantilevers. When these stones are in position the whole of the interstices between the notching and the steel cantilevers should be filled with Portland cement grout by means of the joggles in the joint surfaces, thus ensuring that the weight of the work above is supported by the cantilevers.

Fig. 171 is a section through the centre of the window opening.

Fig. 172 is a section through a stone corbelling supporting a bay over an opening. As an alternative to the construction just given, each course of stone in this case acts as a corbel for the course above, until the projection required is obtained. Sufficient stone is required on the wall to more than counterbalance the overhanging weight.

A sketch showing the construction necessary is given in Fig. 173.

Tracery.—The origin and development of tracery should be of great interest

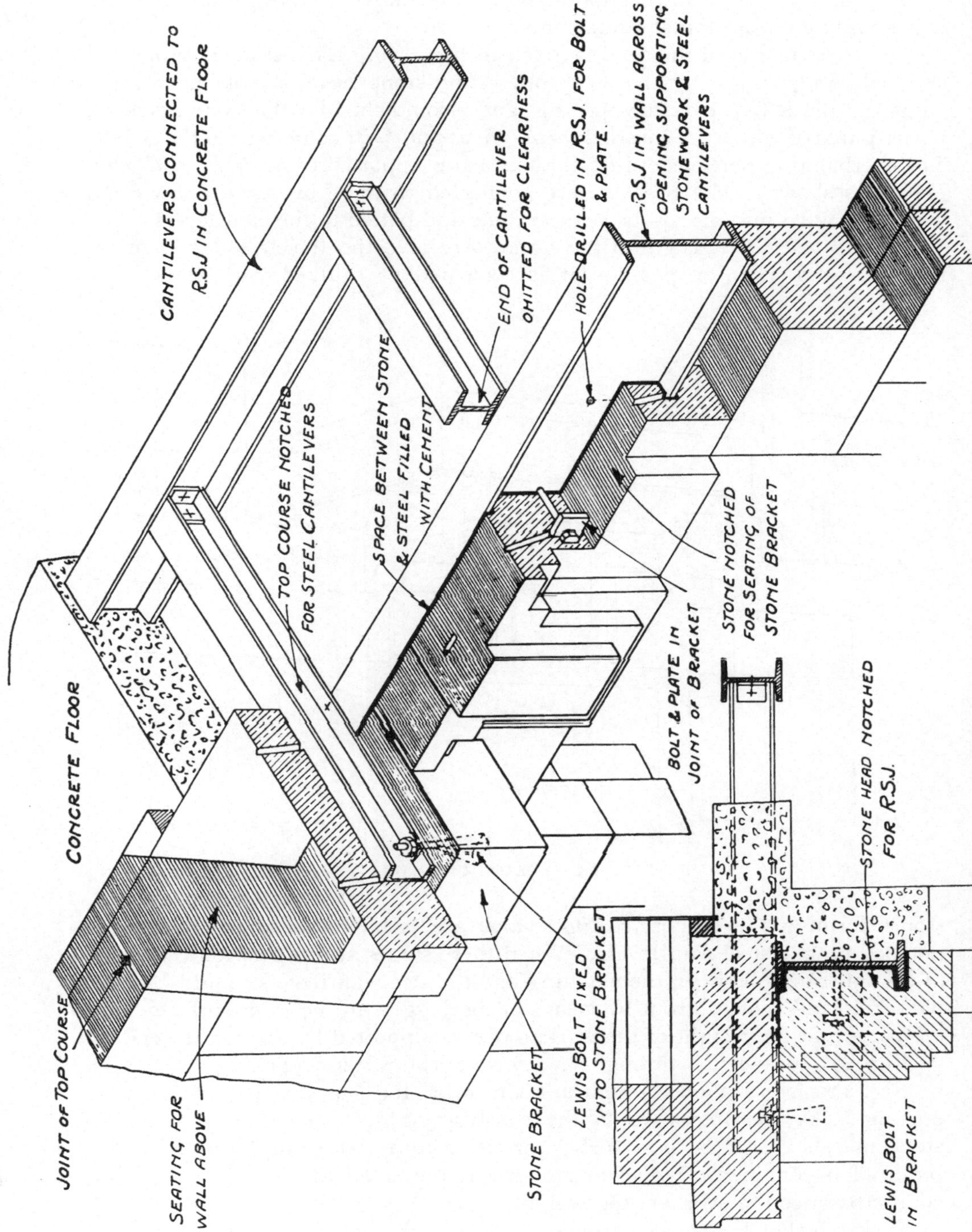

FIG. 170.—CONSTRUCTION OF ORIEL.

FIG. 171.—SECTION THROUGH ORIEL.

to the student of masonry. It is not the author's intention to enlarge upon the subject, a knowledge of which can be obtained by the perusal of various books on Gothic architecture.

Tracery contributes much to the external visible effect of Gothic architecture, and, like most other elements in buildings, has been developed from a simple idea.

Crude in treatment, the germ of development, called "*Plate Tracery*," preceded the tracery proper of the thirteenth, fourteenth, and fifteenth centuries.

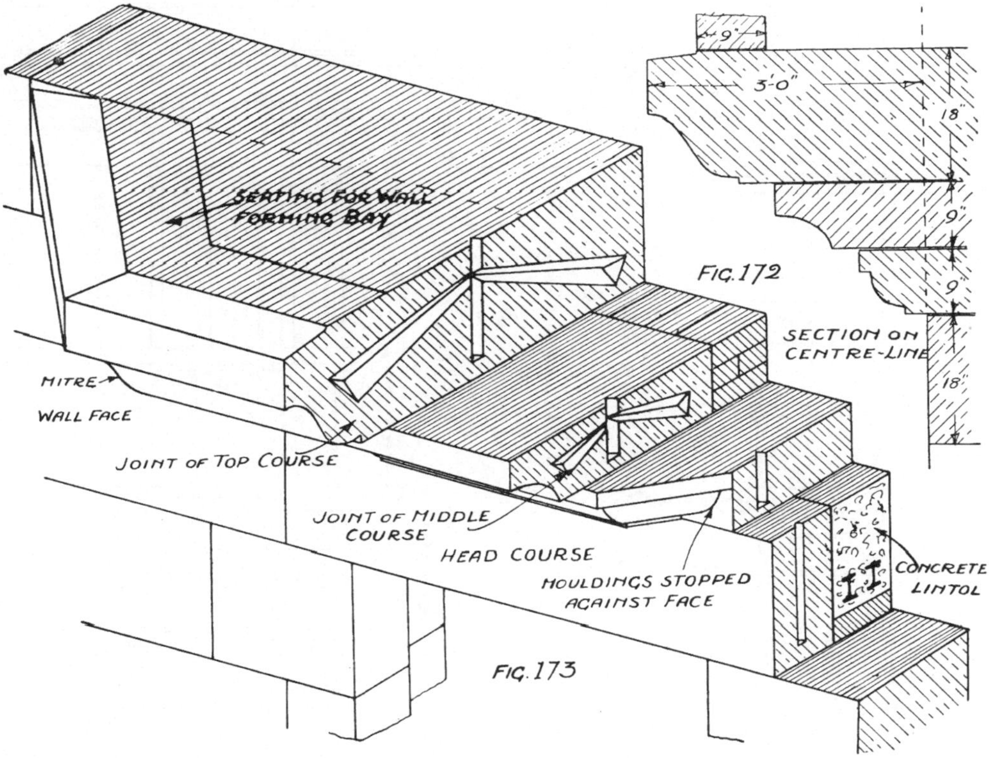

CONSTRUCTION OF AN ORIEL WITHOUT STEELWORK.

Plate Tracery is characterised by perforations in the form of circles, trefoils, etc., in the spandrel slab or plate situated below the *drip stone*.

Tracery Proper or *Bar Tracery* is formed by the piercing of the spandrels in the window head and the continuation of the mullions above the springing of the *window arch* into a consistent design of perforations.

The weight of the superincumbent wall is usually carried by the *window arch* without the aid of the tracery, which might be removed without endangering the stability of the arch.

Foliation was created by the pulling out of the *arch* or bar from its normal course in such a way as to form *projections* or *cusps*, thus making *scallops* called *foils* situated between the *cusps*.

The chief constructional consideration for the masonry student is the arrangement of the jointing, whereby the individual stones are held in position and supported by the other stones comprising the tracery.

The jointing should be arranged to prevent the possibility of any of the stones falling out. The planes of the joints should, in all cases, be normal to the curve from which the joint plane commences.

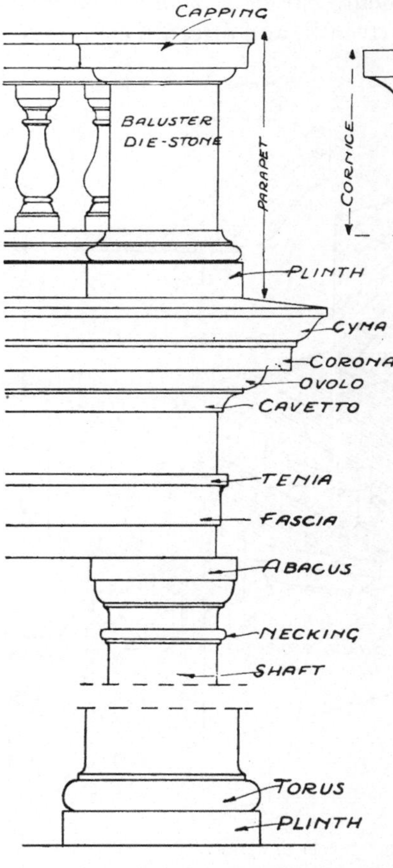

FIG. 174.—COLUMN AND ENTABLATURE OF THE TUSCAN ORDER.

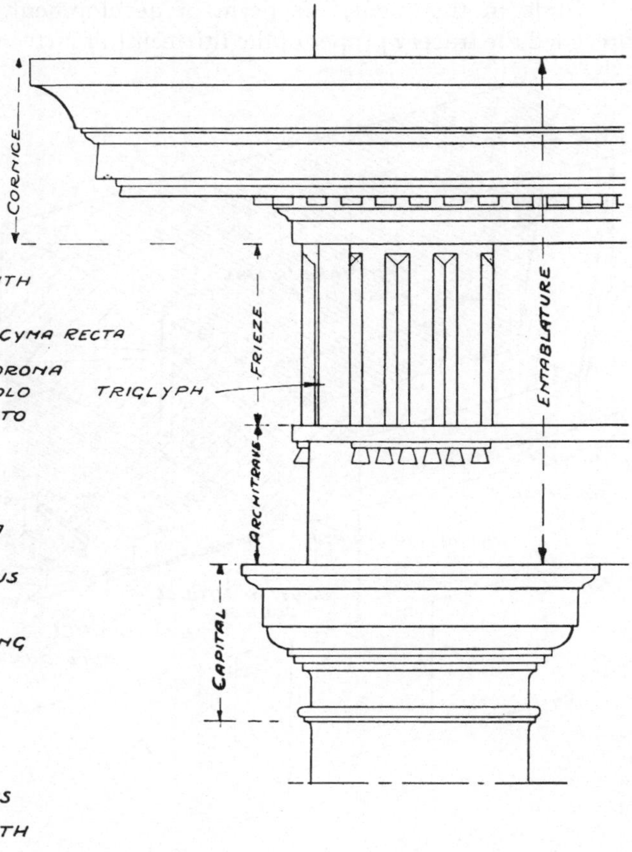

FIG. 175.—ENTABLATURE OF THE ROMAN DORIC ORDER.

The geometry and setting out of tracery is described in the "Geometry of Masonry" section.

Entablature.—In architecture an entablature is the part of an *Order* above the column capitals. It is divided into three parts—*architrave*, *frieze*, and *cornice*.

Fig. 174 illustrates the various parts of the entablature of the *Tuscan Order*, including the column, which is divided into *base*, *shaft*, and *capital*.

Fig. 175 shows the entablature of the *Roman Doric Order*.

Progressive photographs, illustrating the modern construction and fixing of an entablature, are given in Figs. 176, 177, 178, 179.

FIGS. 176 AND 177.—MODERN CONSTRUCTION OF AN ENTABLATURE, VICTORIA HOUSE.

Chas. W. Long, Esq., F.R.I.B.A., Architect.

FIG. 178.—MODERN CONSTRUCTION OF AN ENTABLATURE, VICTORIA HOUSE.
Chas. W. Long, Esq., F.R.I.B.A., Architect.

FIG. 179.—THE COMPLETED ENTABLATURE, VICTORIA HOUSE.
Chas. W. Long, Esq., F.R.I.B.A., Architect.

Cornices.—The term cornice is usually applied to the projecting course which crowns that part of the building to which it is affixed. One important function of a cornice being to protect the surface of the wall from moisture, it should be designed to carry out this purpose effectively. Provision should be made for the proper discharge of rain water from the top surface, also a continuous "*drip*" should be provided as near to the nosing of the cornice as the detail will allow.

A sketch detail of a piece of cornice is given in Fig. 180, illustrating its various parts.

Modillions are the projections under the *corona* of the cornice, resembling brackets.

Dentils are the small blocks in the *bed moulding* of the cornice.

Balustrades.—Sometimes a parapet wall is built above a cornice in the

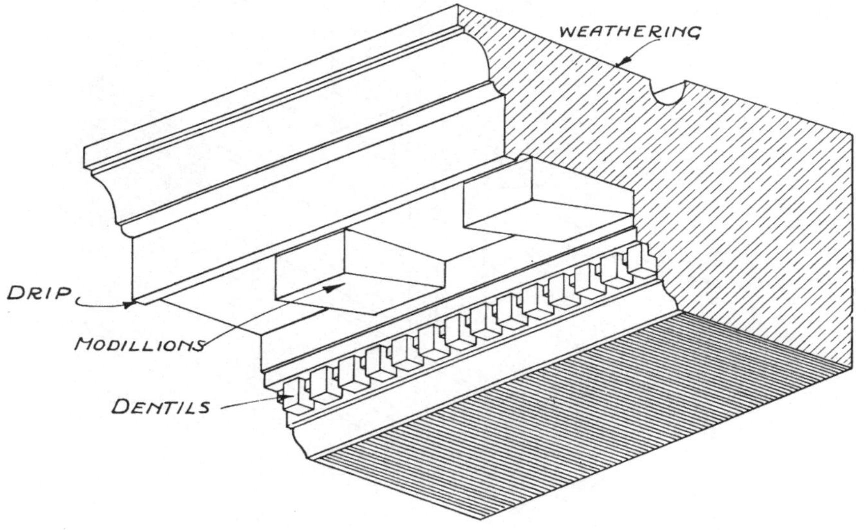

FIG. 180.—DETAIL OF A PIECE OF CORNICE IN ONE BED.

form of a balustrade. Fig. 174 shows a portion of a balustrade in conjunction with the *Tuscan Order*. It is composed of *plinth*, *balusters*, and *capping*. The top surface of the plinth should be worked inclined towards the inside face, whilst horizontal seatings should be provided for the balusters. The interval between the balusters should not be more than half their largest diameter, or less than one-third.

Baluster Die.—Balusters are divided into bays, each bay containing about nine balusters and two half-balusters, which are usually worked on the die-stone of the balustrade at each end of the bay, as in Fig. 174. This stone is called a *baluster die-stone*.

The **Capping**, being a form of coping, should be designed to fulfil that function. Slate dowels should be fitted in the top and bottom joints of the balusters to prevent their displacement.

Construction of Cornices.—With regard to construction, a cornice may be complete in one stone, as in Fig. 181; but large stone cornices are usually built up in a series of courses, each course forming a cantilever for the projecting course immediately above. Sufficient weight on the wall should be provided to more than counterbalance the oversailing portion. The stability of the cornice is often assisted by placing a course of stone immediately above the cornice. This course is called a *blocking course*, the top surface of which should be inclined, and each joint strengthened with *metal cramps*.

When cornices are built up in height with several stones, great care should be exercised in the placing of the beds. *Modillions*, unless supported by

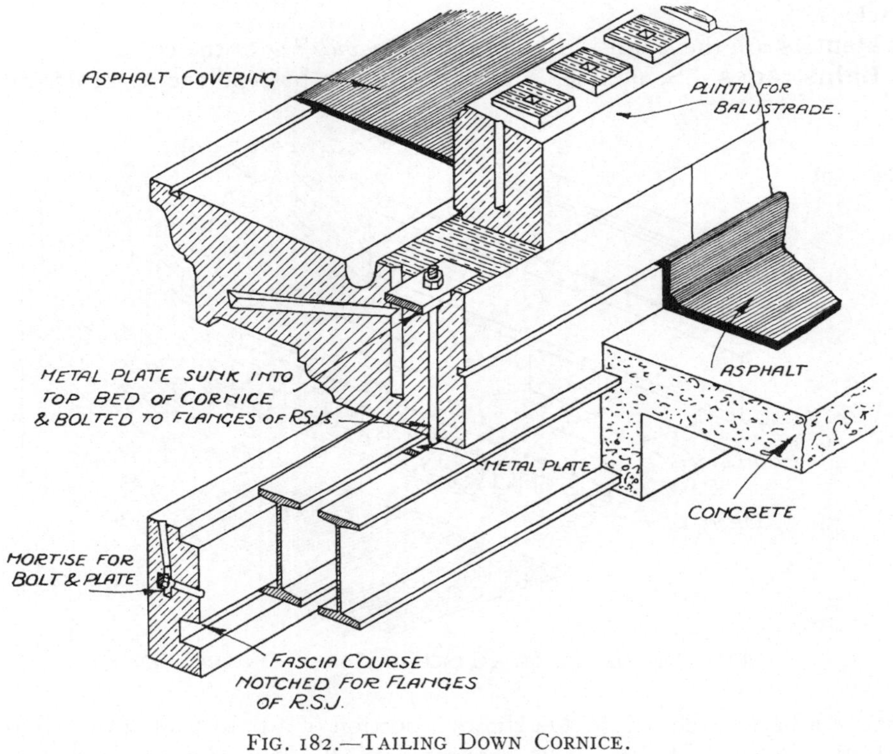

FIG. 182.—TAILING DOWN CORNICE.

steelwork, should be worked out of the solid stone. A bed-joint is often arranged immediately over the *modillions*, but this practice is not to be recommended, although a great saving of labour in working the modillions is gained by placing the bed-joint in this position. Should a settlement of the wall occur, there is a probability of the modillions being broken off. In fixing, the *mortar bed* should be kept clear of the top surface of the modillions, thus reducing the risk of breakage.

Tailing Down a Cornice.—Cornices with large projections often require tailing down to ensure stability. An effective method of accomplishing this is given in Fig. 182. A metal plate is sunk in the top bed surface, a *tailing-down bolt* is threaded on the plate and fastened under the top flanges of the

FIG. 181.—DETAILS OF CORNICE IN ONE BED, VICTORIA HOUSE.

Chas. W. Long, Esq., F.R.I.B.A., Architect.

two R.S.J.'s by means of a second plate, the nut and top portion of bolt being allowed to project into the bottom bed of the plinth course above.

In the modern construction of cornices, steelwork plays a very important part in the stability and the manner of jointing and bedding.

In steel-framed structures, *steel stanchions*, which carry the entire load of the building, are built in the walls. Horizontal R.S.J.'s are connected to these about floor level, to assist in supporting the floors, and are so arranged that the cornice can be secured to them.

Fig. 183 is a section through a built-up cornice, the bottom bed course forming a lintol for the window opening below. The R.S.J. is placed in position and fastened to the stanchion after the bottom course is fixed. The *dentil course* is notched for the flange of the R.S.J. and secured to it by metal plates and bolts placed in each joint, the space between the top bed of the stone and the top flange of the R.S.J. being filled with blue bricks and cement mortar.

Steel cantilevers rest on the top of the R.S.J. and are secured at one end, the other end projecting to within a few inches of the nosing of the cornice. The cantilevers occur at every joint of the top bed of the cornice, which is notched for them.

When the top course is complete, the space between the cantilevers and the stone should be filled with cement grout, so that the top course of stone is supported by the steelwork.

The introduction of steel cantilevers in the joints of cornices with large projections is to be encouraged as a means of minimising risks in the event of fire.

A sketch of the cornice showing the construction is given in Fig. 184, whilst photographs illustrating the same cornice in construction and being placed in position are given in Figs. 185, 186, 187. The finished cornice, showing the detail, is given in Fig. 188.

Supporting Modillion Courses.—Fig. 189 is a section through a large cornice showing the construction for the efficient support of heavy modillions. The modillions in this cornice are independent stones, the spaces between them being filled in with separate stones of the same height as the modillions, thus completing the course. The stones are bolted to a *plate girder* built in the wall. To support the top course of stone, and to hold up the overhanging part of the modillions, steel cantilevers are introduced. A R.S.J. is connected to the ends of the cantilevers, parallel to the face of the wall. *Lewis* or *rag bolts*, fixed in the top surface of the modillions, are fastened to the bottom flange of the R.S.J., whilst the top course, which forms the *corona* and *cymatium* of the cornice, is notched to fit the flanges. This course is supported by the R.S.J. and held in position by metal plates and bolts, and is back-jointed for the soffit stones, which are fixed on the top surface of the modillions as *filling-in stones*. When these stones are in position and grouted, the projecting top surface and horizontal bed for the plinth of the balustrade are made up with concrete. Before this concrete is placed in position, the surfaces of the stones that come in contact with the concrete should be coated with *lime mortar*, or painted with *bituminous paint*, to prevent any stains from the concrete from

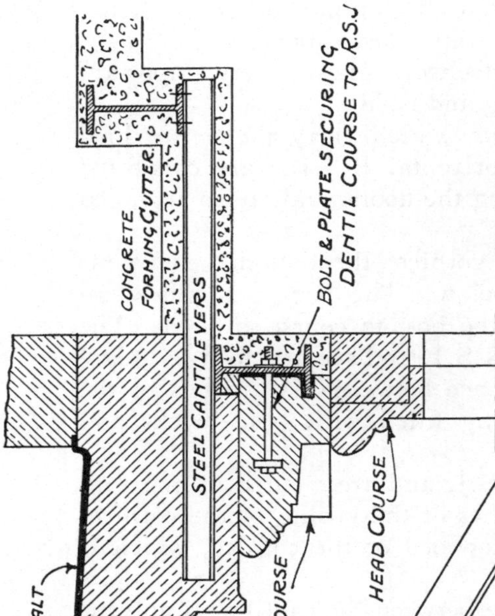

SECTION THROUGH CORNICE
FIG. 183

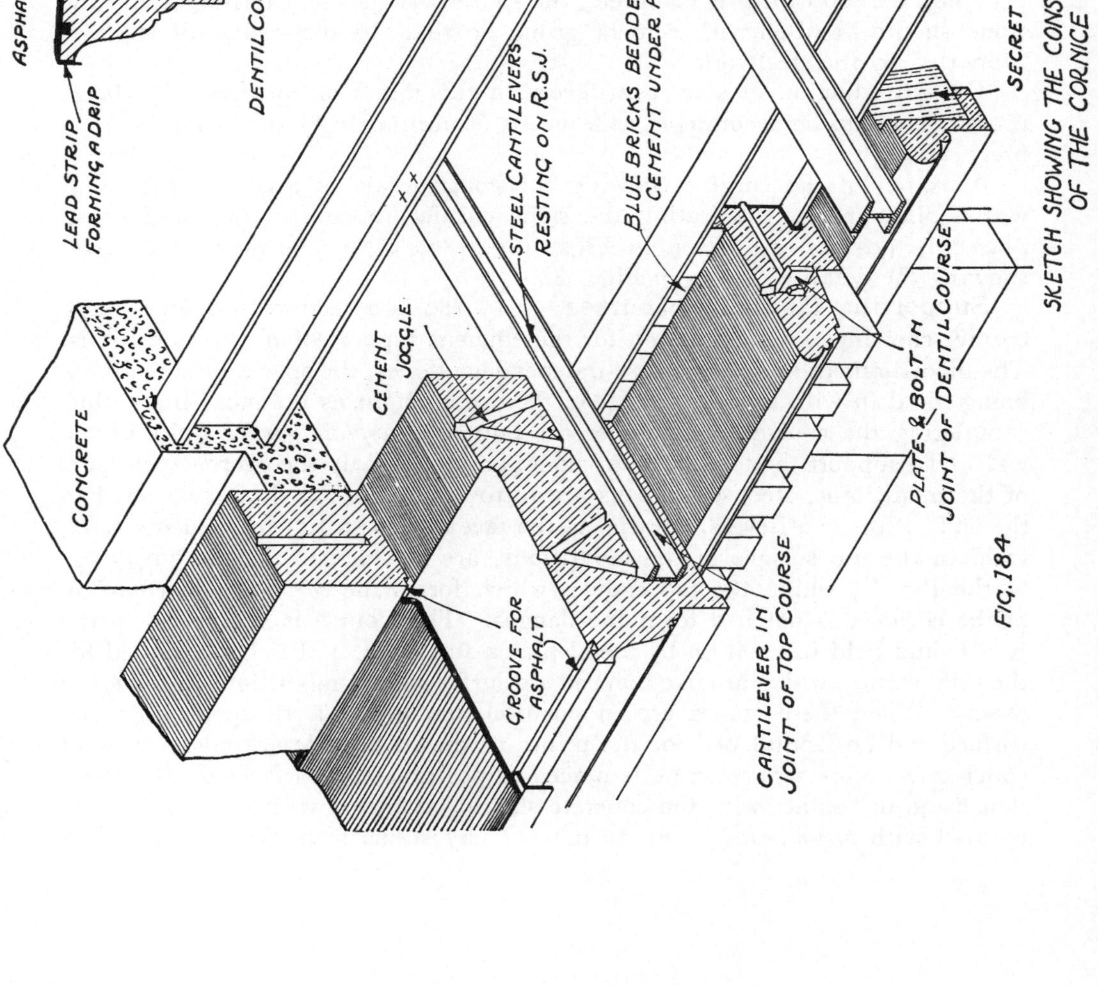

SKETCH SHOWING THE CONSTRUCTION OF THE CORNICE

FIG. 184

Fig. 185.—Bottom Bed of Cornice.

Fig. 186.—Construction of Dentil Course or Middle Bed of Cornice, Victoria House.

Chas. W. Long, Esq., F.R.I.B.A., Architect.

FIG. 187.—CONSTRUCTION OF TOP BED OF CORNICE, VICTORIA HOUSE.
Chas. W. Long, Esq., F.R.I.B.A., Architect.

FIG. 188.—CORNICE COMPLETED, SHOWING DETAIL, VICTORIA HOUSE.
Chas. W. Long, Esq., F.R.I.B.A., Architect.

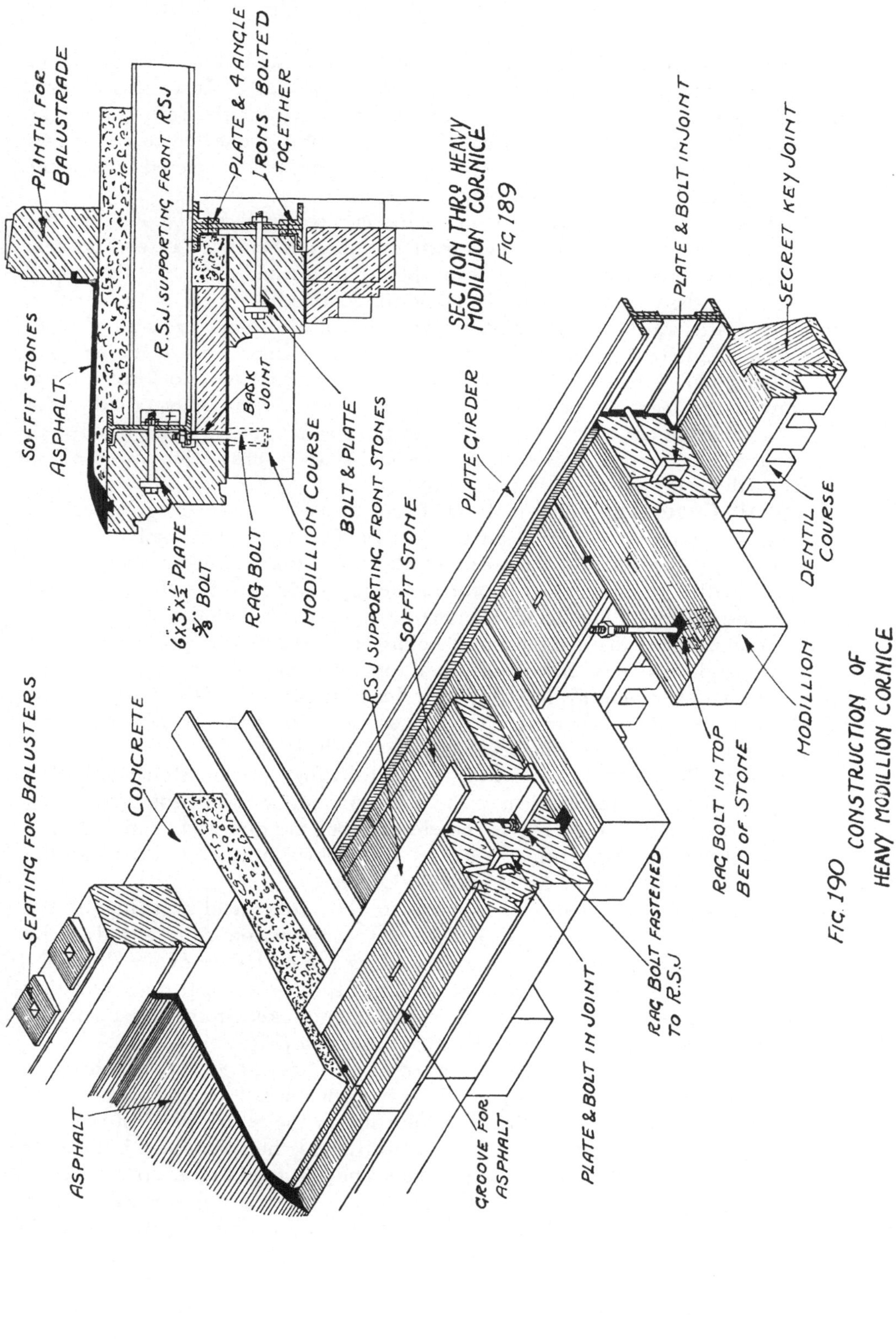

SECTION THRO. HEAVY MODILLION CORNICE
FIG 189

FIG. 190 CONSTRUCTION OF HEAVY MODILLION CORNICE

passing through the stones to the face of the work. The construction is clearly shown in Fig. 190.

Coverings to Cornices.—Exposed top surfaces should be weathered, or worked, inclined towards either the nosing or the wall face. If worked towards the wall face, a channel should be provided to carry off the rain water, the surface of the channel being graduated to outlets and connected to rain-water pipes. The whole of the top projecting surface should be covered with some impervious material, either *sheet-lead* or *asphalt*.

Sheet-lead Covering (Fig. 101).—This is the best form of covering, but it is expensive. It makes a nice contrast in colour when used in conjunction with light coloured stones. The lead should be turned into a "flashing" groove, cut in the vertical face of the stonework immediately above the cornice, the groove afterwards being pointed with cement and sand or caulked with lead. The lead covering should turn down over the front arris of the cornice, and finish about $\frac{1}{4}$ in. below the bottom edge of the nosing, to form a *drip*. To prevent the lead sheets from being lifted by the wind, dovetailed "dot holes" should be cut in the top surface of the cornice, these being filled with molten lead, leaving a raised hemispherical-shaped projection on the top surface of the lead, as shown in the figure.

Asphalt Covering.—This material forms an effective covering, especially when the top surface is partly concrete, as in Fig. 190. The asphalt is laid whilst hot over the surface, and secured along the front edge by being keyed into a dovetailed groove, which is cut in the top surface of the stone, parallel with the nosing line. The disadvantage of using asphalt is that it is impossible to form the required drip, the asphalt being finished along the top arris, thus causing unsightly stains down the front top members of the cornice. This defect may be overcome by introducing a lead strip, which is turned into the asphalt key groove and dressed over the nosing to form the drip, the asphalt being finished on the top of the lead, as shown in Fig. 184.

Masonry Doorway.—A working scale drawing for the stonework of a doorway is given, showing clearly the arrangement for the beds and joints, to ensure good bond. Fig. 191 is the elevation of the doorway, and Fig. 192 is the section on the centre line. Fig. 193 is the horizontal section through the door-jamb, and Fig. 194 is the plan of the cornice.

The joints are arranged with a view to the economical working of the stones, whilst keeping in mind that correct construction is essential. The sizes given for the stones in the *pediment head* are scaled from the drawing. In order that the correct sizes may be assured, a full-sized setting-out of this feature must be drafted, and the sizes of each stone taken from it. A setting-out of this feature is given in the geometrical section (Figs. 440-443).

Figs. 195, 196, 197, 198 are sketches of stones Nos. 2, 23, 34, 14, showing clearly their shape and the labour involved in each stone.

Fig. 199 shows the construction and jointing for the return piece of cornice.

Copings are the covering stones of a wall. As their function is to keep the wall protected from rain water, they should be designed to fulfil this purpose effectively. They should have a weathering, or inclined top surface, also a projection over the front and back surfaces of the wall, and be provided with

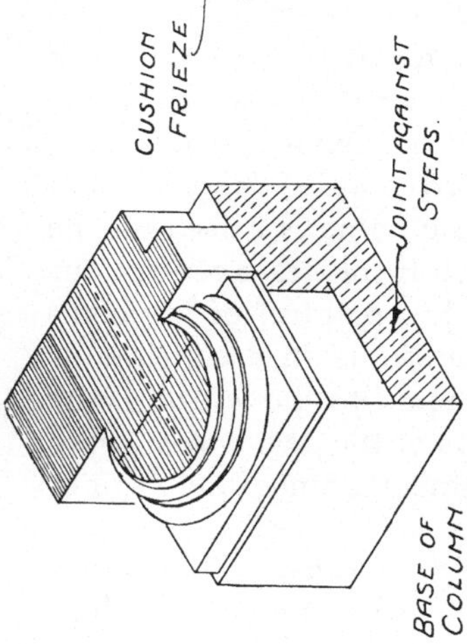

SKETCH OF STONE No 2

FIG. 195.

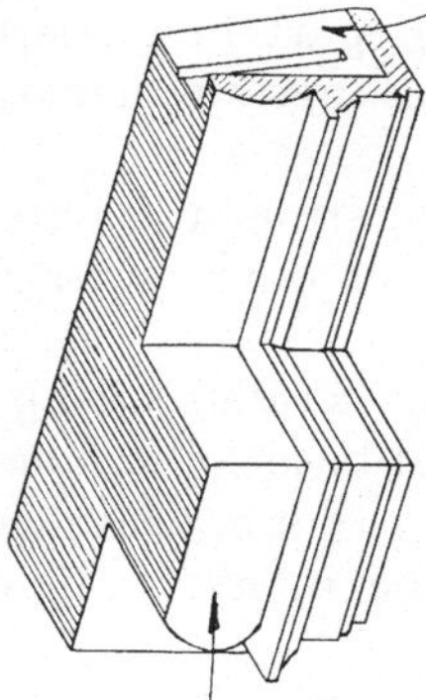

SKETCH OF STONE No 23

FIG. 196.

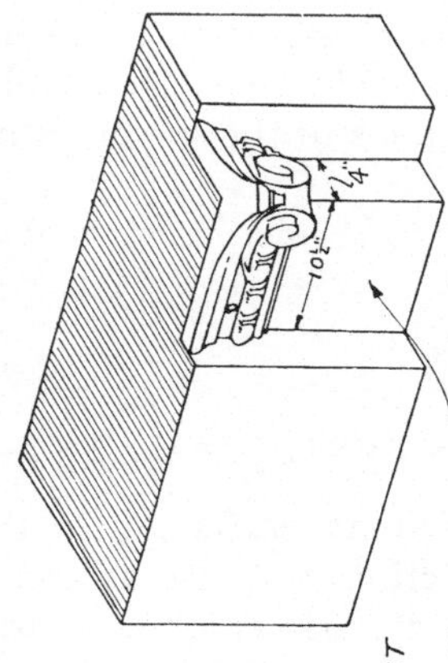

SKETCH OF STONE No 34

FIG. 197.

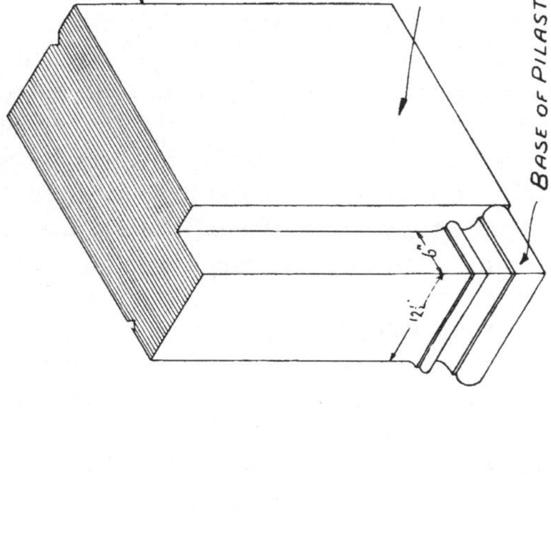

SKETCH OF STONE No 14

FIG. 198.

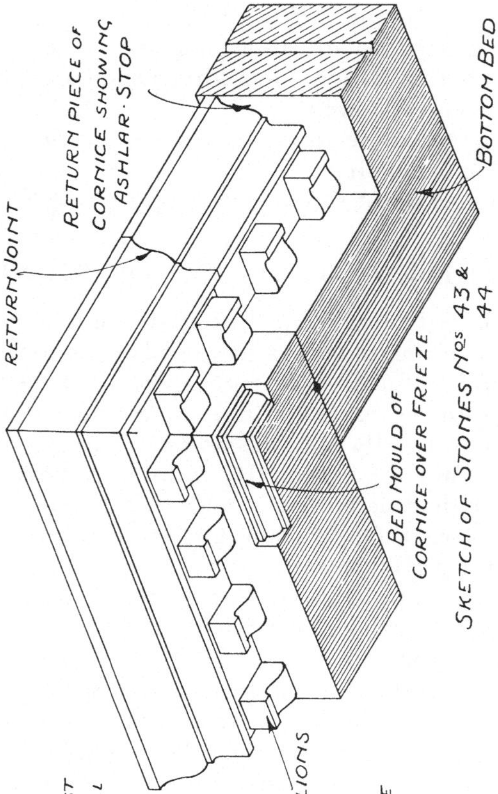

SKETCH OF STONES Nos 43 & 44

SHOWING RETURN END OF CORNICE

FIG. 199.

a "*drip*," formed by the working of a groove on the underside of both projections. The stones should be as long as possible, thus minimising the number of vertical joints, these being the chief cause of weakness. Although the coping sections differ considerably, their fundamental principles should be embodied, additional mouldings and curves being added to suit the surrounding architectural details.

The usual sections are known as *flat*, *weathered*, *saddle-back*, *segmental*, and *Gothic* copings.

A section of a segmental coping is given in Fig. 97.

THE CONSTRUCTION OF STONE DOMES AND VAULTS

There is a wide field of study for the student and young craftsman in the theory of domes and vaults, not only as to their construction, but also as architectural features. The author would advise the perusal of various architectural books on the subject. The stone dome is still in use in modern construction, either internally, as a decorative ceiling, where the stones would be comparatively thin, or externally, forming a finish or crowning feature to a tower. Although concrete construction is playing a very important part in modern dome construction, some fine examples of stone domical ceilings in modern buildings may be seen.

A dome is a spherical or similarly formed roof over a compartment, having a circular, square, or polygonal base. The simplest form of dome springs from a circular base.

A **Pendentive Dome** is one in which the concave surface is intersected by the vertical planes of the wall surfaces of a square or polygonal base equidistant from the axis of the sphere. This form of dome produces a very flat appearance externally, a small saucer-like portion only rising above the intersection of the walls.

Arches are usually formed penetrating the wall surfaces to give light into the chamber. The lower portions of the dome on the inside, which form spandrels between the arches, are called *pendentives*. Some of the stones in the courses of the *pendentives* are bonded with, and worked on, the arch-stones.

Dome and Pendentives.—Whereas in the *pendentive dome* the spherical surface is cut by the vertical wall planes, in this case the dome is raised on the crests of the arches bounding the *pendentives*. In such instances a continuous base for the dome must be formed by corbelling out the spandrel of the compartment in such a manner as to meet the base of the dome in a continuous bed-joint, the diameter at the base of the dome being equal to the width of the compartment, square to the wall faces, whilst the radius of the *pendentive* curve in section is equal to half the diagonal of the compartment immediately under the dome.

Squinch Arches were often formed instead of *pendentives*. These are arches thrown across the angles of a square or polygonal base, thereby developing a figure approximating more nearly to a circle. Small pendentives are then built at the angles if a circular base is required.

The stability of a dome elevated on *pendentives* and *arches* of wide span

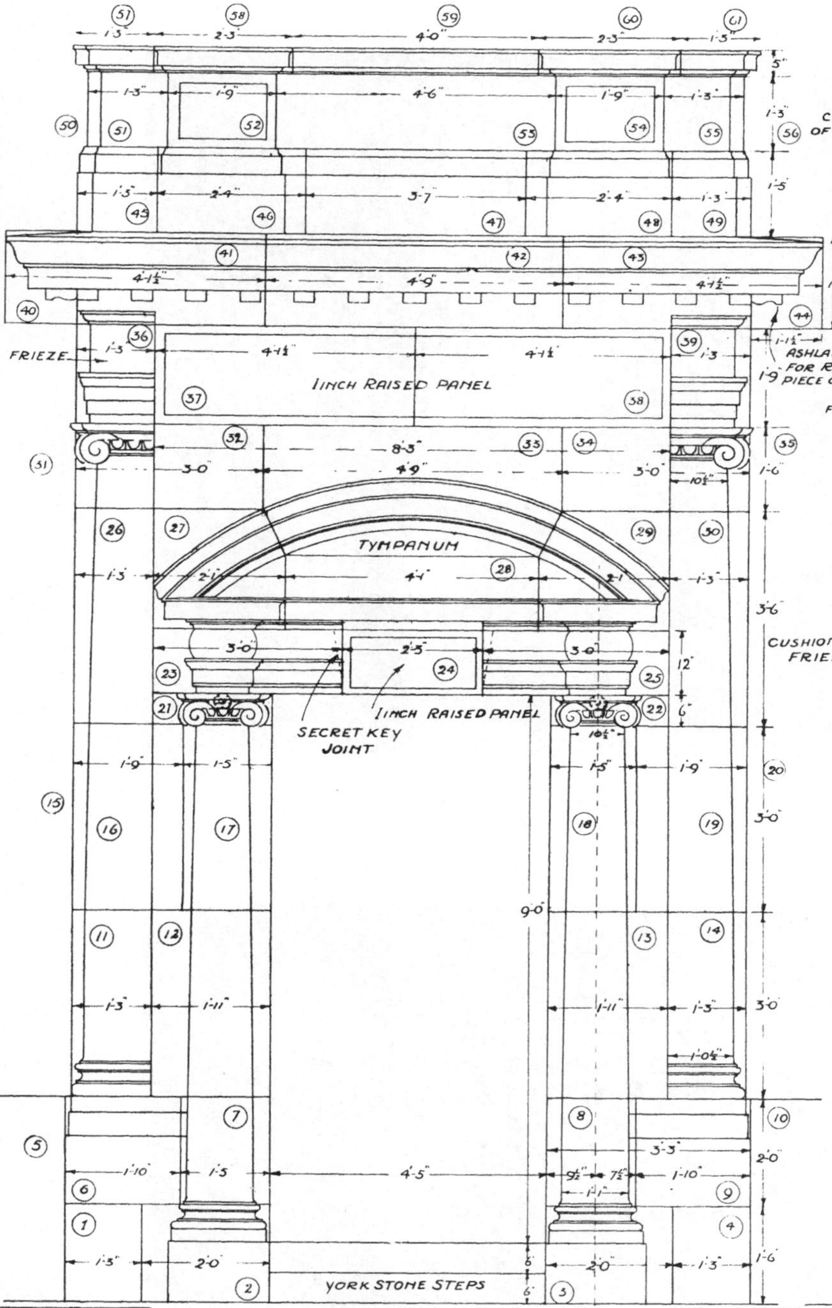

ELEVATION
FIG 191

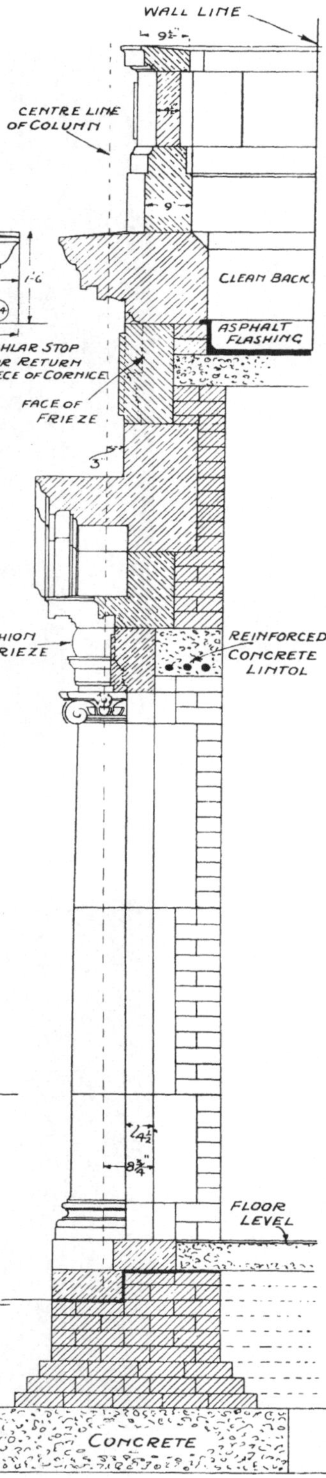

SECTION THROUGH CENTRE
FIG 192

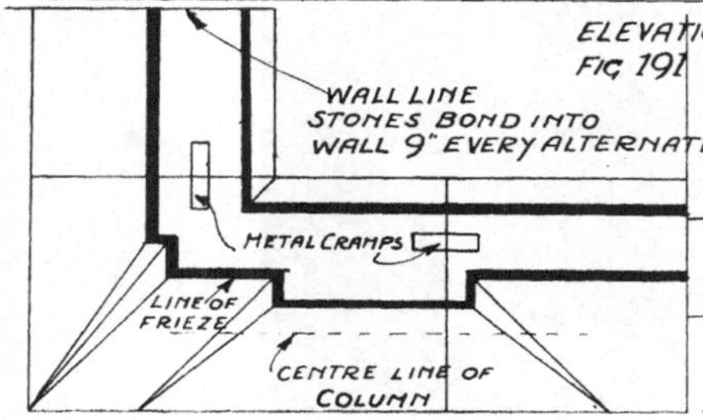

PLAN OF CORNICE
FIG 194

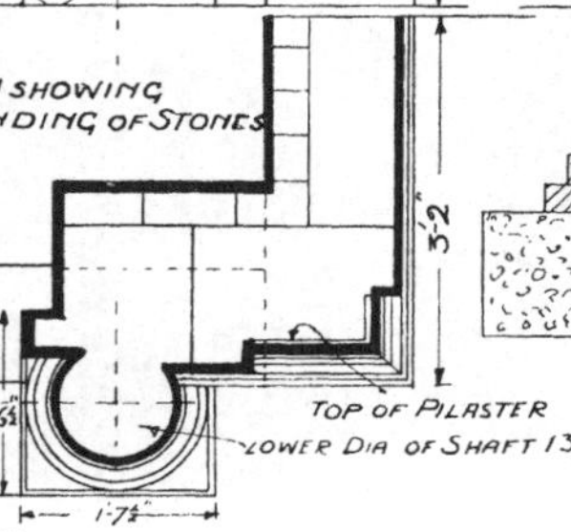

PLAN OF JAMB
FIG 193

SCALE

FIG. 200.—VIEW OF VAULTING AT STRAND ENTRANCE, BUSH HOUSE, LONDON.

Messrs Helmle & Corbett, Architects, New York.

FIG. 201.—BACK VIEW OF THE VAULT DURING CONSTRUCTION, BUSH HOUSE, LONDON.

Messrs Helmle & Corbett, Architects, New York.

can be effectively secured only by an extensive system of abutment applied to the superstructure. It is obvious that in a *pendentive dome* a matter of the first importance is the limitation of its weight. *Hemispherical domes* are generated by the rotation of an arch on its vertical axis, when the section, taken in every direction, will be exactly the same, and the thrust uniform all round. In *polygonal domes* each of its sides is a portion of a separate vault, and exerts a distinct and independent thrust. They lack that coherency and strength which are associated with *hemispherical domes*. The tendency of pointed arches, without surcharge, is to fail by lifting at the crown and falling in at the haunches. This holds good with regard to a dome rising from a polygonal base—that is, a *polygonal dome*—so that, when a dome of this kind is loaded with a lantern, the original tendency to lift is more than counteracted by the superadded weight of the lantern.

The point at which the greatest force is exerted must of course depend on the pitch of the dome. The higher the pitch, the less will be the thrust, and the lower the point of greatest weakness. To counteract the outward thrust of a *hemispherical dome*, *metal ties* in the form of *chains* or *channel sections* may be embedded in the masonry, near the base, to resist the lateral thrust.

The dome differs from an ordinary arch in that it may be left open or unfinished at the apex without prejudice to its power of supporting a lantern, the weight of which should not exceed that of the crowning portion of the dome, which is omitted. A stone dome need not be uniform in thickness throughout. It is best to reduce the thickness of the stone gradually towards the crown by lowering the *striking centre* for the extradosial curve. The dome should be built in horizontal courses of stone, the bed-joints of which should converge to the striking centre of the intradosial curve. The joints between any two stones of any course should be vertical planes passing through the vertical axis of the dome.

One of the difficulties in exterior dome construction is to prevent the rain water, which runs down over the upper surface, from percolating through the joints of the stones. Various methods have been suggested and tried, all with very little success. There does not appear to be any effective method of preventing this leakage through the joints, chiefly owing to the variation of temperature to which these stones, in such positions, are subjected.

A great amount of responsibility for the ultimate success in this respect rests with the craftsman in the erection of the dome. The stones in each course should be bedded solid on a mortar bed composed of 2 parts stone dust and 1 part Portland cement, or 1 part stone dust and 1 part white cement for stones such as Portland stone.

The bed-joint surfaces should be *pitted* with *punch marks*. Great care should be exercised by the fixer in placing the stone in position, so that there may be no need for the readjustment of the stone, thus disturbing the mortar bed. It is this readjustment to make the stones fit the adjacent stones that is often the cause of failure in this respect. Deep joggles should be cut in the vertical joint surfaces with a hammer and punch. When the course is complete and the joints pointed, each joint should be well filled with Portland cement grout, and, if necessary, metal cramps fitted across each joint. It is not advisable to use dowels in the bed-joints of the stones.

VAULTING

Barrel Vaulting.—This vaulting is really an elongated arch or semicircular roof, springing from one wall to another, covering the space between. When two of these vaults intersect, their line of intersection is called a groin line. We term these vaults *groin vaults*.

The simplest form of *groin vault* is the intersection of two vaults of equal span and section forming a straight-line groin in plan, both on the extrados and intrados, as the diagonals of a square. Fig. 200 is a view of the vaulting at the entrance of Bush House, London. The groins in this instance are covered by an enriched form of rib. Fig. 201 shows the vault in construction.

A **Welch Groin** is formed by the intersection of two cylindrical walls, one being of less height than the other. The groin line is then a curve in plan, as shown in Figs. 540-545.

Straight-line groins may be formed over rectangular compartments by making the vertical height of the large vault the same as that of the small vault. The large vault would then be semi-elliptical in section. The method of obtaining these curves and the normal joint planes is clearly demonstrated in the geometrical section.

Gothic Vaults.—The difference between *groin vaults* and *Gothic* or *ribbed vaults* is that, whereas the groins are the source of weakness in *groin vaulting*, in *Gothic vaulting* the *groins* are replaced by ribs, which occur at all intersections, the ribs carrying the weight of the vault, including the in-filling of the bays between the ribs. By the introduction of these ribs, compartments of various forms in plan may be conveniently roofed without producing the awkward intersections which would be necessary in *groin vaulting* covering the same polygonal compartments.

The most troublesome part of the *semicircular vault* is thus eliminated, thereby effectively reducing the thrust, which in a *semicircular vault* chiefly derives from that section. The setting out of a Gothic vault is given in Figs. 581-590.

STONE STAIRS

The design of stairs rarely enters into the province of the craftsman, definite instructions, in almost all cases, being provided by the architect. It is advisable, though, that the student should be conversant with the various principles underlying stair planning, types of stairs, construction, and the terms connected with stairs. *Suitability*, *directness*, and *light* are the chief principles which govern stair planning.

Suitability or *harmony* with the architectural features surrounding the staircase.

Directness for ease in ascending or descending from one floor to another. The maximum amount of light should be provided at the vital points of the stair, such as the beginning, the bends, and landings.

For an easy stair it is necessary to introduce landings at proper intervals in the height. A flight of steps should not contain more than twelve steps. In spiral or geometrical stairs, however, landings should be avoided, as they interrupt the continuity of the helical curves. The width of the treads should be equal throughout, this condition holding good when both *winders* and *fliers* are in combination in a flight, the width of the tread being measured on a *walking line*, 18 in. from the free ends of the steps.

The *rise* of the steps should be uniform throughout the flight. For ease in ascent or descent, the dimension of the steps, or the relation of the width of tread to height of riser, must be considered. For instance, it is found by experience that $5\frac{1}{2}$ in. rise and 12 in. tread produces an easy stair. The following rule is often adopted: the width of tread multiplied by the rise equals approximately 66 in.; or another formula may be adopted, viz., 2 risers + 1 tread = 23 in., the minimum width of the tread being 9 in.

Terms.

The **Staircase** is the chamber or compartment containing the stairs.

Flight of Steps.—A series of steps without a landing.

Landings (Fig. 210) are the flat resting-places between any flight of steps. They are named according to the space they occupy in the staircase, such as **Half-space Landing**, which extends across the width of the staircase, or **Quarter-space Landing**, occupying the same width or slightly more than the width of the stairs.

It is essential that all the steps in a flight should have the same rise and the same width of tread. The **Rise** is the vertical height between two treads. The **Tread** is the horizontal upper surface of the step, whilst the **Going** is the horizontal distance from the face of one riser to the face of the next riser, not including the nosing, which is the amount of projection of the tread over the front of the riser. The **Total Rise** is the height from one floor to another.

Fliers (Fig. 211) are steps of uniform and parallel width of tread, and are used in the straight portion of stairs.

Curtail Step (Fig. 211) is the bottom step of a stair, with one or both ends projecting beyond the ends of the steps in the flight, the projecting portion being in the form of a semicircle.

Bull-nosed Step is the same as a curtail step, but the projection is in the form of a quadrant.

Well-hole is the clear central space round which the stair turns, as in Fig. 211.

Winders are steps of triangular form on plan, and are used for changing the direction of a flight of steps, either through a right angle or round the curved portion of a staircase, forming a continuous flight. A sketch of a *winder* step is given in Fig. 202.

Methods of Supporting Steps.—The simplest method that can be adopted is building the steps in the wall at both ends. If a headway under the stairs is not required, the steps may be rectangular in section, as shown in Figs. 203 and 204.

Steps Built into the Wall at One End and Supported at the Other.—These steps are free at one end, and supporting a hand-rail or balustrade.

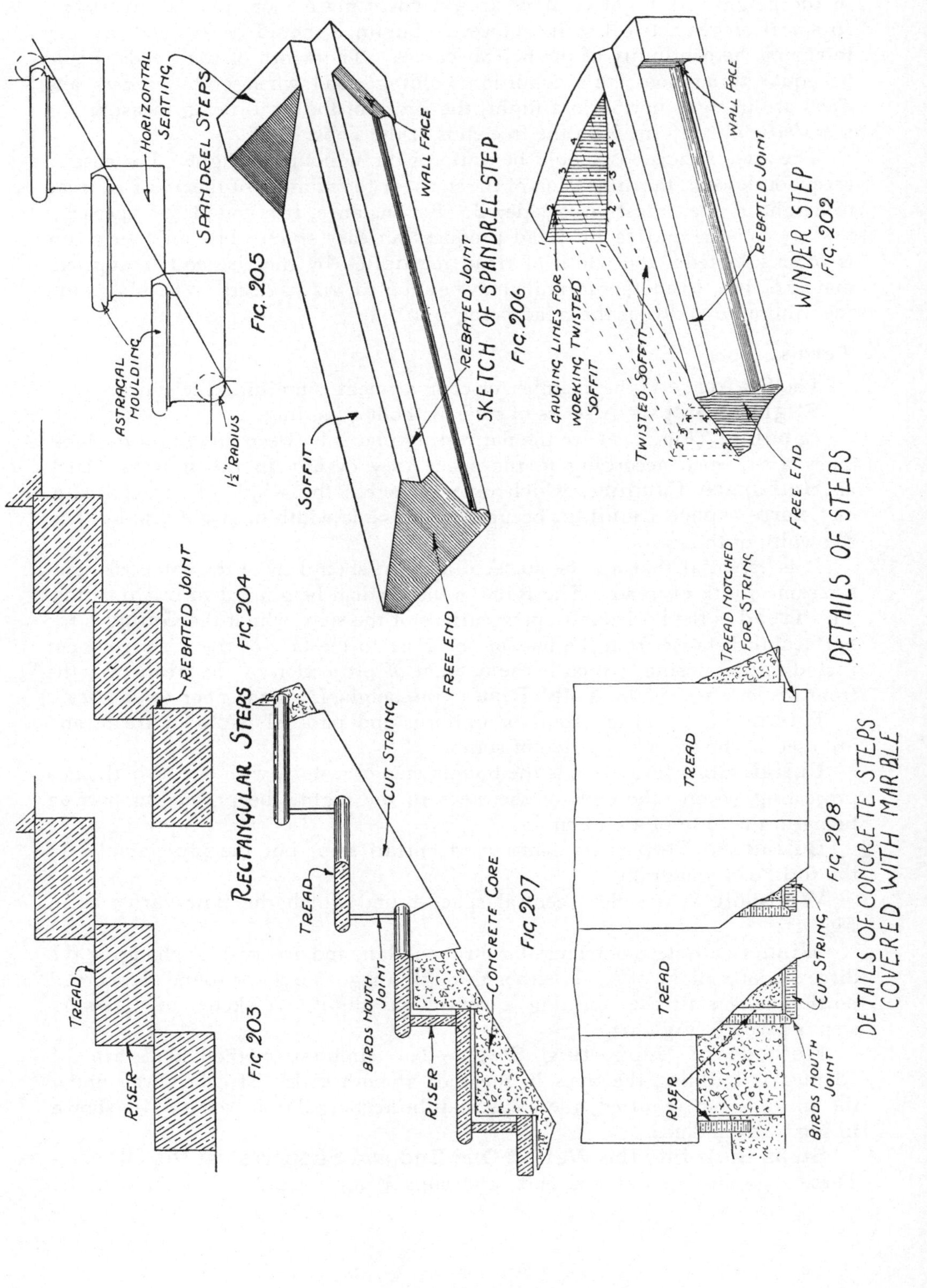
FIG 203
TREAD
RISER
RECTANGULAR STEPS
REBATED JOINT
FIG 204
ASTRAGAL MOULDING
HORIZONTAL SEATING
1½" RADIUS
SPANDREL STEPS
FIG. 205
SOFFIT
REBATED JOINT
WALL FACE
FREE END
SKETCH OF SPANDREL STEP
FIG. 206
GAUGING LINES FOR WORKING TWISTED SOFFIT
TWISTED SOFFIT
REBATED JOINT
WALL FACE
FREE END
WINDER STEP
FIG. 202
TREAD
CUT STRING
BIRDS MOUTH JOINT
RISER
CONCRETE CORE
FIG 207
TREAD NOTCHED FOR STRING
TREAD
FIG 208
CUT STRING
BIRDS MOUTH JOINT
RISER
DETAILS OF STEPS
DETAILS OF CONCRETE STEPS COVERED WITH MARBLE

The free end of these steps usually rests on a wall built underneath. It is in these walls that *rampant arches* occur. These steps should be rebated to fit the step below.

The London Building Act provides that all hanging or free-end steps should be supported at the free end by rolled steel joists, thereby minimising the risk of collapse in the event of fire.

Where head-room is desired, and also to obtain a good appearance from the flight below, the lower surface of the step is usually worked to the pitch of the stairs. This splayed surface should only be worked to the wall face, the remaining part of the surface forming a horizontal seating on the wall. The bearing surface of one step upon another should be formed so as to transmit the pressure from one step to another, and to prevent any lateral or sliding movement of the steps. The bottom step, or landing, should form a secure abutment for the other steps in the flight, as shown in Fig. 209.

Form of Steps.—The rectangular step is used chiefly for garden-terrace work, or in positions where the undersides of the steps are not exposed.

Spandrel Step (Fig. 205).—This form of step is used where head-room is required, the underside of the step being worked to form a raking soffit, and moulded if desired, the portion built into the wall being left rectangular. A sketch of a spandrel step is given in Fig. 206.

Concrete Steps covered with Marble.—Stone stairs are not in general use to-day, preference being given to concrete stairs built *in situ*, and covered effectively with marble.

The steps are built up with thin slabs of marble, thus forming the treads and the risers, as shown in Figs. 207 and 208.

In these circumstances it becomes necessary to cover the ends of the concrete steps with a marble *cut string* similar to that used in wood stairs. The construction and jointing for the step and string is shown, whilst the setting-out of this type of stair is given in the geometrical section.

FORMS OF STAIRS

Stairs with Return Flights or Dog-legged Stair.—Fig. 209 shows the sectional elevation through a dog-legged stair, Fig. 210 being the plan. The free ends of one flight are immediately over those of the flight below.

Open Newel Stairs (Fig. 211).—These stairs conform to the shape of the staircase in plan, whilst newels are employed at the turnings and an open space or well-hole is provided between the successive flights.

The Central-newel Turret Stair (Fig. 212) is now seldom used. A portion of the central newel is worked on the narrow end of the steps, the wide end being built into the wall. Whilst the construction of this type of stair is sound, a great difficulty arises when repair is needed.

Geometrical Stairs (Fig. 213).—The characteristic feature of these stairs is the continuity of the strings and the hand-rail from floor to floor, the formation of these stairs depending on geometrical principles. They may be arranged in either rectangular, polygonal, circular, or elliptical staircases in plan.

The setting-out of a geometrical stair is given in the geometrical section.

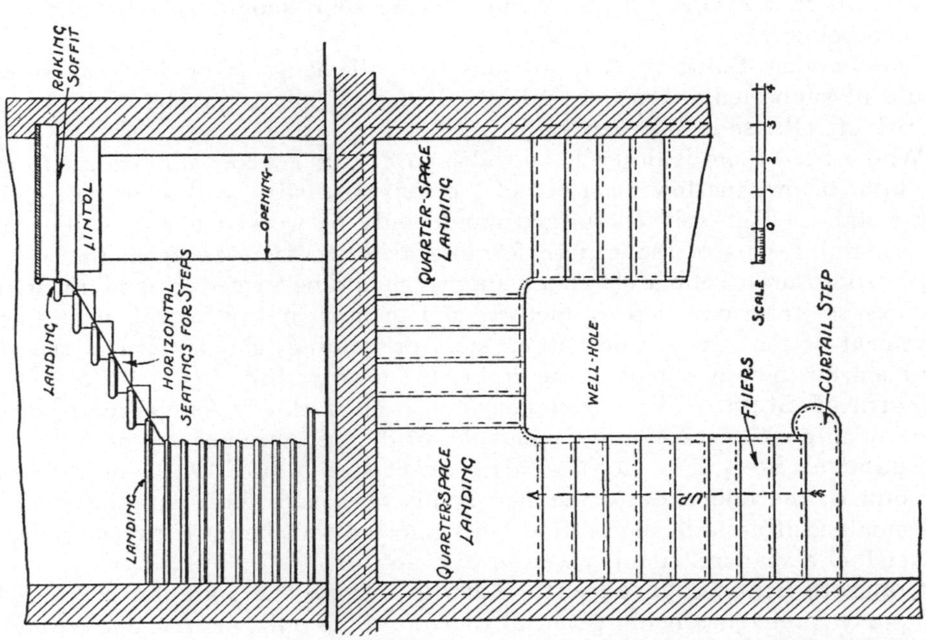

FIG. 211.—OPEN NEWEL STAIR.

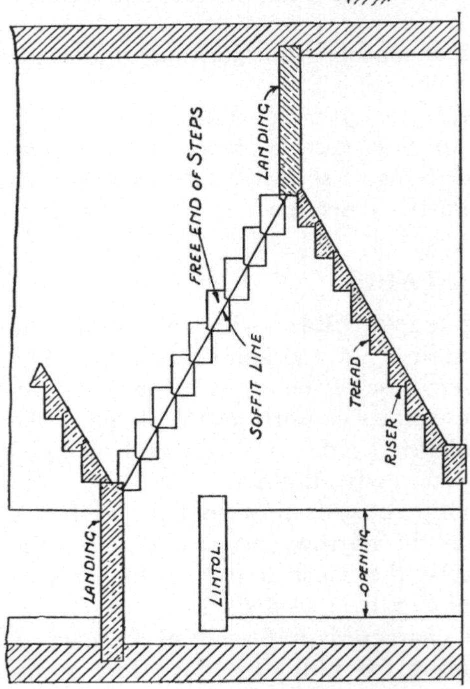

FIG. 209.—SECTIONAL ELEVATION.

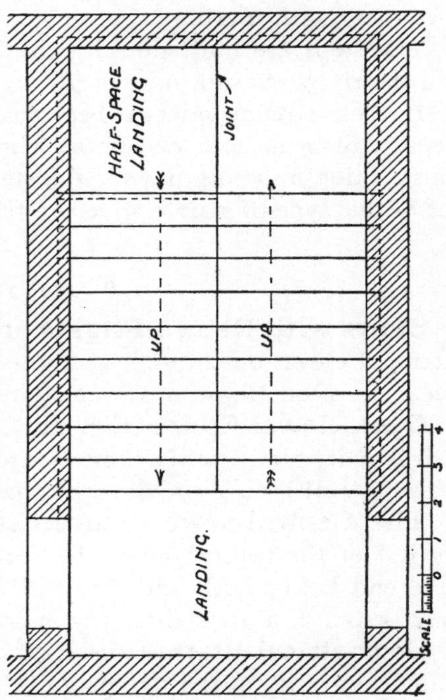

FIG. 210.—PLAN OF DOG-LEGGED STAIR.

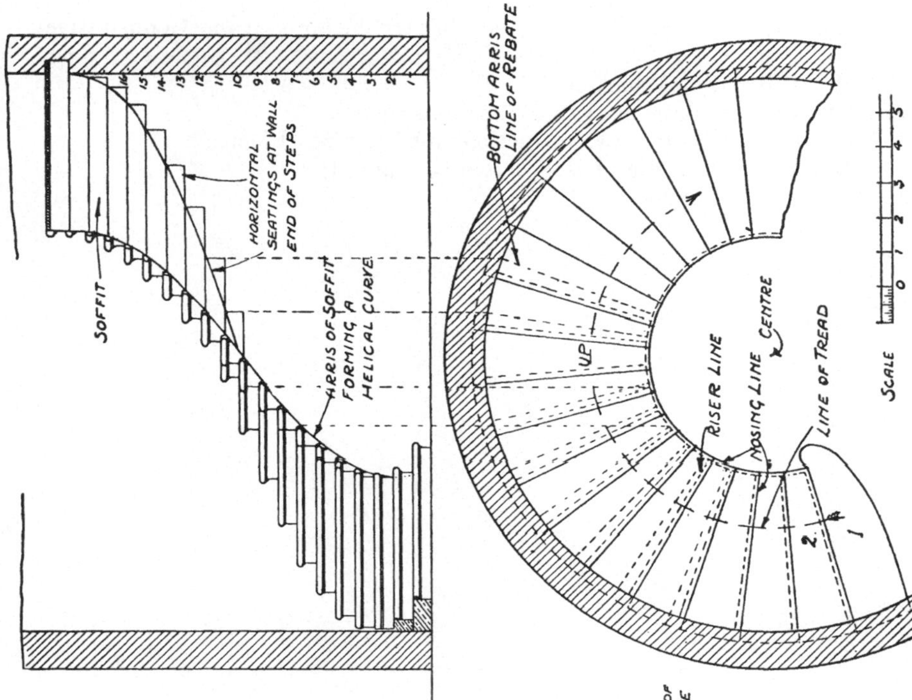

FIG. 213.—GEOMETRICAL STAIR.

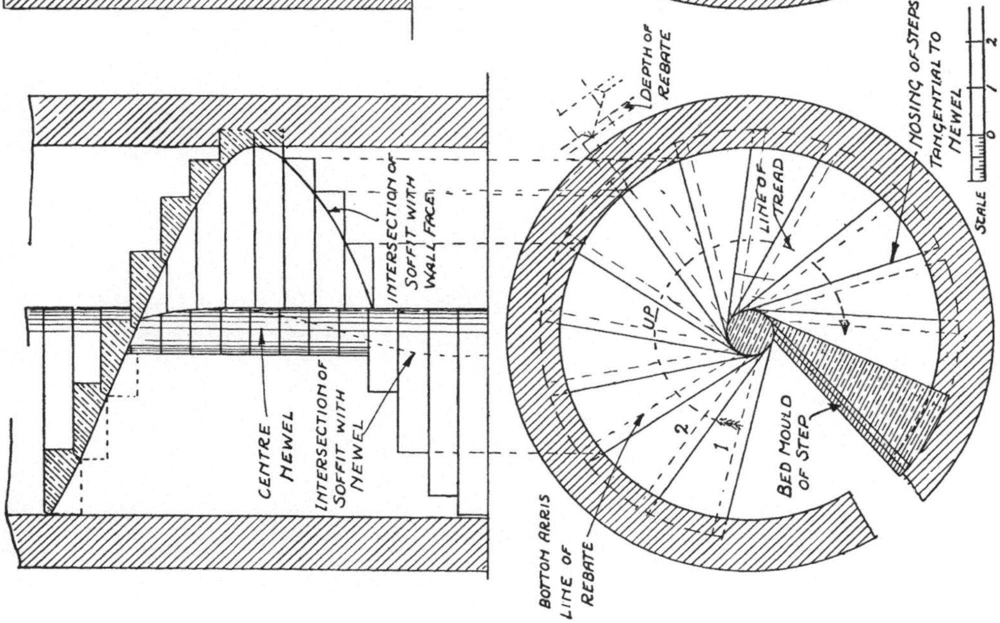

FIG. 212.—TURRET STAIR.

BOOKS ON THE SUBJECT

BOND, FRANCIS, " Gothic Architecture in England."

FLETCHER, Sir BANISTER, F.S.A., " A History of Architecture on the Comparative Method."

GREENHALGH, R., " Building Educator."

JAGGARD and DRURY, " Architectural Building Construction."

MITCHELL, GEORGE F., " Building Construction and Drawing (Elementary)."

MITCHELL, GEORGE F., " Building Construction (Advanced)."

SHARPE, E., " Decorated Windows."

CHAPTER IV

HOISTING AND SETTING OF STONEWORK

Scaffolding—Hoisting—Appliances for Lifting—Centering—Mortars—Limes and Cements.

Scaffolding.—In order that walls and structures may be built, it is necessary to provide scaffolding, or stagings, to enable the craftsman to place the stone on the wall at the heights required. There are various forms of scaffolding, each constructed to meet the requirements of the particular class of work to be erected.

Stone-faced buildings generally require the erection of more substantial scaffolding than brick walls, because of the heavy individual stones that are usually necessary in this class of work. Scaffolds for the erection of brick walls are partly supported by the wall itself.

For brickwork the platform upon which the craftsmen work is laid on *putlogs*, which span the distance between the wall and the outside upright poles, which are called *standards*. One end of the *putlog* is placed in a recess in the face of the wall, which is called a *putlog hole*, whilst the other end is allowed to rest on a *ledger pole*, which is fastened in a horizontal position to the *upright standards*.

Masons' Independent Scaffold.—Generally speaking, it is not possible to allow recesses for *putlogs* in the faces of the walls faced with wrought stonework, so that it becomes necessary to erect scaffolding independent of the wall. This type of scaffolding is called a "*Masons' Independent Scaffold*," and is usually similar to that shown in Fig. 214.

Steel Tubular Scaffolding, which is now considerably used for large building work, requires a modification of the above arrangement, but the general principles are the same. Fig. 215 shows this type of scaffolding.

Suspended Stages.—This type of staging is suitable for some classes of work, but its scope is rather limited. It may be used with advantage where large ashlar surfaces and slight projecting courses form a casing to steel-framed structures, but its use usually entails readjustment when large projections occur, which often means reversion to some form of rigid scaffolding. Figs. 216 and 217 illustrate work proceeding on a suspended staging.

Cantilevered Scaffolds.—This type of scaffolding is frequently used where the erection of walls commences at various levels, or when the lower part of a building is required to be free, in order to allow for the completion of the lower floors whilst the erection of the upper floors is still proceeding.

Baulk timbers or *rolled steel joists* are used as cantilevers, projecting through the window opening to the full width of the required staging. These are counterbalanced by being securely fastened between the floors on the inside of the building, the other ends supporting the staging or scaffolding.

In some districts the walls are erected almost entirely from the inside of the building, this being termed *overhand fixing*. The fixer places the stones in position by leaning over the wall from an inside staging, which is erected on trestles from floor to floor. This method cannot be recommended either for economy or efficiency. Almost without exception it is necessary to resort to *cantilevered scaffolds* for the completion of some portion of the work. The above statement is based upon the author's own experience of this type of work.

Hoisting.—Methods of hoisting or lifting stones vary according to the class of work and type of building. For the erection of large buildings, mechanical devices or cranes are in general use for raising the materials. These cranes are placed on "*tower gantrys*," the level of the platform upon which the crane is erected being higher than the roof of the building.

On steel-framed buildings cranes are often erected on the top of the steelwork or connected to the front of the steelwork for hoisting the stonework and other building materials. The various forms of cranes are too numerous for details of them to be given in this section. Where cranes are in use, the stones are usually lifted direct out of the vehicle in the roadway, and either stacked on a staging specially provided for them or placed direct on the wall in the position required. The latter method is not to be recommended, but *hand jib cranes* or other lifting apparatus, such as *chain tackles* fastened to the steelwork above, should be provided for assisting in placing the stones in their correct position on the wall.

Where *derrick cranes* are not being used, it is necessary to hoist the stones either by some other mechanical or hand device.

As a means of storage, and to assist in the erection of the stonework, for buildings in main thoroughfares, a staging known as a *gantry* is usually built of baulk timber, over the footway, the front scaffolding being erected from this. The stones are then lifted out of the vehicle in the roadway and stacked on the gantry by means of a *running tackle*.

Masons' Hoist.—Provision must be made for the lifting of heavy stones where cranes are not in use. On buildings other than *steel-framed structures*, an arrangement of poles similar to that shown in Fig. 218 is erected sufficiently above the scaffold staging to permit the stones being lifted and handled on the wall. The horizontal *ledger poles* are carried up at least two stages above the platform upon which the workmen are engaged. *Transome poles* are then placed to span from the inside to the outside ledgers, whilst on top of these, and parallel with the face line of the building, is placed a scaffold pole termed a "*head-tree*," which forms a beam to which the lifting tackle is secured. This hoist is lifted scaffold by scaffold as the work proceeds. The steelwork is utilised as a means of securing the lifting apparatus on *steel-framed structures*. This is done by forming a series of cantilevers, to the underside of which are fastened rolled steel joists, parallel to the face line of the building. *Chain tackles* provided with runners are fitted over the bottom flanges of the rolled

Fig. 214.—Erection of Masons' Independent Scaffold.

Fig. 215.—Steel Tubular Scaffolding.

FIGS. 216 AND 217.—SUSPENDED STAGINGS.

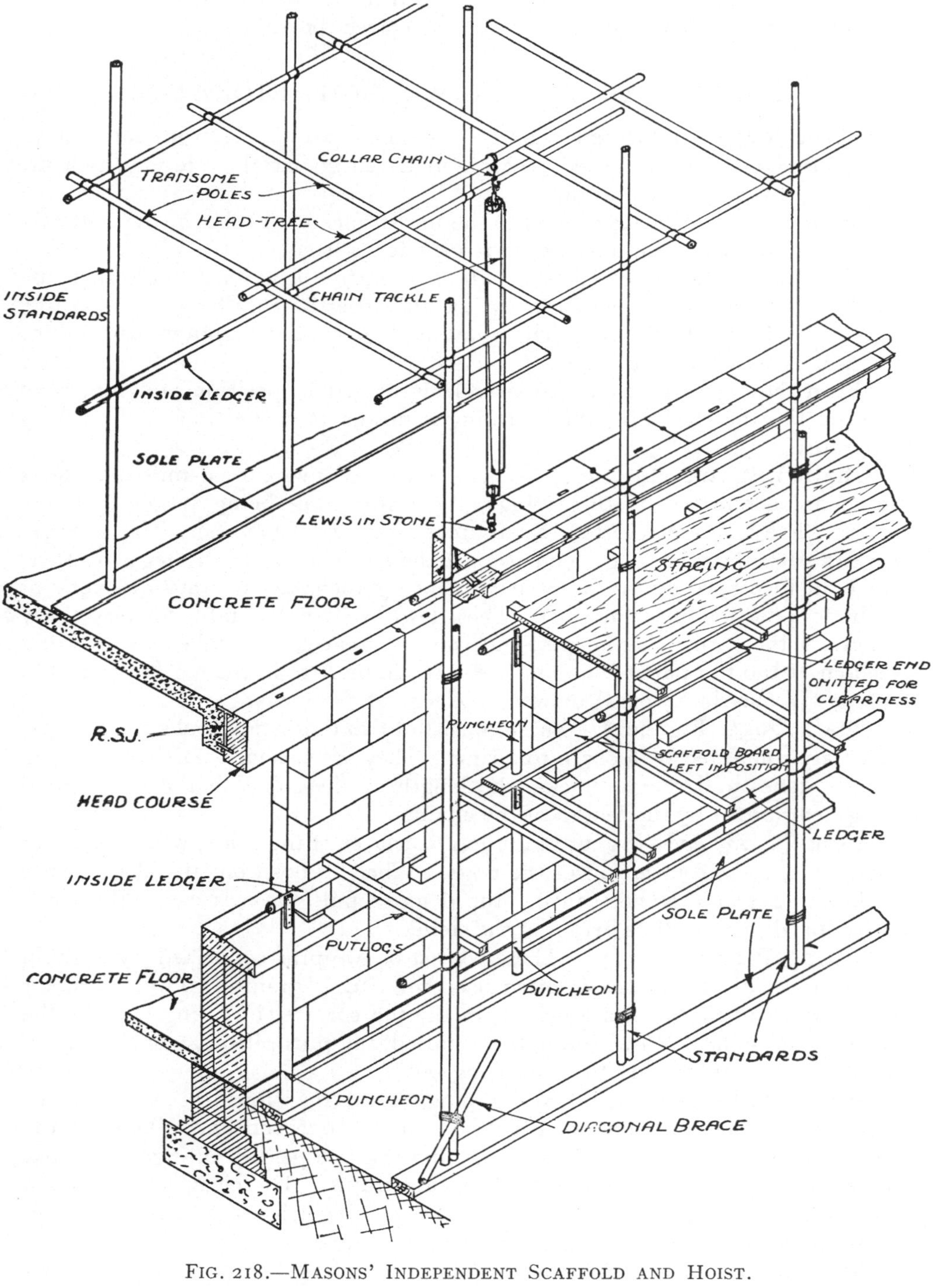

FIG. 218.—MASONS' INDEPENDENT SCAFFOLD AND HOIST.

steel joist, thus permitting the stone, when lifted, to be moved longitudinally, or parallel with the face of the wall, to its correct position.

APPLIANCES FOR LIFTING WROUGHT STONEWORK

Three-legged Lewises (Fig. 219).—These are made from mild steel, and consist of two dovetailed pieces and a centre parallel piece, which are connected by a shackle and pin.

An undercut mortise is formed in the top surface of the stone, the undercut portion being made at a less angle than the lewis legs.

The mortise should be such that the two outside legs may be inserted into the mortise together, thus allowing for the insertion of the centre parallel leg. The lewis should fit fairly tight, and grip at the bottom part of the mortise, as shown in Fig. 220.

When the legs are fitted, the shackle is placed in position and connected by inserting the steel pin, thus forming a means of connection to the lifting apparatus.

Chain Lewises (Fig. 221).—These consist of two curved mild steel legs, which are connected by three rings, the centre ring being larger than the others. The legs are inserted in an almost square-cut mortise. The arrangement is such that the rings tend to pull the top portion of the legs together, which in turn spreads the lower part, causing them to bite into the stone at the bottom portion of the mortise. These lewises are not suitable for very heavy stones, as they are liable to pull out, unless great care is taken in allowing freedom at the top part of the mortise. If the mortise is cut too large, a thin *iron wedge* may be inserted between the legs of the lewis.

Chain Dogs (Fig. 222).—These consist of two mild steel hooks called *dogs*, which are connected by a chain to a ring. They are specially suited for lifting jointed stones and rough blocks, the length of the stone that can be lifted being determined by the length of the chain.

A small "*dog hole*" is cut in the joint surfaces of the stone, with a hammer and punch, for the insertion of the dogs. When in position, the chain forms a triangle over the stone, causing the dogs to pull towards each other in a horizontal direction, thus gripping the stone.

Lifting Pins (Fig. 223).—These consist of two pins connected by a chain to a ring. This apparatus is often used for lifting granite. Two holes, for the insertion of the pins, are drilled in the top surface, inclining towards the centre of the stone. When the pull is applied, the pins bite against the side of the holes, thus preventing them drawing out.

Sling Chains are used for wrapping round the stones as a means for lifting. The chain is fitted with a large ring at one end and a hook at the other, and when placed round the stone, the chain is passed through the hook, whilst the ring is attached to the lifting apparatus. The clean arrises of the stone must be protected from damage by placing pieces of wood in such positions as will prevent the chain injuring the stone.

Skip Chains consist of four chains about 3 ft. in length, grouped together by a large ring and terminating with hooks. They are used for lifting *skips* or *trays*.

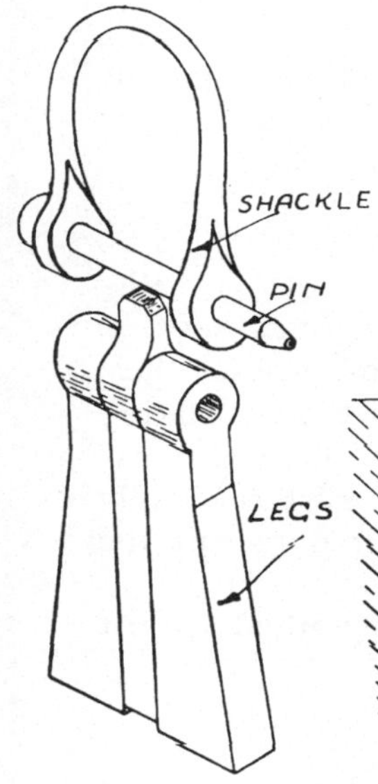

FIG. 219. THREE-LEGGED LEWIS.

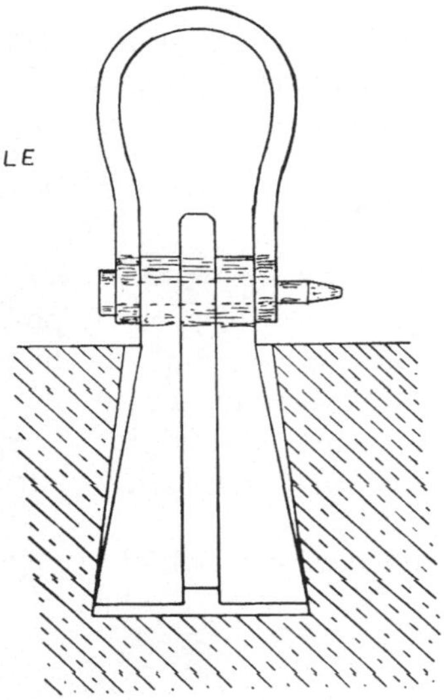

FIG. 220. LEWIS IN POSITION.

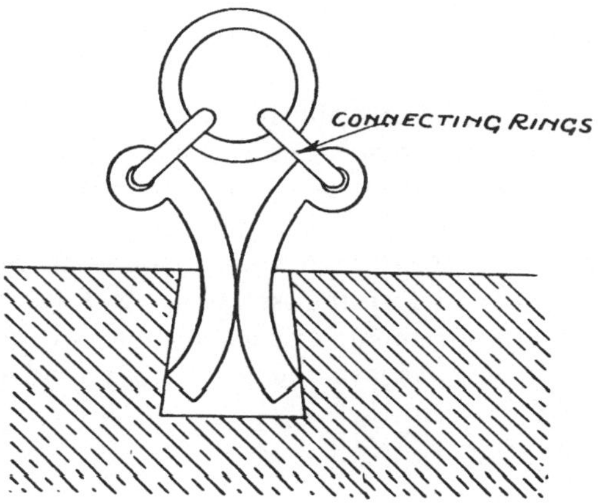

FIG. 221.—CHAIN LEWIS.

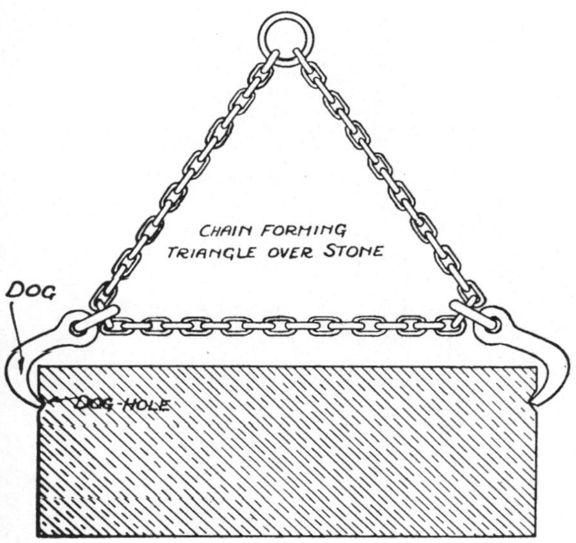

FIG. 222.—CHAIN DOGS.

FIG. 223.—LIFTING PINS.

Skips are timber boxes specially adapted for hoisting numerous stones in one lift.

Rope Tackle comprises a combination of two and three sheaf blocks, through which is wreathed either a *hemp* or *steel rope*.

Endless Chain Tackle.—There are several patterns of lifting apparatus which come under this heading, being chiefly patents, such as *Morris* or *Western Patterns*. The weight to be lifted determines the size of the apparatus.

CENTERING

The erection of suitable centering for the support of *arches*, *fascia courses*, *domes*, *vaults*, etc., during their construction is of great importance for their ultimate success. The fixer should be acquainted with the methods of fixing centering, and be able to state what is required for any given piece of work. The centering should be sufficiently strong and rigid to resist effectively any deforming stresses, its form and construction being varied to suit the requirements of each individual case.

Owing to the increasing load of the *voussoirs* which are gradually built

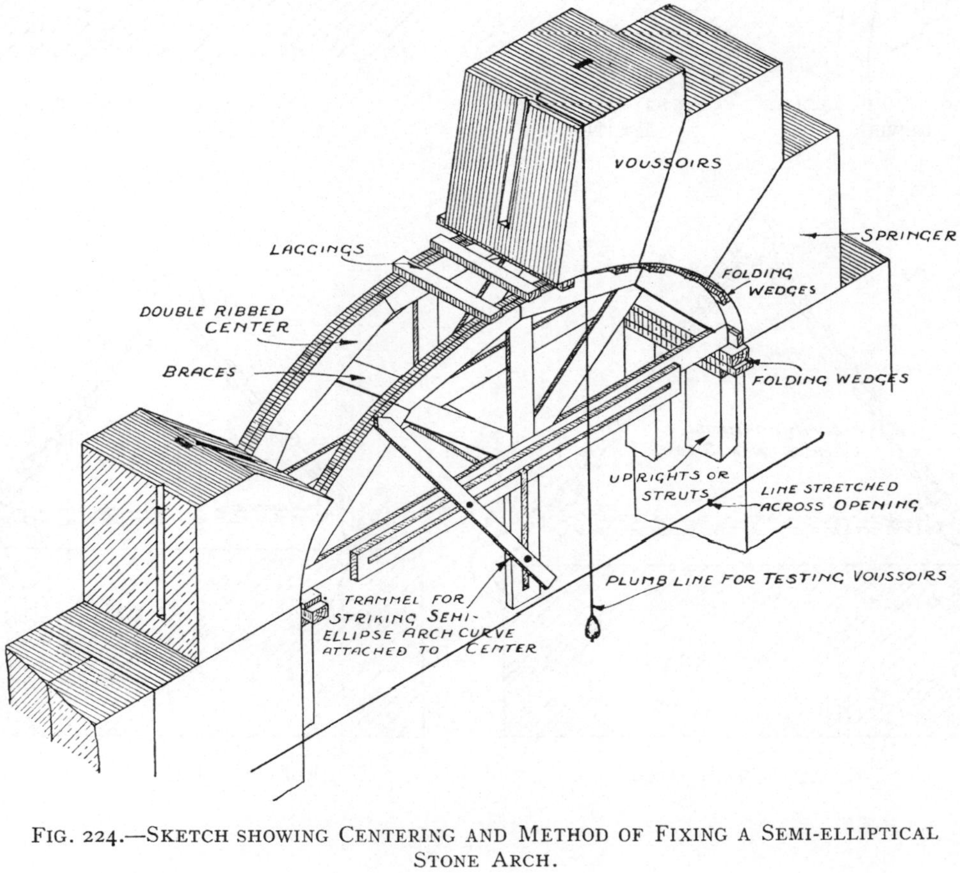

FIG. 224.—SKETCH SHOWING CENTERING AND METHOD OF FIXING A SEMI-ELLIPTICAL STONE ARCH.

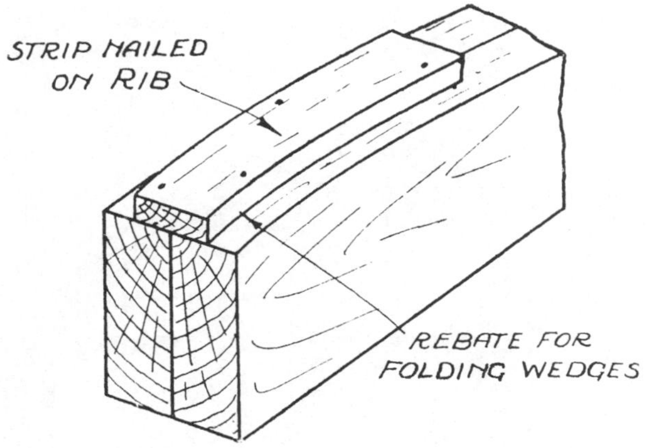

FIG. 226.—SECTION THROUGH CENTERING FOR THE SUPPORT OF DOME.
(See Fig. 228.)

FIG. 225.—CENTERING FOR SUPPORTING A LARGE ARCH DURING CONSTRUCTION.

FIG. 227.—CENTERING FOR LARGE ARCH IN FRONT OF THE DOME CEILING, BUSH HOUSE, LONDON.

Messrs Helmle & Corbett, Architects, New York.

FIG. 228.—CENTERING FOR THE SUPPORT OF THE STONEWORK OF CEILING, BUSH HOUSE, LONDON.

Messrs Helmle & Corbett, Architects, New York.

FIG. 229.—CENTERING FOR SUPPORTING A FASCIA COURSE, VICTORIA HOUSE.
Chas. W. Long, Esq., F.R.I.B.A., Architect.

FIG. 230.—FIXING ON CENTERING, VICTORIA HOUSE.
Chas. W. Long, Esq., F.R.I.B.A., Architect.

up from the *springers*, there is a tendency for the pressure to raise the crown, and to depress the haunches of the centering, therefore they must be *braced* to resist this tendency effectively. They should also be rigid when fixed, and provision should be made for the *striking*, thus minimising the risk of damage to the work when the centering is removed.

Centers for Arches.—Fig. 224 is a sketch illustrating the method of fixing a stone arch. The center should rest upon folding wedges, placed on the top of the upright, so that when the arch is complete, and the joints grouted, the wedges may be loosened, thus allowing the center to drop clear of the arch soffit. This applies chiefly to centers fixed to the correct outline of the arch curve, as in the case of centering for *Gothic arches*. Fig. 225 shows a center in position ready for the fixing of a large arch.

Cross pieces or *laggings* should be nailed on the centering at intervals to suit the size of the individual stones, preferably two laggings for each *voussoir*. The curve outline of the *centering*, including the *laggings*, should be made to 1 in. less radius than the *arch line*, thus allowing room for *folding wedges* between the top surface of the *laggings* and the soffit of the arch. A *radius rod* or *trammel* should be attached to the *centering* for testing the accuracy of the curve of the arch.

In fixing the arch, each *voussoir* should be tested for "*plumb*" by using a *plumb line*, as shown.

Centering for Gothic Arches.—These are best made and fixed to follow the correct curve line of the arch, so that the stones may be laid direct on to the surface of the *laggings*, which should be placed only a few inches apart.

Centering for Groined Ceilings, Domes, Vaulting, etc.—Centering for these classes of work should be built up in a series of ribs and not boarded over, so that the fixer is able to get at all the surfaces of the individual stones. The centering should be built to give the correct curve for the masonry, so that the stones would require only slight adjustment. A good method is to form the ribs in section, as shown in Fig. 226, so that whilst the centre portion of the rib gives the correct outline of the vault, space is allowed at the sides for wedging. This method minimises the risk of *flushing* the stones during their adjustment. Photographs showing the centering for supporting the *domed ceiling* at the entrance of Bush House, London, are given in Figs. 227 and 228.

Centering for Fascia Courses and Architraves.—The successful fixing of these courses requires great care and skill on the part of the craftsman. Rigid timber head-trees or supports should be erected as a temporary staging, both to rest the stones upon and to assist in placing them into their correct position. The sizes of the timber and the number of *struts* or *uprights* depends upon the width of the opening to be spanned and the weight to be carried. They should be made sufficiently strong to carry the weight of the stones without sagging.

The top surface of the head-tree should be fixed about 1½ in. below the soffit line of the course, and folding wedges provided for the adjustment of the stone, as shown in Figs. 229 and 230. The centering for *arches*, *vaulting*, *domes*, etc., should be struck after sufficient time has been allowed for the abutments to be built, but before too much weight has been concentrated above the work.

The centering for *fascia courses*, etc., should remain intact until the erection of the building has advanced sufficiently to warrant their *striking* or removal. The wedges should be slackened slowly, thus preventing any sudden weight coming on to the stones. The bedding material for the course above should be kept clear of the top surface of the *fascia stones*, so that the weight of the work above is transmitted back to the top surface of the steelwork.

Mortars used for masonry vary according to the kind of stone and structural requirements. The careful mixing of the ingredients of mortar is an important factor in its ultimate success.

Owing to the fact that the strength of *rubble walls* depends to a very large extent upon the setting properties of the mortar, and also because of the large quantity required to fill all the interstices between the rough stones, it is advisable to use a binding material, having sufficient setting qualities, to enable the work to withstand the tendency to crush, due to concentrated loads on the wall. The admixture of four parts sand to one part Portland cement, or three parts sand to one part hydraulic lime is suitable for this class of work.

Wrought stonework does not depend to any great extent upon the *bedding material* for its strength. The arrangement of the *bonding* and the large true surfaces obviates the necessity for a hard setting mortar.

Masons' putty, which is a combination of three parts *stone dust* and one part *lime putty*, is chiefly used for the bedding of such stones, the joints being grouted with either neat *Portland cement* or a combination of *cement* and *stone dust* in proportion of three parts *stone dust* to one part *cement*. *Keene's cement* is often used for grouting instead of *Portland cement* for stones such as *Portland*, but in ordinary circumstances it is advisable to use a *white Portland cement*, thereby minimising the risk of staining at the joint surfaces.

Where the strength of the work is almost entirely dependent upon the cementing material, for instance, where courses of stones are supported by steelwork, it is advisable to fill all the interstices between the stone and the steel with neat Portland cement.

For setting *polished granite*, an admixture of two parts washed sand to one part Portland cement makes the most suitable bedding material, although two parts *granite dust* to one part *Portland cement* forms a good mortar for this purpose.

For setting *solid marble* of the *white* variety, "*Atlas White*" *cement* makes an excellent bedding material, but for marble wall linings, etc., *plaster of Paris* or *Keene's cement* is most suitable.

LIMES

The term lime is applied to several materials, all of which are produced by calcining or burning some form of limestone, the product being named according to the composition of the limestone burned. All limes contain free lime, *calcium oxide* (CaO), and slake more or less quickly on the addition of water.

Quicklime is a term applied to lime which slakes readily and which is in the state in which it comes from the kiln.

Fat Lime is a term applied to a pure calcareous lime, which forms a *plastic paste* or *putty* when slaked with a suitable proportion of water. *Fat limes* are made by burning limestone or chalk containing at least 96 per cent. of calcium carbonate.

Lean Lime is the term applied to any lime which *slakes* on the addition of a suitable proportion of water, but does not form a highly plastic putty. Such limes are formed when any limestone containing more than 5 per cent. of *silica* is over-burnt. It is almost impossible to over-burn a pure limestone, or one that is free from silica.

Dolomitic Limes are made by burning *dolomitic* or *magnesian limestones*. They are suitable for the manufacture of mortar, where the presence of magnesia is not objectionable.

Hydraulic Lime is the term applied to any impure lime which possesses the property of *setting*, or forming a hard, rock-like mass when mixed with a suitable proportion of water, and at the same time contains sufficient *free lime* to enable it to slake readily. It contains also a considerable proportion of combined *silica* and *alumina*.

The most powerfully *hydraulic limes* contain approximately 65 per cent. *lime*, 10 to 13 per cent. *alumina*, and 23 to 25 per cent. *silica*. According to their foreign content they may range from *eminently* to *feebly hydraulic* in properties. It will be noticed that hydraulic limes in essential constituents are somewhat similar to Portland cement. They are chiefly made from a natural mixture of limestone, sand, and clay, such as that constituting the greater part of the *lias formation*. When an artificial mixture of limestone with a suitable clay is burnt, the product is known as *Portland cement*.

Hydraulic lime is burnt in the same manner as *fat lime*, but a higher temperature is required. Both are sold in powdered form.

Selenitic Lime is a *feebly hydraulic lime* containing about 5 per cent. of *plaster of Paris*.

Slaked Lime.—When a lump of lime is left in the air and falls to powder, it is said to have slaked. This change may be effected by the addition of water to the lime, when a large amount of heat is developed, and a compound of lime and water known as *calcium hydrate*, or *hydrate of lime* ($CaOH_2O-Ca$), is produced. Lime may be either *dry slaked* or *wet slaked*. In the former the product is a powder, whilst the product of the latter is a paste or putty.

Lime Putty is a soft plastic paste made by mixing *fat lime* with about three times its weight in water, and allowing the mixture to stand until it solidifies. This putty does not readily recarbonate, and may be kept for many months in the open air. It has the advantage of being completely slaked, whereas a dry-slaked lime, which is merely mixed with water and used at once, may contain unslaked particles. This is one of the chief reasons why lime putty should be used for the bedding material for stones such as Portland stone.

Portland Cement is an artificial cement, produced by the *calcination* or *burning* of certain materials yielding *silicates* and *aluminates* of lime, which form the chief constituents. It is a carefully and scientifically manufactured product, obtained from the natural or mechanical combination of carbonate of lime with silica and alumina in the form of *shale* or *clay*. These materials, in the proportion of approximately 75 per cent. limestone and 25 per cent. clay, are mixed in a *wash mill* or *mixer* until the lumps are broken up and a slurry formed by the addition of water. This slurry is passed through a mesh 180 to square inch, and is afterwards conveyed from the *wash mill* to a series of *agitators*, then to long tubular *rotating kilns*, which are slightly inclined to the horizontal and heated to a temperature of about 1,500° C. at the bottom and 426° C. at the top. The *slurry* enters the kiln at the top end, and gradually works its way down, first becoming dry, then fused, and finally leaving the kiln in the form of *clinker*, which is cooled in a rotating cooler. The final stage in the process of manufacture is the grinding of the *clinker* to very fine powder in *rotating tube mills*, by the percussive and abrasive action of *steel balls* which are placed inside.

The atmosphere of the mill is kept moist by the introduction of a jet of steam, the pressure and quantity of which can be altered at will. Thus every particle of cement in the mill is subjected to hydration, thereby eliminating any free lime which might exist in the clinker. During this process, the setting-time of the finished cement is regulated by the addition of about 1 per cent. of gypsum. Provided that the materials are suitably selected, and the various stages of manufacture properly carried out, the test of the quality of the cement is in the *fineness* to which it is ground. The finer the powder, the greater are its adhesive or cementing qualities. For a rapid-hardening cement the fineness is such that it will pass through a sieve having 180 × 180 meshes per square inch, leaving a residue of only 3 per cent.

CHAPTER V

PRINCIPLES OF STONE-CUTTING

Coping or Splitting Stones—Preparation of Surface of Operation—Shaping the Stone—Working Concave and Convex Surfaces—Working Moulded Jamb-stone—Working Circular Mouldings—Working Return-end of Cornice—Working Moulded Window-sill—Working Voussoir for Arch—Working Keystone with Projecting Sloping Face—Working Voussoir for Semi-elliptical Skew Arch—Working a Column Stone by Hand—Working a Pediment Springer—Working a Moulded Base for Attached Column—Working Voussoir for Semicircular Arch in Cylindrical Wall—Working Niche Stone—Working Dome Stone—Working Stone for Pendentive Dome—Working Stone for Intersecting Vaults—Working Stone for Welch Groin—Working Stone for Interpenetration of Dome—Working Stone for Ramp and Twist Work.

THE methods and principles of stone-cutting described in this section are placed before the young craftsman or student as a foundation upon which to erect a superstructure of higher technique of the craft.

The conversion of rough quarry blocks of stone into wrought stonework is now to a very large extent done by machinery.

The surfacing described in this section may seem superfluous to those engaged in modern masonry works, but it must be remembered that it is essential for the young craftsman to have an intimate knowledge of the principles described. All the operations necessary for the completion of the stone by the mason, after the machines have fulfilled the purpose for which they have been installed, are dependent upon the mason's knowledge of these fundamental principles. It is for this reason they are described in detail in this section.

The working of various stones taken from the geometrical plates is dealt with, and it is hoped that the student will use these plates in conjunction with the methods here described.

COPING OR SPLITTING STONES

Stones are usually cut to size or slabbed under the "*frame saw*" or the "*diamond saw*," or, in the case of Bath or similar soft stones, cut to size with "cross-cut" saws by hand labour.

Sawn slabs are often supplied to masonry firms whose works are not equipped with machinery, or where the stone is worked by hand on the site of the building being erected. Under these conditions it becomes necessary to "*cope*" or "*split*" the stones roughly to the size required, as in Fig. 231, which shows the methods of *coping*, suitable for Portland or similar stones. When the position of the "*cope*" is determined, a line is marked across the top surface, as at A, B. From points A and B lines are squared down the side surfaces of the stone. If the stone is *quarry axed*, the line from point A should be marked approximately square from the top surface, the line on the opposite surface from point B being "*boned*" by adjusting two straight-edges, as explained for surfacing. The stone should be allowed to rest on an *iron bar* placed immediately under the cut: one end of the stone should be left free, wedges being placed under the other, to keep the top surface of the stone horizontal. A V-shaped groove is then cut in the surfaces of the stone, so that the centre of the groove corresponds with the cutting line already marked on the stone. Sometimes the line is stabbed

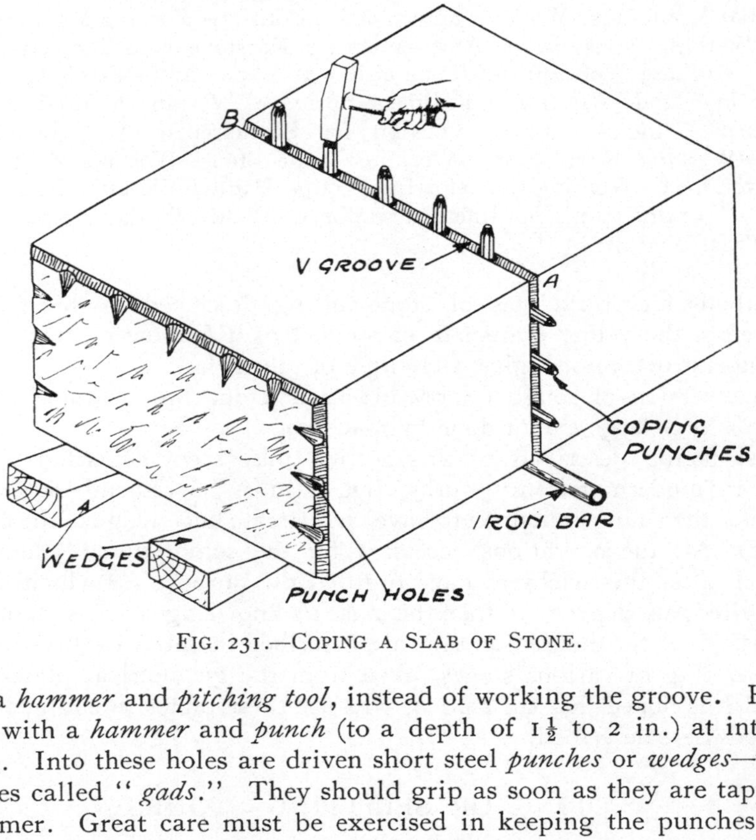

FIG. 231.—COPING A SLAB OF STONE.

in with a *hammer* and *pitching tool*, instead of working the groove. Holes are now cut with a *hammer* and *punch* (to a depth of $1\frac{1}{2}$ to 2 in.) at intervals of 4 to 5 in. Into these holes are driven short steel *punches* or *wedges*—these are sometimes called "*gads*." They should grip as soon as they are tapped with the hammer. Great care must be exercised in keeping the punches parallel to the direction of the cut, or the surface of the cut will be very irregular. The punches should be gently struck in rotation, whilst water poured into the holes often assists in the splitting of the stone, especially if allowed to remain for a while. Thin slabs of stone or marble are readily split by *nicking* the cutting line with a *hammer* and *chisel*, and tapping the back surface of the material with a *hammer*.

The harder stones, such as granite, etc., are split by drilling holes about 3 in. deep and $\frac{5}{8}$ in. diameter at intervals along the cutting lines. These holes are either drilled by means of a *pneumatic drilling machine* or with a "*jumper*."

Into these holes are placed two thin pieces of iron called "*feathers*," the wedge being placed between the *feathers* and struck with a hammer until the stone splits.

SURFACE OF OPERATION

For the conversion of a rough block of stone to a stone of definite shape, that is, to fit correctly into a position in the building predetermined by the drawings, it is necessary that the mason should select one surface of the stone as a *surface of operation*.

This surface differs according to the ultimate shape of the stone to be cut. In some instances it is best to select the face of the stone, whilst at other times it is most suitable to select one of the beds.

It is important that the *surface of operation* when worked should conform to the definition of a plane. It should coincide in every direction with a straight line, therefore it should be out of "*twist*," so that any surface may be squared true from it.

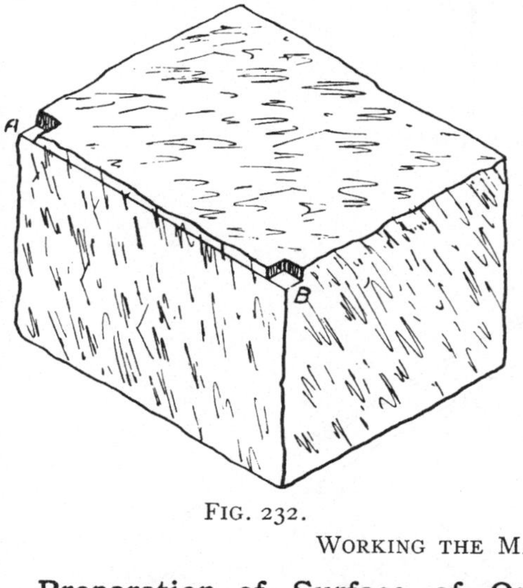

FIG. 232.

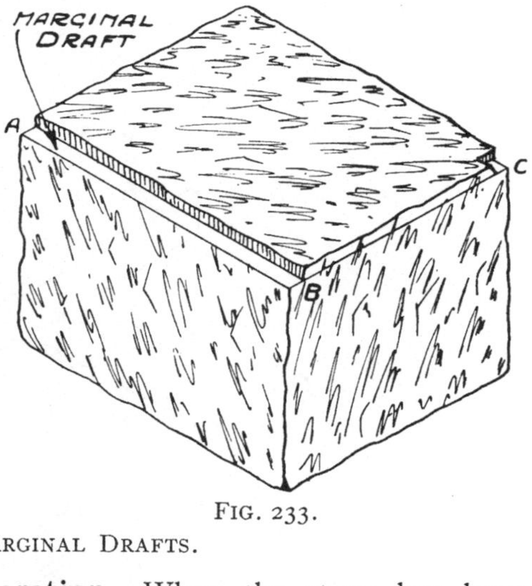

FIG. 233.

WORKING THE MARGINAL DRAFTS.

Preparation of Surface of Operation.—When the stone has been chosen, to clear the size of the finished stone, and the surface of operation selected, a guiding line is marked on one of the side surfaces, as at A B, Fig. 232. This line is cut up with a *hammer* and *pitching tool*, and the corners A and B cut in as shown, with a *mallet* and *drafting chisel*. A marginal draft about $\frac{3}{4}$ in. wide is now worked between A and B and tested for straightness by applying a straight-edge. Next a point C is selected, which should be approximately square with the rough surfaces of the block. Cut in corner C, and work the draft between B and C as previously described (Fig. 233).

It is now necessary to obtain the exact position of point D. To do this, proceed as shown in Fig. 234. Place a straight-edge on the draft A B, and hold another straight-edge on the opposite surface of the block, so that the edge coincides with the position of corner C. These straight-edges are now sighted through, and whilst keeping the edge correct with corner C, the other end of the straight-edge is raised or lowered until the edges of both are parallel.

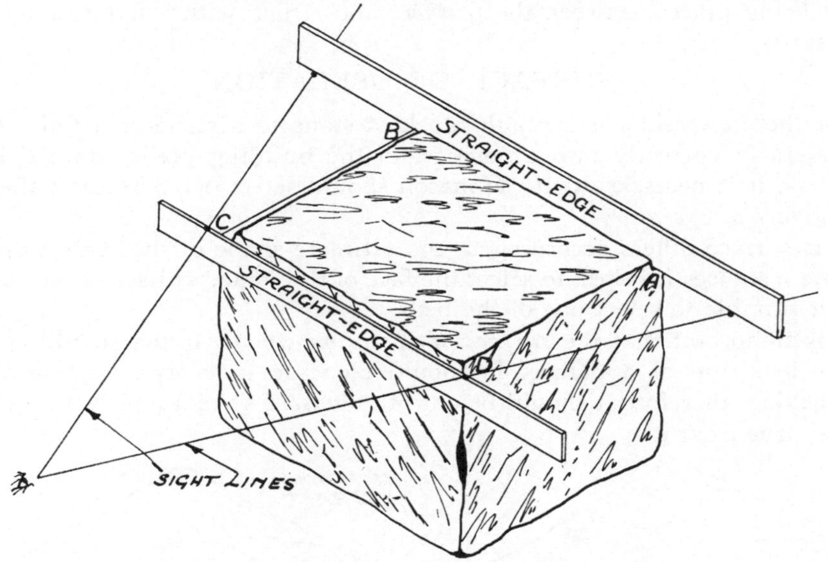

FIG. 234.—BONING THE LINE.

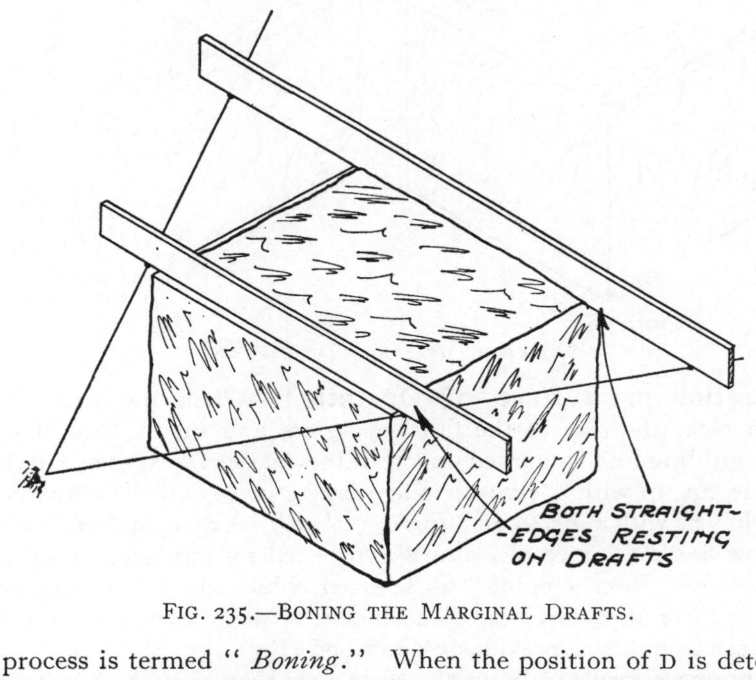

FIG. 235.—BONING THE MARGINAL DRAFTS.

This process is termed " *Boning*." When the position of D is determined, a straight draft is worked between C and D and tested with both straight-edges until the draft is correct, as shown in Fig. 235. Next work a draft between D and A, which will complete the four marginal drafts.

Boning blocks are often used for surfacing granite and similar stones. These blocks are made of hard wood, about $1\frac{3}{4}$ in. cube. The corners of the stone are cut in with a chisel and tested by sighting through two straight-edges resting on the top surface of the blocks, as in Fig. 236.

To remove the superfluous stone, enclosed by the marginal drafts, a series of *furrows* is worked along the surface with a *hammer* and *punch* parallel with the draft A B (Fig. 237). These *furrows* should be fairly close to the surface required. The surface is then "*clawed*" in parallel drafts by using a *mallet* and *claw chisel*. The drafts should be worked in the same direction as the furrows and tested by applying the straight-edge (Fig. 238). A series of drafts is now worked with a *mallet* and *boaster*. These should be worked parallel to the marginal draft A B, each one being tested and reworked until it is correct before the next draft is *boasted*. Each draft is a guide for the working of the next, and so on across the surface. The straight-edge should then be applied

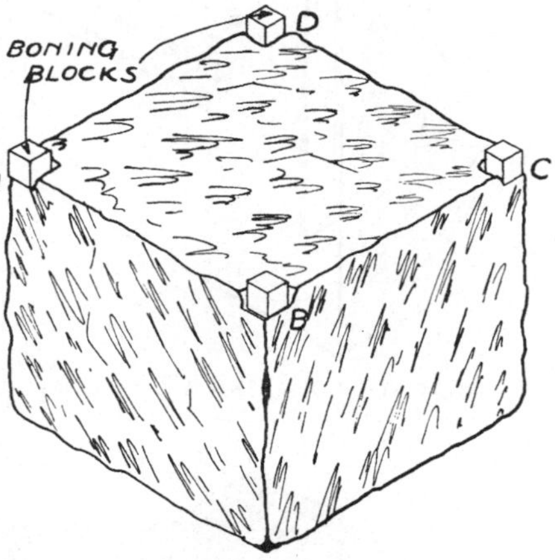

FIG. 236.—METHOD OF USING BONING BLOCKS.

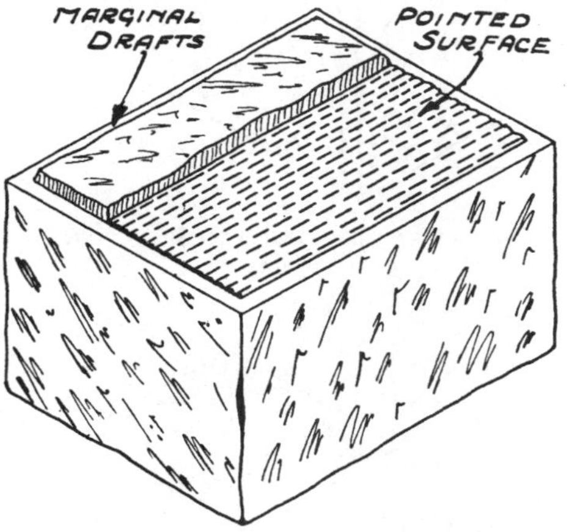

FIG. 237.—POINTING A SURFACE.

diagonally across the surface. If the stone is round in one direction and hollow in the other direction, it is proof that the surface is *twisting*. It is then necessary to rework the marginal drafts, and repeat the processes already described.

Shaping the Stone.—After the *surface of operation* has been worked true, the *templets* or *moulds* should be applied to the stone, the outline of the mould being *scribed* or marked on the surface by means of a *scriber*, after which a pencil may be used to define the lines clearly. The various surfaces may now be worked square from the *surface of operation*, in stages, as described above, the drafts being tested by applying a steel square (Fig. 239). The *stock* of the square should be held when being applied to the stone, and the blade gradually brought into contact with the draft. The stone should be made to fit the square and not the square to fit the stone. The square should be applied with the *blade* and the *stock* at right angles to both surfaces.

Circular Surfaces.—These are worked in a manner similar to that described for straight surfaces, except that the first draft is cut to the curved outline of the mould marked on the stone and tested with the reverse templet of the curve.

To Work the Stone to the Required Shape.—Commence by working one of the beds square to the sawn surface X (Fig. 240), and then apply the

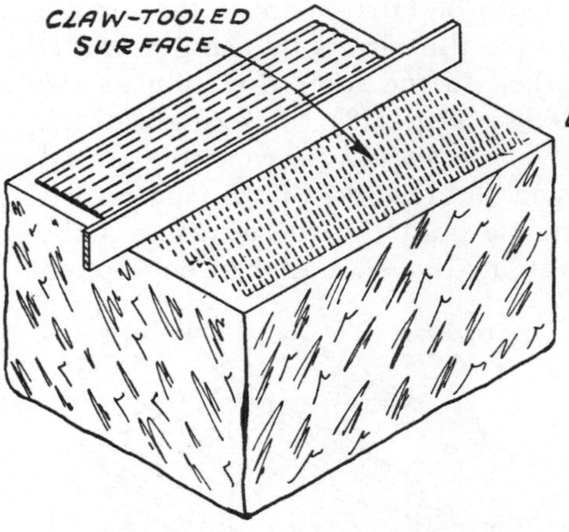

FIG. 238.—CLAWTOOLING AND BOASTING A SURFACE.

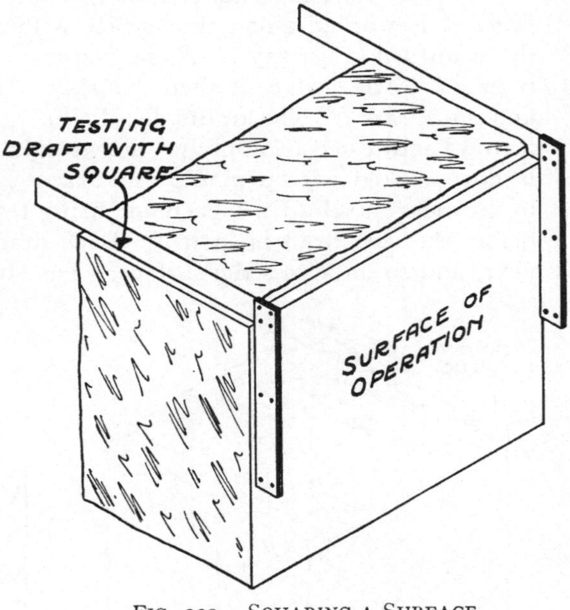

FIG. 239.—SQUARING A SURFACE.

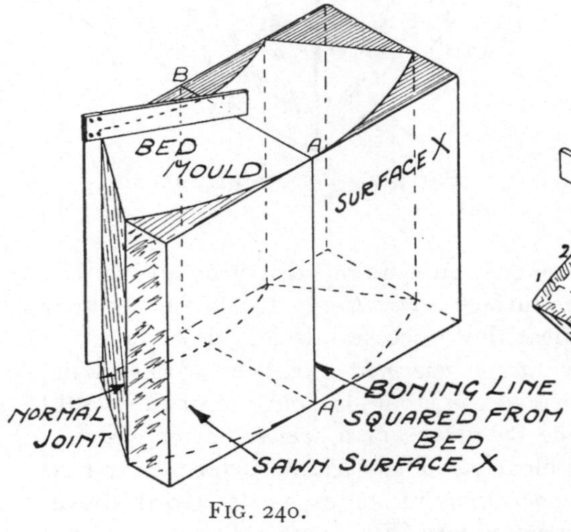

FIG. 240.

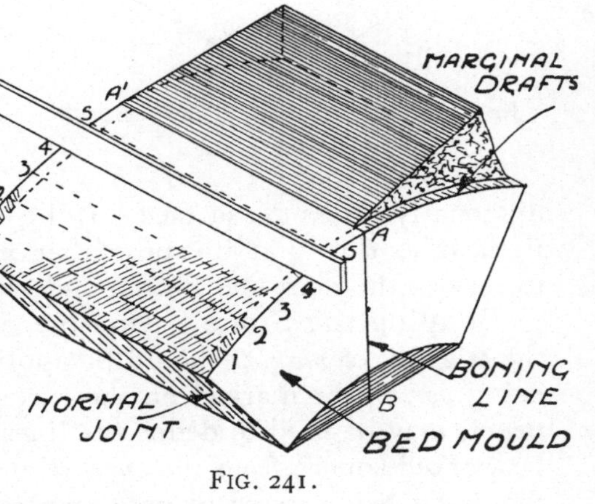

FIG. 241.

WORKING CIRCULAR SURFACES.

bed mould, giving the curve and shape of the stone required. The *boning line* A B marked on the mould should be transferred to the stone, and squared down the vertical sawn surface from point A to the other bed, which should be worked parallel to the first bed and to the height required. Obtain the true position of the *boning line* on this bed from point A′ by the process of *boning*, as before described. The *bed mould* should now be applied to the surface, *lines down*, and adjusted to the *boning line*.

Now work the normal vertical joints, testing them with a square from the *bed surface*, as shown in Fig. 240. The circular face of the stone may now be worked. First work marginal drafts to the curve of the bed mould applied on each bed, and connect them across the surface by straight marginal drafts, at the extreme ends of the stone, thus forming the joint arrises. Mark a series of points on the curved marginal drafts, as at 1, 2, 3, 4, etc. (Fig. 241), and work the surface of the stone in a series of straight drafts, testing each draft with a straight-edge, held parallel to the *vertical boning line*, or to the points 1, 1, 2, 2, etc.

To Work a Moulded Jamb-stone.—Assume that the stone "bankered" is a rough block of stone. Select one of the beds for the *surface of*

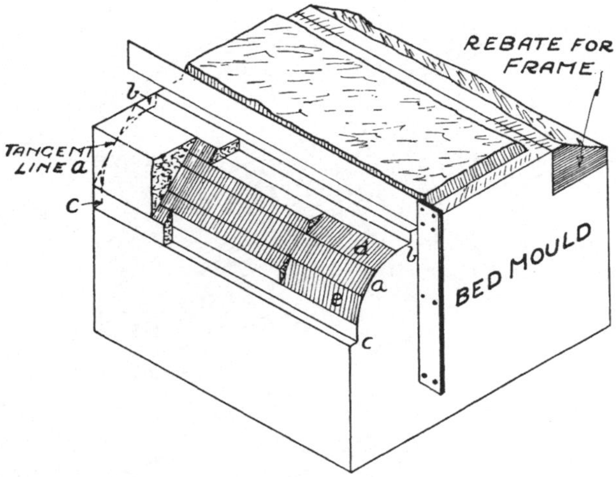

FIG. 242.—WORKING A JAMB-STONE.

operation, work this surface and mark on the *bed mould*, which will give the outline of the jamb. Work the face of the stone square to the bed, and then the reveal square to both the face and the bed (Fig. 242). Next measure on the faces the height required, and work the other bed square to the face and the reveal. Now scribe the *bed mould*, *lines down* on this surface, adjusting it to the arris of the face, and the reveal. Next work the vertical joint. The working of the moulding is the next operation. Mark a tangent *a* to the ovolo moulding, *boning* it on the opposite surface so that the point of contact with the curve is in the same relation at each end of the stone. Connect the moulding at each end with lines, also the chamfer lines. Now work the chamfer. Next cut in the fillets *b* and *c* at each end of the stone, and work them true, in drafts from end to end. A series of small tangents *d* and *e* should now be marked on and worked. By repeating this process, a curve may be worked to a fine degree of accuracy.

The *check* or *rebate* for frame should be worked next. It is better to leave this until the moulding has been worked, so that the stone may be worked back if required.

Working Mouldings by Hand.—Mouldings and cornices are usually run through by machines, but it is essential that the processes necessary to execute them by hand be explained.

No definite method for the working of mouldings can be stated, as each craftsman evolves a method peculiar to his own style. The difference in the texture of the various stones calls for different procedure: the general principles are, however, the same. For instance, mouldings on Bath stone are worked differently from mouldings in Portland stone, whilst marble calls for a treatment differing from either. It is not advisable to cut marble direct into sharp angles, because the material is liable to fly owing to its brittle nature. The

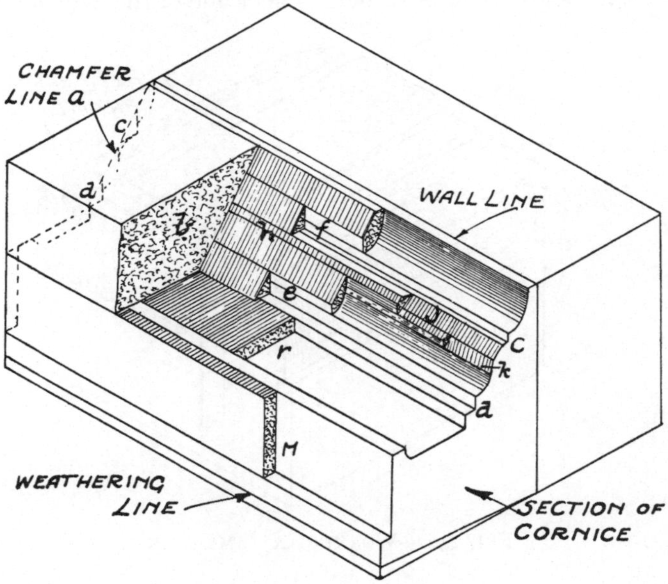

FIG. 243.—WORKING CORNICE BY HAND.

mouldings in Bath stone may be *sawn* direct into the fillets with a *fillet saw*. Fig. 243 shows the method of working mouldings in Portland and similar stones by hand.

The group of mouldings forming a cornice must be worked in stages, or a series of sinkings, each sinking preparing the way for the finishing of the group.

Commence by marking on the chamfer line *a*, and work out the check *b* down to the level of the drip. Now produce the fillet lines at *c* and *d* to intersect the chamfer line *a*, and work the small checks as at *e* and *f*. Mark on the arris lines for the *cavetto* mould and also the hollow portion of the *cyma recta* (ogee), and produce this curve to meet the chamfer line *a* in *h*. Complete the outline of the *cyma recta* by marking and working small tangents, as at *j* and *k*. Next work the front fillet *m*, and lastly the throat *r*.

To Work Circular Mouldings (Fig. 244).—Assume the circular stone already described in Figs. 240 and 241 to be a moulded *voussoir* or *arch-stone*, irrespective of the direction of the *bedding planes*. The *bed mould* would now be the *face mould*, the convex surface and the joints being already worked. The concave surface may now be worked to the outline of the mould, in a manner similar to that described for the convex surface.

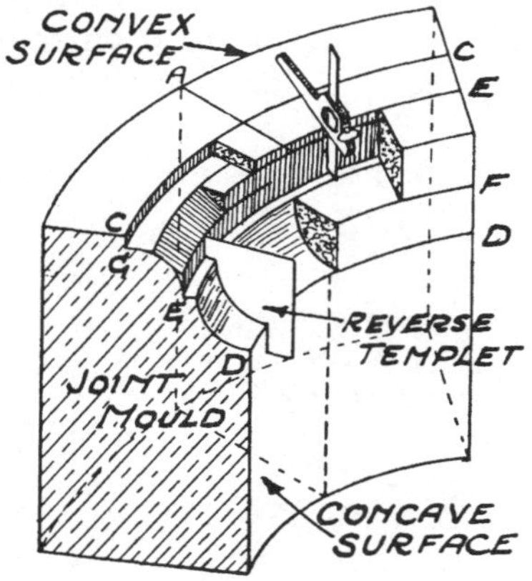

FIG. 244.—WORKING CIRCULAR MOULDINGS.

The joint moulds, giving the section of the arch moulding, are now marked on the normal joints. Trammel the lines of the moulding at C C, round the face of the stone from the convex surface, also trammel the arris line D along the concave surface, parallel to the face of the stone. Now square the angle of the fillet E to the face of the stone, and extend it out to the concave surface at F. Trammel these lines as before, and work out the check to these lines, testing it with a *sinking square*, from both surfaces. Next repeat the process for fillet G. A series of tangents marked to the curve of the *ovolo* moulding, and worked as shown in the figure, will complete the contour of this member. The *cavetto moulding* may now be worked by trammelling the arris line of the fillet E parallel to the concave surface, the exact shape of the hollow being obtained at each joint by working a draft to the section marked on the stone, and being tested with a reverse templet of the hollow.

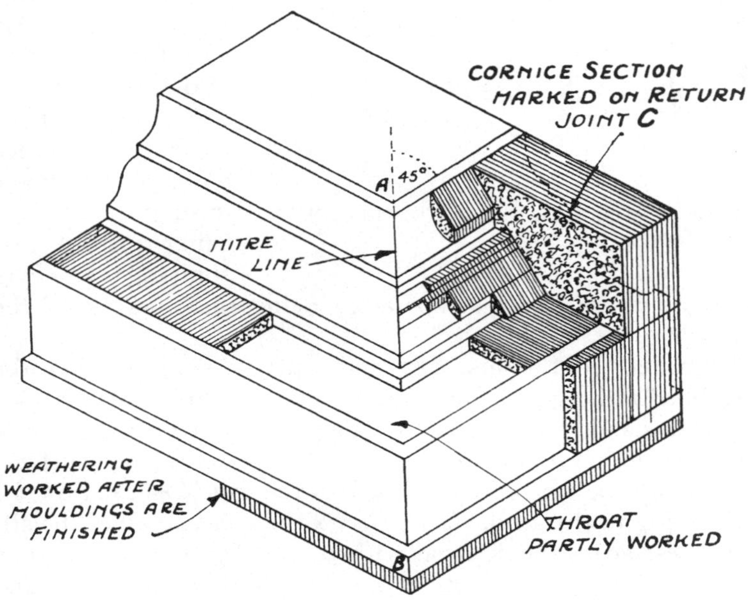

FIG. 245.—WORKING RETURN PIECE OF CORNICE.

Working the Return End for a Piece of Cornice (Fig. 245).—Assuming that the mouldings have been worked through the front of the stone, with the exception of the throating, the first operation is to mark the *zinc section* on the return joint C. Now scribe the *mitre line* from A to B, which, if drawn correctly, would give the correct outline of the return mouldings. To scribe this mitre line, a wide straight-edge is held in the position shown in Fig. 246, the exact angle to hold the straight-edge being determined by placing a square on the bed surface of the stone and keeping the flat surface of the straight-edge tight to the blade of the square.

If the return moulding is 90° to the face, a *mitre square*, comprising an angle of 135°, may be held on the *nosing*, and the straight-edge adjusted to it, as Fig. 247. A long pencil is then held flat on the wide surface of the straight-edge and moved whilst the point of the pencil is in contact with the stone, until the outline of the mitre is marked distinctly on the stone.

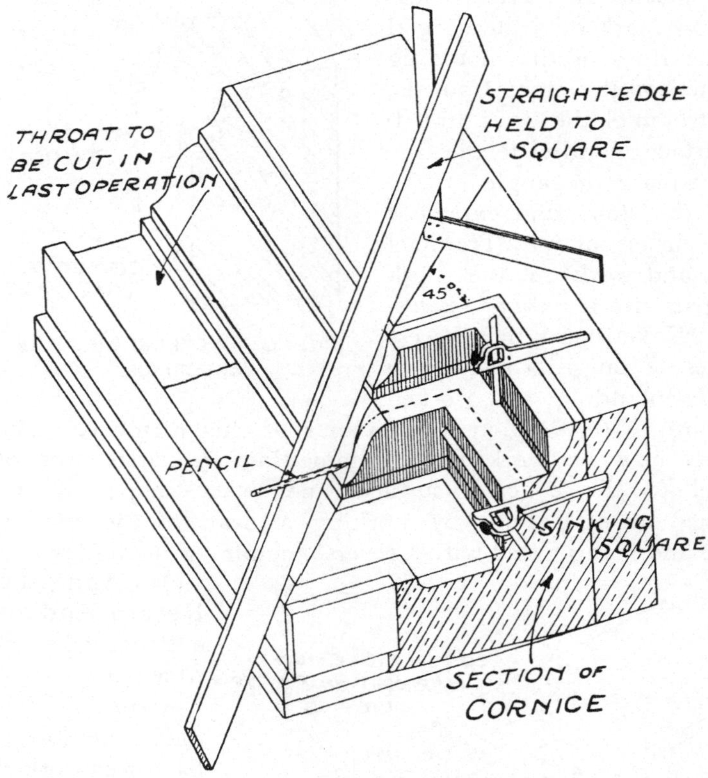

FIG. 246.—SCRIBING A MITRE LINE AND WORKING AN INTERNAL MITRE.

The mouldings are now worked step by step, as explained before, the *throat* and *weathering* being worked after the mouldings are completed.

To Work a "Break" in a Piece of Cornice (Fig. 246).—Mark on the *mitre line* as explained, also mark the *zinc section* of the cornice to the correct depth of the *break*. Then work the various members in stages, each stage being tested by applying the *sinking square*, as shown in the figure, the last operation being the working of the *throat* and the *weathering*.

To Work a Moulded Sill (Figs. 247 and 248).—Assuming that the stone is sawn, or slabbed to height, before being bankered, first mark the *bed mould* on the top surface, and work the *nosing* along the front edge, square to the top bed. Now work the joints square from the *nosing* and the bed surface. These joint surfaces should be worked clean, back to the *wall line* to allow for the return of the moulding to the ashlar face. Mark the *zinc section* of the sill on to the joint surfaces, adjusting it flush with the *nosing* and the top bed surface. The bottom bed line and the depth of the *nosing* should now be marked on the stone from joint to joint. The moulding can now be worked in stages, as at A, B, C (Fig. 247). Next scribe the *mitre lines* on each end. The position for these is obtained by measuring the distance of the projection, along the *wall line*, from the joints. Return the mouldings on to the ashlar face, by working the members square from the front moulding and parallel to the beds. The return of the *throat* together with the front throating can now be worked.

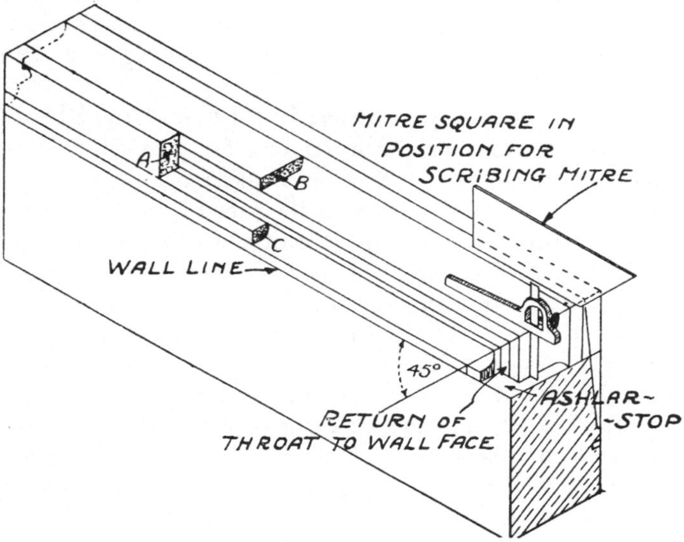

FIG. 247.—WORKING A MOULDED SILL.

The top surface, including the *weathering* and the *stooling*, should be worked next. Transfer the lines from the bed mould, indicating the width of *opening* and *reveals*, and work the *weathering* as suggested in Fig. 248.

First cut up the nosing line and work a straight draft from joint to joint. When working this draft, the chisel should be inclined at the slope of the *weathering*. Next work the drafts A and B correctly to the slope of the *weathering*, and dress the surface between A and B as described for working a plane surface. The *stooling* and *intersection* may now be worked, and the sill completed by working the *groove for water bar*, either through the sill from joint to joint or stopped just beyond the reveal line, as desired.

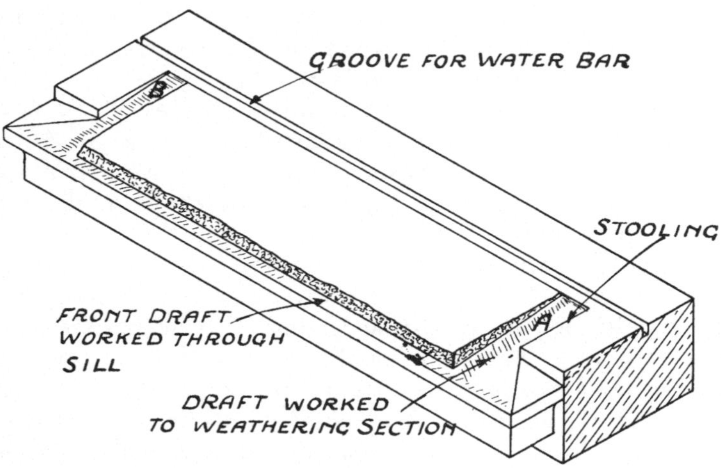

FIG. 248.—WORKING WEATHERING FOR SILL.

To Work a Voussoir or Arch-stone from Fig. 340.—The moulds required are: *face mould* and *section*. Select a piece of stone of suitable size, and assume one surface only as sawn. True up this surface, making it the *face* of the stone and also the *surface of operation*. Mark on the *face mould* (Fig. 249), and work the top bed surface, the vertical joint, and the two normal joints all square from the face, and to the correct outline of the *face mould*. Next work the clean soffit, which should be tested by applying a reverse templet of the curve to the stone (Fig. 250). If desired, the *face mould* may be marked on the back vertical surface to assist in working the stone to shape. The *vertical joint mould*, or *section*, should now be marked on the joint, giving the outline for working the detail of the rustication (Fig. 251).

To Work the Keystone with a Projecting Sloping Face.—This stone can either be worked from a vertical face, in which case the stone should be cut to the shape of the face mould taken from elevation, or it may be worked from a face sawn to the slope of the face of the keystone.

This would necessitate the use of a *developed face mould* and *bevels* to work the top bed surface and the soffit. The joint should be squared from the sloping face, by holding the square in a horizontal position, or parallel to the top bed surface. The former method is the better to adopt, and is here described. As explained for the voussoir, apply the *face mould* to the *surface of operation*, and square from this surface the top bed (Fig. 252). Mark on the centre line A B, and square a line on the top bed from point A to A′. "*Bone*" a line down the back vertical surface, from point A′, and apply the *face mould*, lines down, to this line as a guide to subsequent working.

Next work the normal joints, which should be worked square from the *surface of operation*, and complete the outline of the face mould by working the bottom clean surface.

Now apply the joint mould to the joint surfaces, which will give the correct position for working the slant face of the keystone (Fig. 253). The rebate, or check for frame, is worked last.

Working a Voussoir in Semi-elliptical Skew Arch (Figs. 444-449): **Stone No. 2.**—The moulds required are: *face mould*, or *section of prism*, *bed mould*, *developed joint mould*, and *developed soffit mould*.

The size of the stone required is the shape of the *prism section* and the length of the *bed mould*.

Select the top bed as the *surface of operation*, and square from this the vertical joint H″ D″, H′ D′. Next work the end surfaces of the prism fairly true and apply the *face mould*, adjusting it to the arrises of the top bed and the vertical joint. Now complete the working of the prism (Fig. 254). Next mark on the *top bed mould*, adjusting it to point H″; then mark on the *developed joint mould*, adjusting it to point G, also the *lower developed joint mould*, to point D″, and the *developed soffit mould*, which should be placed correct with points E and F (Fig. 255). The inside and outside faces may now be worked to the lines marked on the stone. If the arch is enriched by mouldings, the outline on the members would be provided on the sections, and worked parallel to the line of the arch curve. Usually the above stone would be worked to the shape of the prism in a planing machine, in which case it is only necessary for the mason to mark on the various moulds and work the faces to the correct angle of skew.

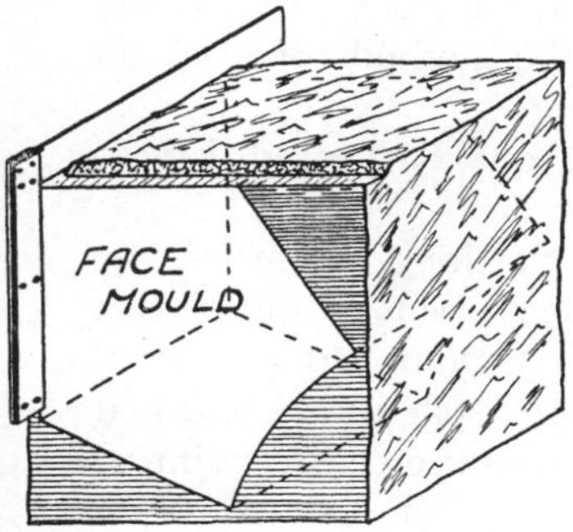

FIG. 249.

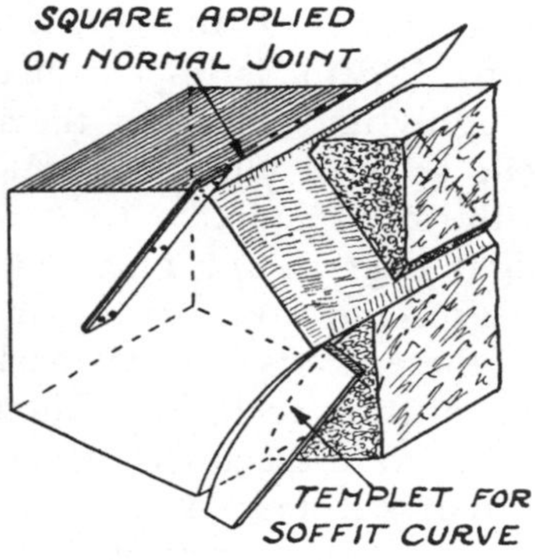

FIG. 250.

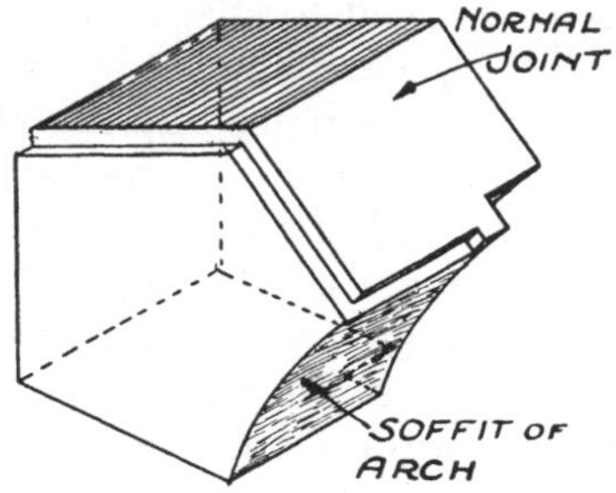

FIG. 251.

WORKING A VOUSSOIR.

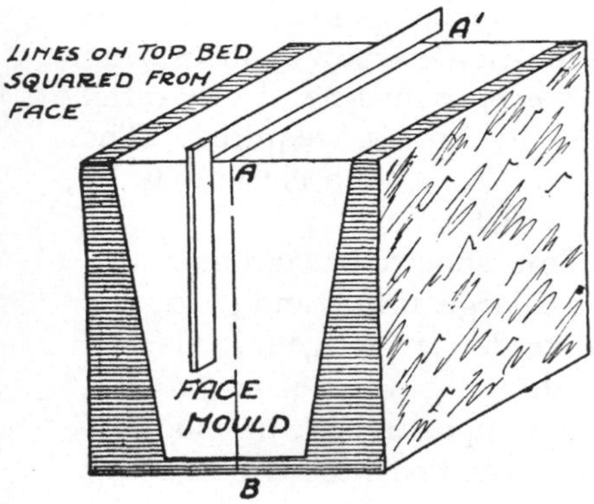

FIG. 252.

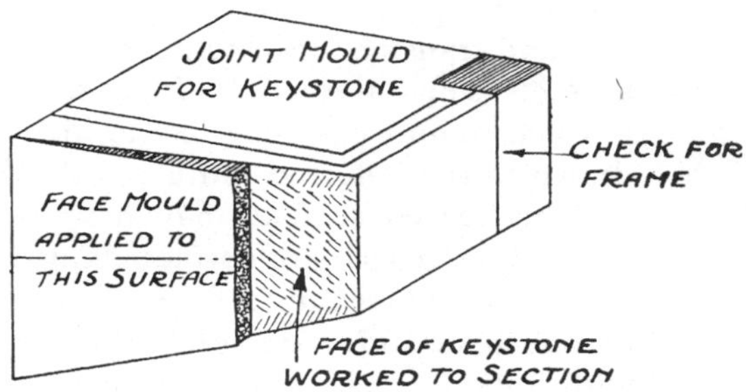

FIG. 253.

WORKING A KEYSTONE WITH SLOPING FACE.

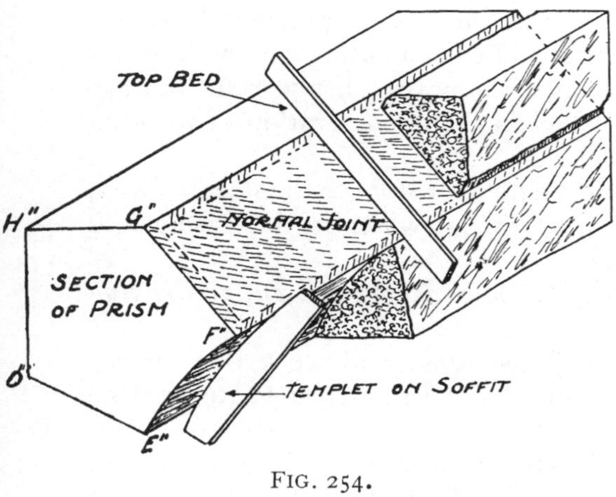

FIG. 254.

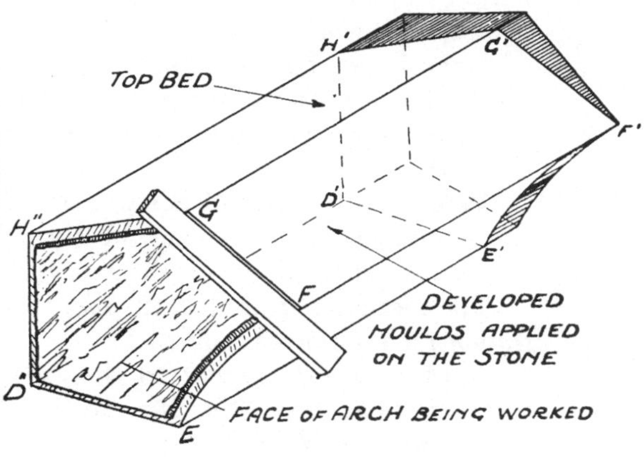

FIG. 255.

WORKING VOUSSOIR FOR SKEW ARCH.

To Work a Column Stone by Hand.—Although these stones are usually turned to shape in a *centre lathe*, it is important that the methods of working by hand should be thoroughly understood by the student.

Assume that the stone selected for this is a rough quarry-dressed block, selected to comply with the size required.

First work two straight *angle drafts* along one of the long arrises of the stone, as shown in Fig. 256, and work the top and bottom beds to the correct height, and square from these drafts. Then mark the bottom bed mould in a suitable position on one of these surfaces. Diameter lines should be marked on the moulds, and transferred to the stone, thus obtaining the centre point C. Produce these diameter lines X X and Y Y to the rough surfaces of the stone.

It must be remembered that the centre of the column stone on the top bed must be exactly above the centre on the bottom bed, so that the circles on the top and bottom bed are parallel, and have their centres on the same perpendicular axis. To ensure this, measure the distance A Y (Fig. 257) and transfer it to the other bed at B Y′. Then *bone* a line Y′ Y′ on one bed parallel to the line Y Y on the opposite bed. Repeat this process with the diameter X X by measuring A X, and transferring it to the other bed at B X′ and *boning* the line X′ X′ parallel to X X. The intersection of the lines X′ X′, Y′ Y′ is the centre C′ required. The top bed mould may now be applied to the diameter lines, or a circle may be drawn to the required radius, with C′ as centre, as in Fig. 256.

The curved surface may now be worked. This should be accomplished by working a series of surfaces tangent to the curve, large at first and gradually becoming smaller, until the required shape is produced. If a tangent is marked on one bed, and a straight-edge held correct to the tangent line, and *boned*, with another straight-edge held to the curve on the opposite bed, the point of contact will be in the same relation to the curve, on both beds, the circles being concentric, although of different radii. By a repetition of these methods the whole of the curved surface may be worked, as in Fig. 258. When an *entasis* is required on the shaft, these tangent surfaces must be worked true to a reverse templet of the *entasis curve*. If the column stone is to be *fluted*, it is not necessary to work the whole of the surface clean, but only the portion coinciding with the bands between the flutings, or just sufficient to allow for the lines for the fluting arrises to be marked on the stones, the portion between the flutings being left approximately rough, and finished in the one operation of working the flutings.

Working Pediment Springer.—Select a piece of stone to the size required, and commence by preparing the *surface of operation*, which in this case would be the face of the stone on the *nosing line*.

The moulds required are: *face mould*, *bed mould*, *true section*, and *raking section* for the normal joint.

Apply the *face mould* to the *surface of operation* and work the various surfaces square from the face, until the stone has been shaped to the outline of the *face mould* (Fig. 259).

To do this, first work the bottom bed surface and apply the bed mould, adjusting it to the position of the *face mould*. Then work the vertical and normal joints, and mark on these surfaces the sections mentioned.

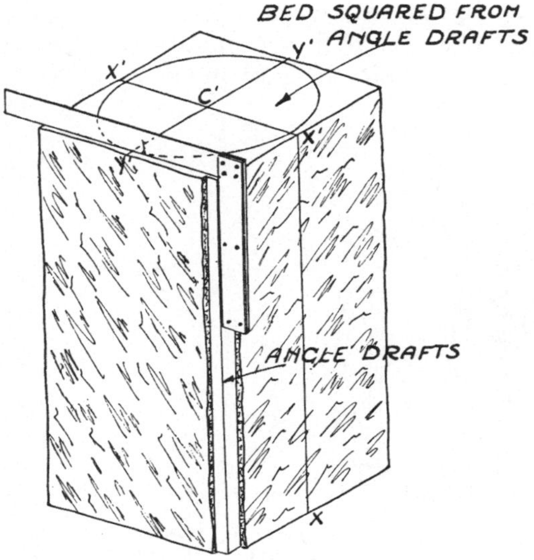

FIG. 256.

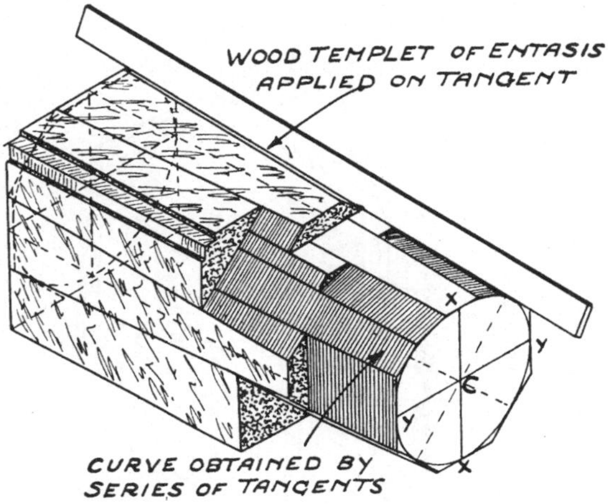

FIG. 258.

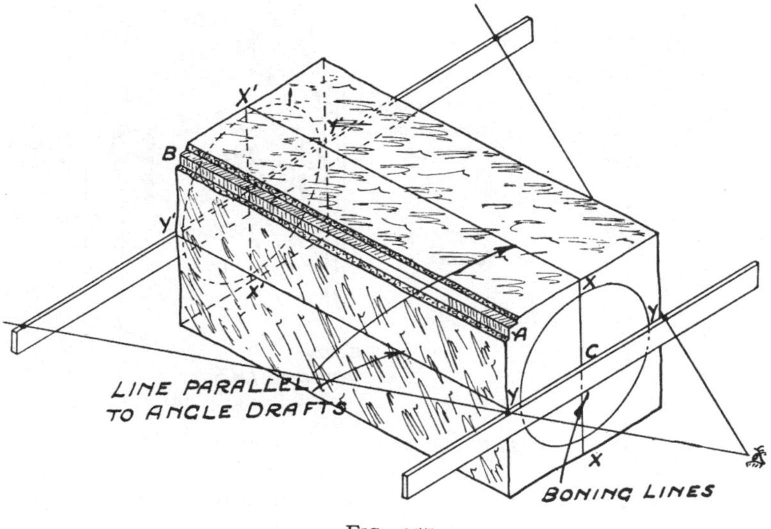

FIG. 257.

WORKING COLUMN STONE BY HAND.

Now square a wide draft down from the nosing at A. This draft should be wider than the *nosing*. Mark the horizontal outline of the nosing on the draft and work the top raking surface of the springer, to the lines on the stone, testing it by *boning* the marginal drafts. The return horizontal mouldings should now be worked to the outline transferred from the face mould. This

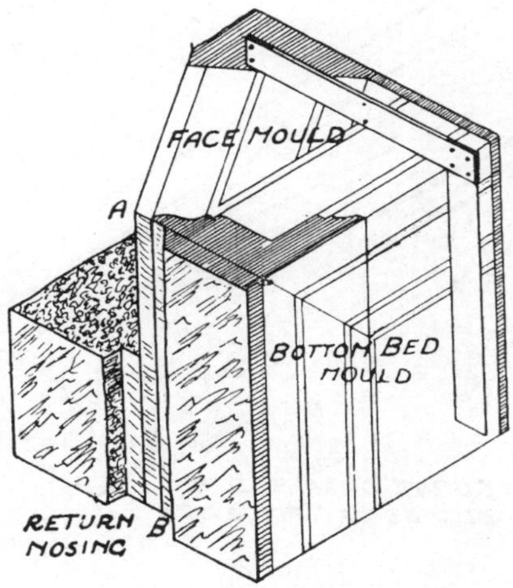

FIG. 259.

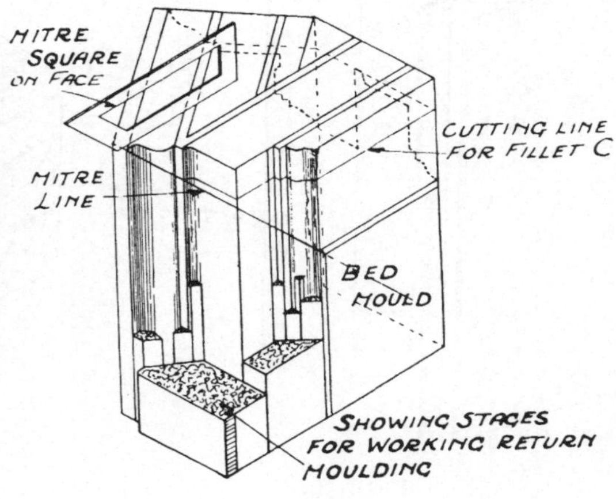

FIG. 260.

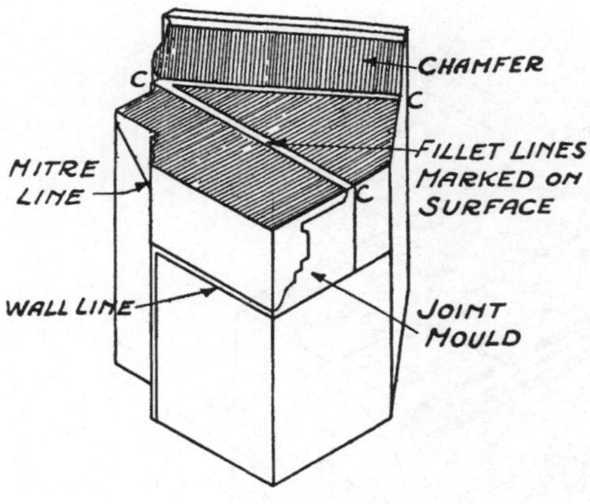

FIG. 261.

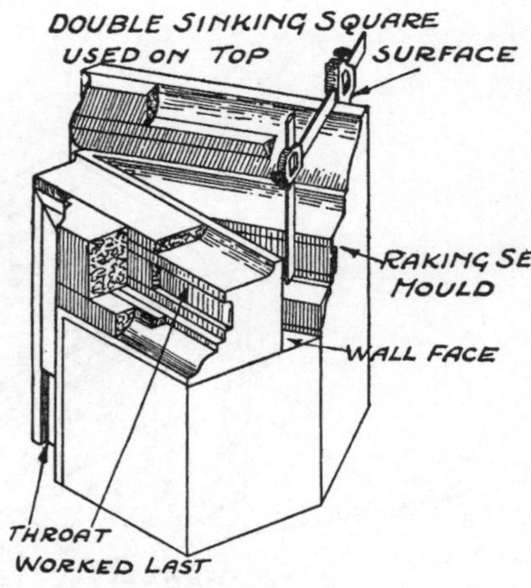

FIG. 262.

WORKING A PEDIMENT SPRINGER.

moulding should be worked in stages, as previously described. Scribe in the *mitre line* (Fig. 260) and commence working the front of the stone.

First work a surface to coincide with the fillet C at the base of the *cyma recta*, and a chamfer forming a tangent to the curve of the *cyma recta*. Now mark on the fillet line C (Fig. 261), connecting it with the fillet C on the vertical joint and C on the raking joint. Next work the horizontal front moulding, and complete the working of the ogee moulding. Finally, the lower members of the raking cornice may be worked, these being tested during the various stages by applying a *double sinking square* from the top raking surface, as in Fig. 262. It is important that these members should be worked parallel to this surface, in a series of checks, to die on to the top horizontal or weathered surface of the horizontal cornice. The throating must be worked in the last operation.

Working a Moulded Base for Attached Column (Figs. 263 and 264).—The lower member of the base marked A and B is usually left square, whilst the members above follow concentrically the curve of the shaft.

The moulds required are: *bed mould* and *section*.

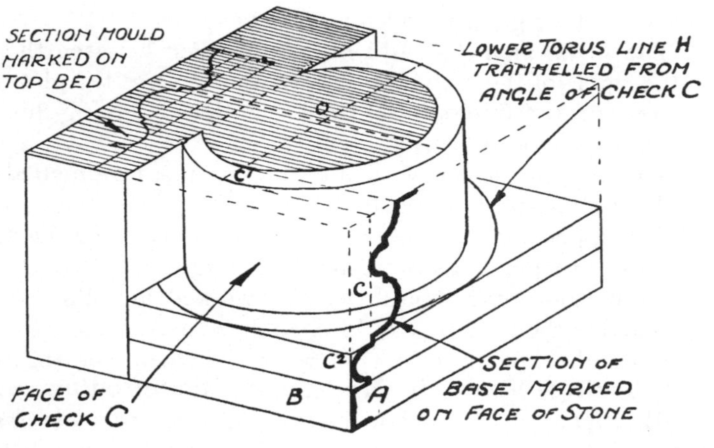

FIG. 263.—WORKING MOULDED BASE.

Assume that the top and bottom beds have been sawn. Square the stone to the shape of the *bed mould*, making all the surfaces square to the beds. Mark the *section* on the front vertical surface, adjusting it so that the front of the section is flush with the side surface B, and the top true with a line parallel to surface B, drawn from the circumference of the shaft, transferred from the bed mould. To work the circular part of the base, the members of the moulded section must be worked in a series of checks, each check representing the definite shape of the moulding at that part of the section.

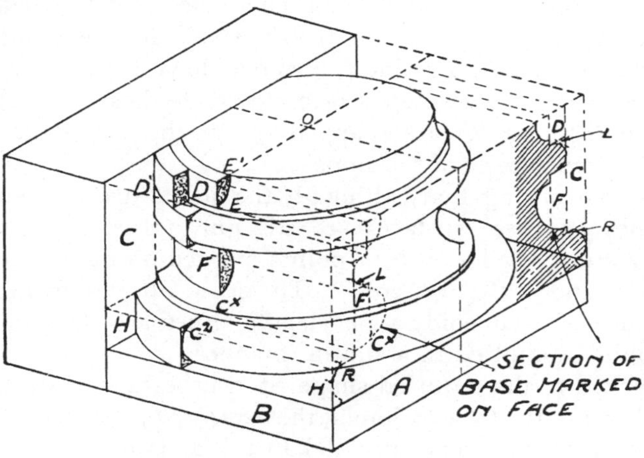

FIG. 264.—WORKING MOULDED BASE.

First mark check C (Fig. 263) down from the top surface, draw a parallel line to surface B along the top surface, to intersect the centre line of the shaft at C^1. With centre O draw the circle representing the plan of this check; also a line at C^2, parallel to the bed, round the three vertical faces. The check should now be worked true to these lines, by squaring down from the circular line on the bed surface, and in from the vertical faces (Fig. 263). Repeat this operation with check D (Fig. 264), except that in this instance the line D′ must be trammelled parallel from the top bed, because the check is circular in plan. Next mark on the fillet line E and work the top hollow section between E and E′, which should be tested with a reverse cut to the curve of the member in section. Mark on the lines for the check at F, and sink the check to the level of the fillet by gauging with a *sinking square.*

The *scotia mould* should now be worked between the fillet lines F and C^x to the shape of the section by applying a reverse templet of the *scotia section.*

Next draw a line parallel from the fillet C^x, giving the plan of the projection of the torus mould at H.

This portion can be worked to the line just trammelled, and the line already on the vertical face of the stone.

The upper and the lower *torus members* can now be completed by working a series of tangents, as indicated in the figure at R and L. The lower *quirking* of the bottom *torus* should next be worked and the *torus section* completed. In practice the section mould should be marked on the top bed surface with checks and tangents set out for reference when working, as in Fig. 263.

Working Voussoir No. 3 for Semicircular Arch in Cylindrical Wall (Figs. 450-459).—The moulds required are: *developed outside* and *inside face moulds*, with vertical *boning lines* marked on them; *bed mould*, with horizontal *boning lines* marked on; and *developed joint moulds*. The developed joint moulds are required because a moulded section runs round the arch.

First work the stone as described for the working of a convex and concave surface (Figs. 240 and 241). When this operation is completed and the bed mould marked on the stone, together with the *boning line* $4^x\,4^{x\prime}$, square a line down the convex surface from point 4^x to point 4, also one down the concave surface from point $4^{x\prime}$. Now adjust the developed face moulds to these vertical lines, as in Fig. 265. These *face moulds* are wrapped round the cylindrical surfaces, thus determining the true shape of the stone.

It must be remembered in working that, although the joints are twisted surfaces, all horizontal lines on these surfaces are straight, therefore they may be tested by applying a straight-edge to *level* points marked on the inside and outside drafts, as at 1 1′ and 2 2′ (Fig. 266). The same applies to the soffit, which, being the surface of a cuneoid, all horizontal lines are straight, hence the straight-edge should be applied in a similar manner.

The stone being now worked to the required shape, it is necessary to work the moulded section. Apply on to the twisted joints, the *developed joint mould*, and trammel the fillet lines along the stone, parallel to the arris of the arch curve. The moulding is worked in a series of square checks, as at C and D (Fig. 267), before described. The ovolo mould should be finished by tangents as at E and F, the *boning lines* being left on the stones. Fig. 268 shows the stone worked to shape, and illustrates the curve for the cuneoidal soffit.

To Work a Niche Stone No. 4 (Figs. 490-495).—The moulds required are: *face mould*, *bed mould*, and *joint mould*. Assume that the stone has been sawn to the correct height before being bankered.

True up the top bed surface and use this as the surface of operation. Mark on the *bed mould* and work the face square to the bed and to the front edge of the *bed mould*, then apply the *face mould* to the surface, adjusting it to the *bed mould* (Fig. 269). Now work the vertical joint and the vertical normal joint to the outline of the *bed mould*, marking the joint mould on the normal joint surface. The curve between E N should now be worked. This could be

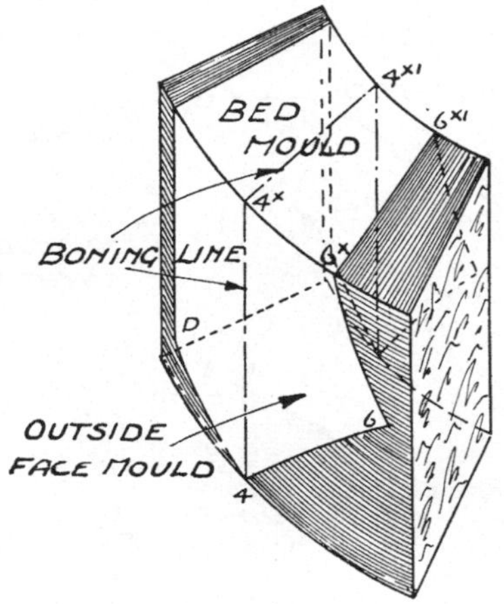

FIG. 265.

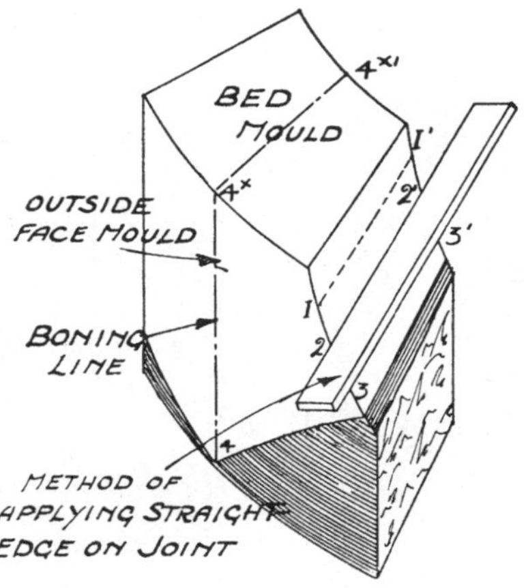

FIG. 266.

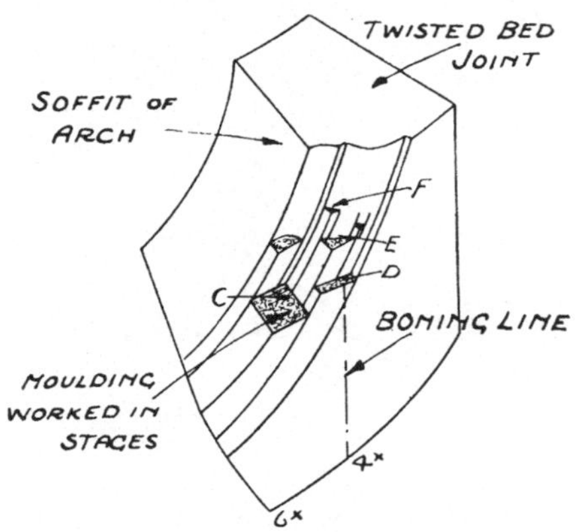

FIG. 267.

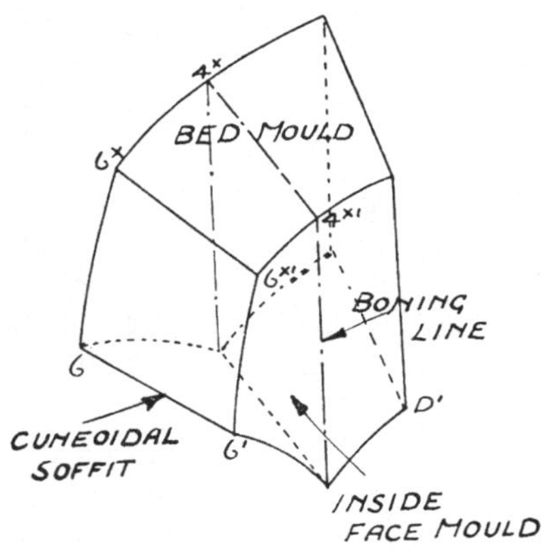

FIG. 268.

WORKING VOUSSOIR FOR SEMICIRCULAR ARCH IN CYLINDRICAL WALL.

accomplished by squaring down from the line marked on the bed, or by working a draft to the curve between E and N, the curve being tested with a *reverse templet* of the curve. A pliable steel straight-edge can be squeezed into the curve, and the horizontal arris line E N marked on the stone, to the straight-edge.

Next work the conical bed joint between the arris line J J and E N, the surface being tested by applying a straight-edge normal to the curve. Mark on this surface the outline for the horizontal seating, as in Fig. 270. The bottom horizontal and conical bed surfaces should be worked next. The true angle and intersection of these surfaces can be obtained by working to a *reverse templet*, cut to the angle made by the curve line E C and the inclined line C D, and applied at intervals on the spherical surface between C and M. As an alternative, the line D D on the bed surface, representing the plan of the line of intersection, may be squared down from the bed surface, with the aid of a *sinking square*, as shown in Fig. 271, the conical surface being worked between the arris line C M and the intersection line D D. Now work the clean

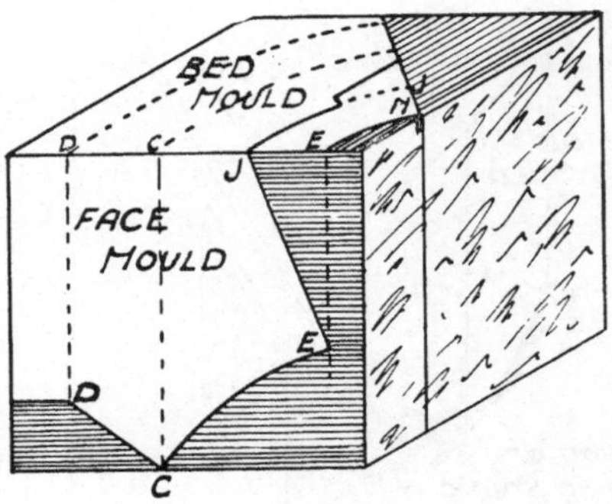

FIG. 269.

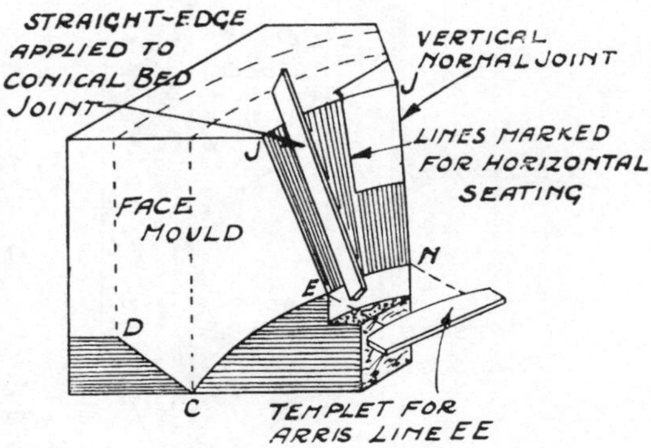

FIG. 270.

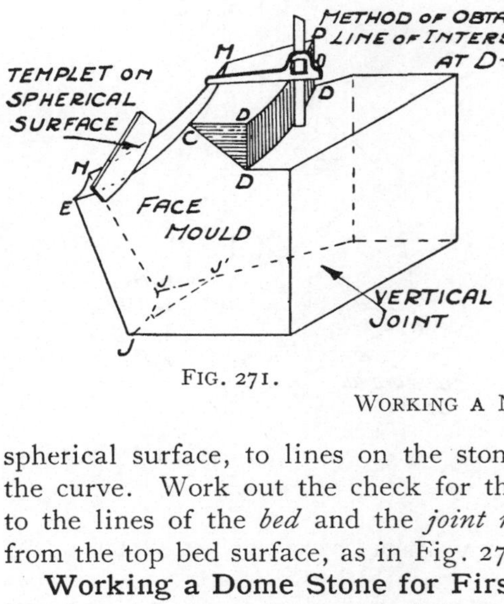

FIG. 271.

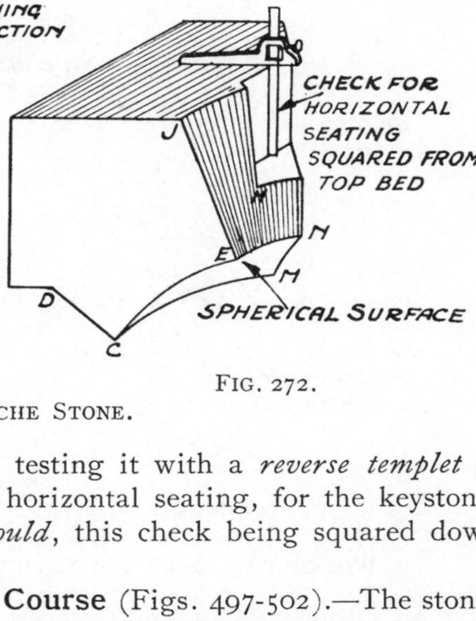

FIG. 272.

WORKING A NICHE STONE.

spherical surface, to lines on the stone, testing it with a *reverse templet* of the curve. Work out the check for the horizontal seating, for the keystone, to the lines of the *bed* and the *joint mould*, this check being squared down from the top bed surface, as in Fig. 272.

Working a Dome Stone for First Course (Figs. 497-502).—The stones for this course are best worked from a rectangular block, contained in the length of the stone in plan, the overall width of the *bed mould*, and the vertical height of the section.

The moulds required are *bed mould* and *section*, templets for the horizontal arris line B on the top bed and the horizontal arris line C on the bottom bed; also reverse templets for the inside and outside spherical faces.

Assuming that the stone has been sawn to the correct height, mark the *bed mould* on the surface chosen for the bottom bed (Fig. 273). Prick through the mould the outline of the arris line C C. Now work the vertical normal joints square from the bottom bed, to the outline of the *bed mould*, and apply the *section mould* to these joint surfaces, adjusting it to the lines marked on the bottom bed. Mark the *bed mould* on the top surface of the stone, at the same time pricking through the mould the outline of the arris line B B. In order

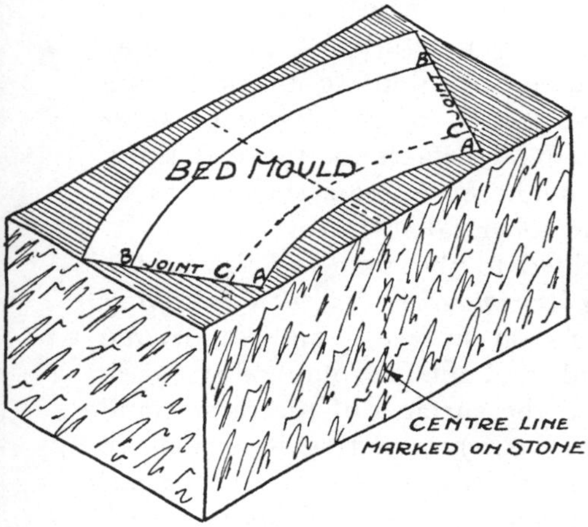

FIG. 273.

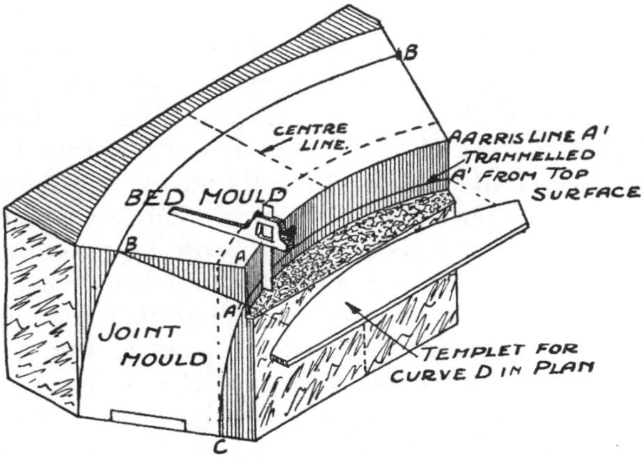

FIG. 274.

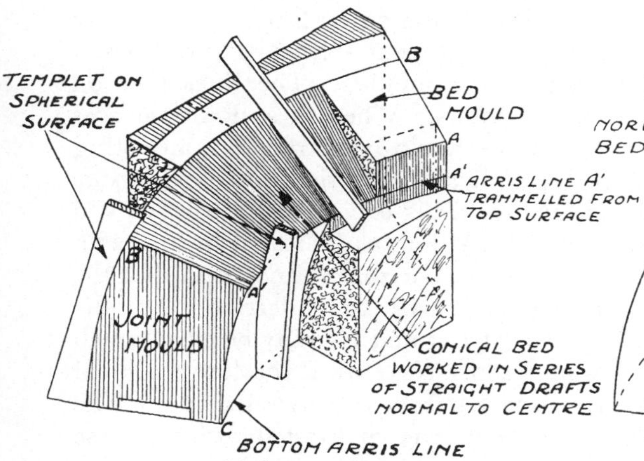

FIG. 275.

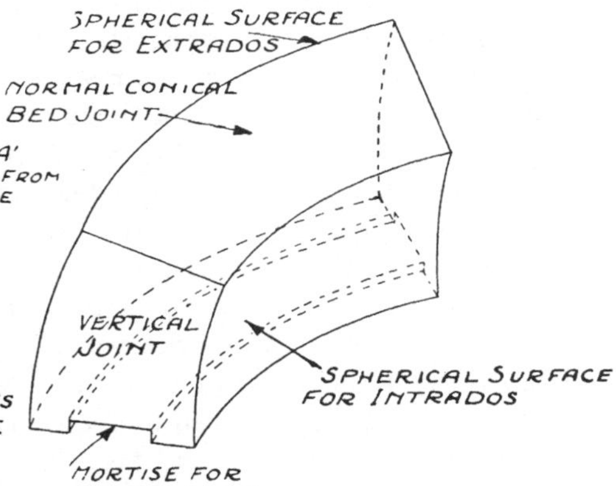

FIG. 276.

WORKING DOME STONE.

to obtain the correct position for the horizontal arris line A′ A′, it is necessary to work a portion of the inside surface to the outline of the *bed mould* between A A. This may be accomplished by working down from the plan to the curve marked on the stone from the bed mould, and testing it by applying a *sinking square*. A draft may be worked along the stone at the level of the arris line, in which case the curve should be tested by applying a *reverse templet* of the curve (Fig. 274). Trammel the line representing the arris line along this curve from the top surface. The conical bed may now be worked between the line A′ A′ and the outside top arris line B B (Fig. 275). This surface should be tested with a straight-edge held normal to the curve of the spherical faces. Now work the outside and inside spherical surfaces to the lines on the stone, testing them with a *reverse templet* of the curves. The mortise for the *table joint* could be worked last. Fig. 276 shows the finished stone.

Working Dome Stone from a Prism.—The stones for the other courses may be worked in a similar manner, but material and labour may be economised, especially if machines are used, by adopting the following method. The stones may be worked to the section of the *prism* or *cover mould* (Fig. 499), thus providing the horizontal surfaces for the bed lines, as at A and B (Fig. 277). A minimum amount of waste stone has to be removed when working the inside and outside spherical surfaces.

Assuming that the stone has been machined to the shape of the prism, the moulds required are: *inner* and *outer plane moulds*, *bed moulds* giving the top and bottom horizontal bed line curves on surface A and B; *section* through dome and *reverse templets* for the inside and outside spherical surfaces.

Commence by marking a centre line round the surfaces of the prism, and apply the *outside plane mould* to surface X (Fig. 277), adjusting it so that the straight edge of the mould coincides with the arris line $4^y 4^y$ of the prism, and the centre line marked on the stone. Next mark on the *inside plane mould*, adjusting it as before, to the centre line and to the arris line $2^x 2^x$ (Fig. 278). Mark on the *bed mould* for the horizontal bed line on surface A and the *bed mould* on surface B, adjusting them also to the centre line. Draw lines C and D, and E and F connecting up the various moulds. The lines now on the stone give the position for working the joints. When these joint surfaces are worked, the true section through the dome should be marked on, adjusting it to points C, D and E. Next work the top conical bed to the lines of the *inside plane mould* and the *bed mould* on surface A. This conical surface is worked true by applying a straight-edge in a normal position, as indicated in Fig. 279.

Remember that the arris lines C C and E E are horizontal lines, and that the width of the conical beds is parallel; then it will be understood that the outline of the arrises for the spherical faces at D D and F F may be trammelled from these arris lines, as in Figs. 279 and 280. Next work the lower conical bed, as Fig. 280, and trammel the arris line F F from the arris line E E. The inside and outside spherical faces may now be worked and tested with a reverse templet of the curve, as in Fig. 281.

Working Stone No. 3 for Pendentive Dome (Figs. 504-513).—The moulds required are: *face mould*, *bed mould*, *inclined joint moulds* for stones 2 and 3, and 3 and 4, *vertical joint mould* and *reverse templets* for inside and outside spherical surfaces, and *templet* for curve 4^x.

Commence by selecting a *surface of operation*, for which in this case the face of the stone is best. Assuming that the surface is sawn, mark on the *face mould*, and work the top bed surface square to the face. Now apply the *bed mould*, adjusting it to the position of the *face mould* (Fig. 282).

Next work the vertical normal joint square from the top bed surface, to the outline of the bed mould, adjusting it to lines squared down the surface

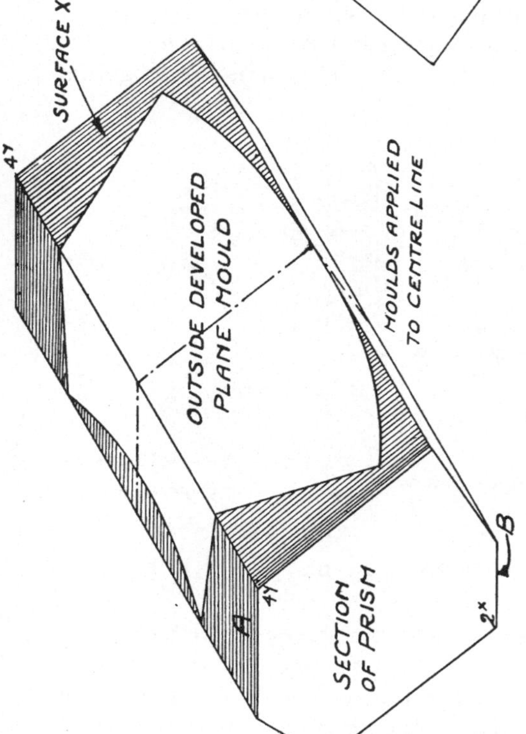

FIG. 277.

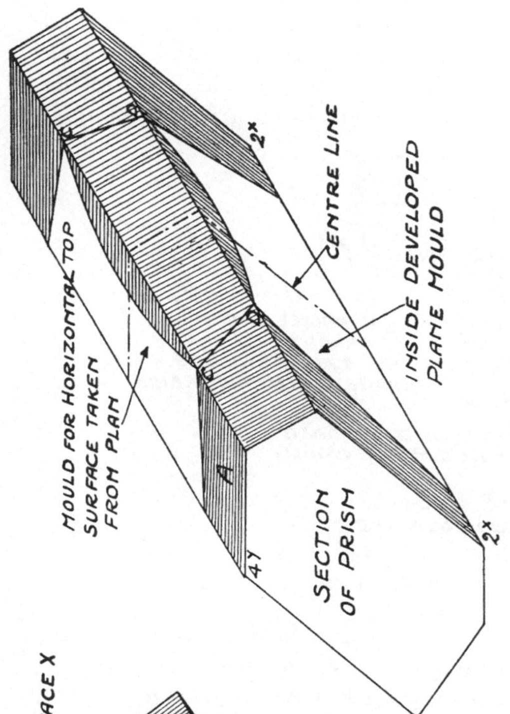

FIG. 278.

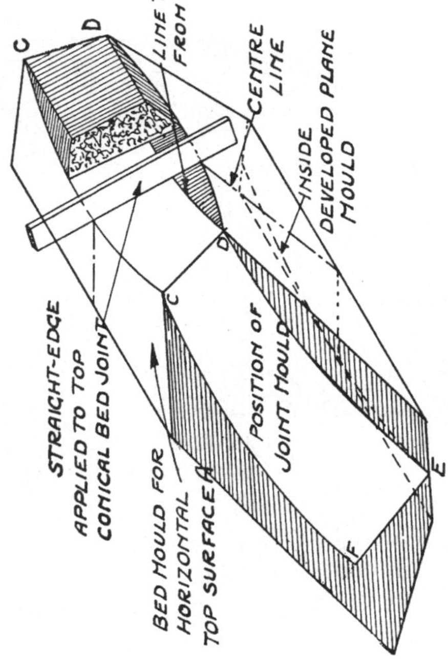

FIG. 279.

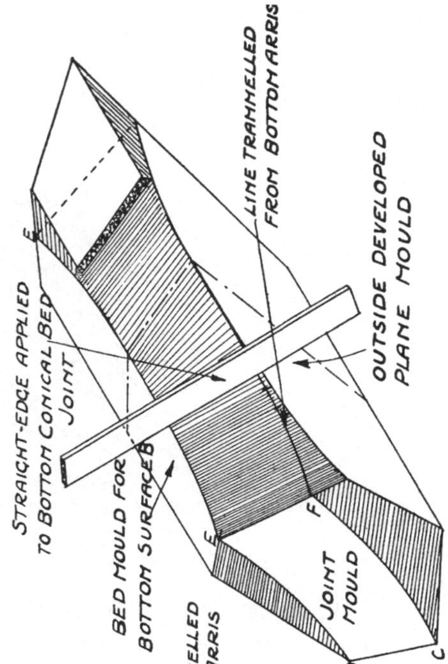

FIG. 280.

WORKING DOME STONE FROM PRISM.

from points 3^y, 4^y, 4^x. The point 4^y, being on the same level as the bed surface, governs the position of the mould in vertical height. The soffit portion of the arch between 3 and 4 may now be worked square from the face, to the outline of the arch curve, and tested, if required, by applying a *reverse templet*

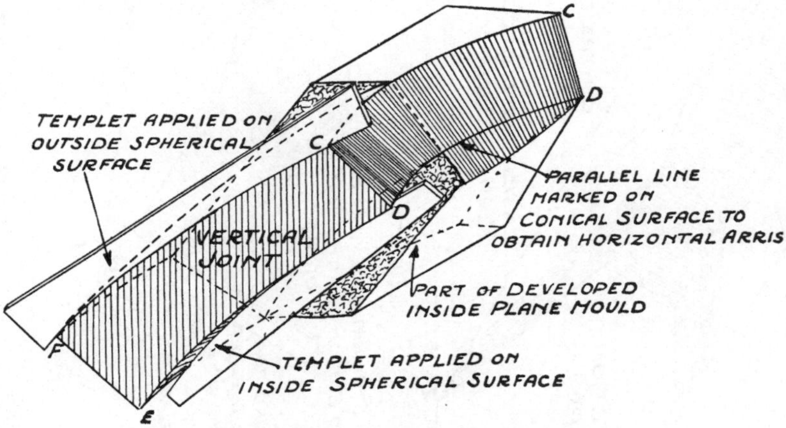

FIG. 281.—WORKING DOME STONE FROM PRISM.

of the arch curve. Next obtain the arris line for the horizontal bed joint $4^x 4^x$. This may be accomplished by working down square from the top bed surface, to the line of the curve $4^x 4^x$, and testing with a *reverse templet* of the curve and trammelling the line $4^x 4^x$ from the top bed. This arris line, together

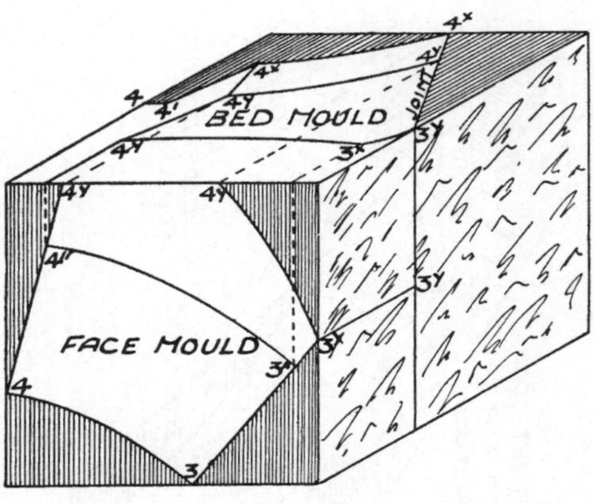

FIG. 282.

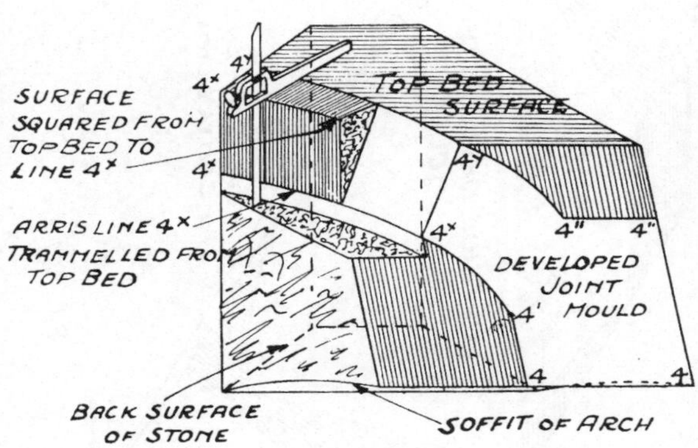

FIG. 283.

WORKING A STONE FOR PENDENTIVE DOME.

with the curve bed line $4^y 4^y$, determines the outline of the conical bed joint, which can now be worked and tested with a straight-edge, as in Fig. 283. Now work the normal arch joints square to the face of the stone, and apply on these surfaces the developed joint moulds. Next work the outside spherical face, the outline of which is determined by the lines already on the stone; the intersection of the extradosial curve and the top surface of the archivolt arch being worked into, by squaring and gauging from the face of the arch, along the curved line marked on the stone from the *face mould*, and the working of the spherical surface to the *reverse templet*, as in Fig. 284.

To work the inside spherical surface, first obtain the width of the inside face of the arch at 3′ 4′ by squaring from the soffit and applying a *sinking square*, which at the same time gauges the width required for the face, and produces the line of intersection with the inside spherical surface, as in Fig. 285. When this line of intersection has been obtained, the inside spherical face may be worked and tested with a *reverse templet* of the curve.

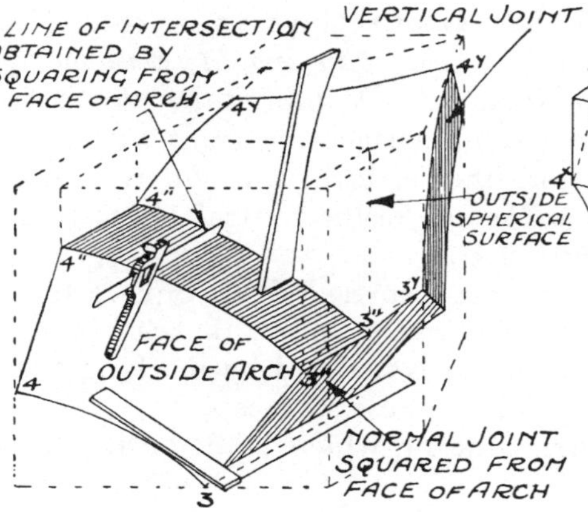

FIG. 284.

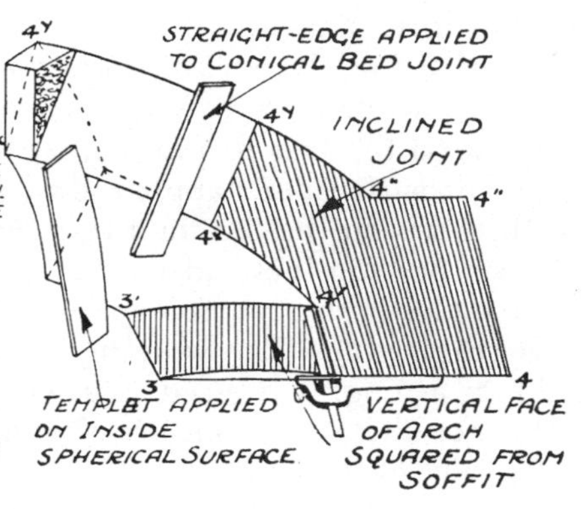

FIG. 285.

WORKING A STONE FOR PENDENTIVE DOME.

Working Stone No. 2 for Intersecting Vaults (Figs. 532-539).— The moulds required are: *bed mould*, *face mould*, and *joint moulds*. Select a piece of stone to the size of the *bed mould* and to the full height of the *section mould* for the small vault.

Commence by working the *surface of operation*, which would be the top bed surface. If the stone is supplied as a slab sawn to the required height, make sure that the *surface of operation* is true and out of *winding* by *boning*. Mark on the *bed mould* and work the joints square from the bed, to the lines transferred from the *bed mould*. Apply the *joint moulds* to these surfaces, adjusting them to the lines marked on the top bed surface (Fig. 286). Next work a surface square from the top bed, approximately to the arris line D D′′. Mark on this surface the *section mould* for the small vault (Fig. 287).

Next work the inclined normal bed joint J D through the stone, and mark on this surface the *developed joint mould*. The inclined bed joint between J^2 D′′ and J^2 and D may also be worked. The curved soffit for the small vault should now be worked, and tested by applying a straight-edge from the front to the back of the stone, or by applying a *reverse templet* to the curve.

It is now necessary to determine the point of intersection for the *intradosial groin line* on the bottom bed at C, by squaring from the joint at C″, along the bottom surface, to intersect the line C C′ in C (Fig. 288). Scribe a *mitre line* on the curved surface between points C and D. This will represent the *intradosial groin line*. The curved soffit for the large vault can now be worked as explained.

The inclined bottom bed joints should now be worked. To do this, work to the bottom arris lines C C′, C C″, and the lines on the joints. The inclined surfaces meet in a straight intersecting line between H^2 and C. These surfaces may be worked true by squaring in from the joints, as Fig. 288, and boning with the arris lines C C′, C C″. The *developed inclined joint mould* is now applied to the lower joint at H C, thus completing the working lines for the extradosial surfaces. These are worked into the extradosial groin line and tested frequently by applying the reverse templet of the extrados curve (Fig. 289).

Working Stone No. 2 for Welch Groin (Figs. 540-548).—Moulds required are: *face mould*, *bed mould*, *vertical joint mould*, *developed joint moulds Nos.* 1 *and* 2, and *developed soffit mould*. Assuming that the stone has been sawn to the size required, make the top surface the *surface of operation* and mark on the bed mould. Next work the vertical joint square to the face and the top surface, and apply the *section mould* for the vertical joint, adjusting it to the lines on the face and bed (Fig. 290).

Now work the inclined normal joints between H and 2. Square the arris line 2 2 along this surface from the face, and mark on the *developed joint mould* (Fig. 291). The working of the stone is assisted by marking the face mould on a roughly dressed or sawn inside vertical surface.

Mark on the bottom surface the arris line 1 1, which is the horizontal arris for the soffit of the arch.

The soffit surface can now be worked straight through from the face to the back surface, and tested with a *reverse templet* of the curve.

Mark on this surface the *developed soffit mould* for the stone No. 2, keeping it correct to the arrises of the surfaces already worked. This mould gives the curve for working the inside cylindrical surface and also the groin line on the intrados (Fig. 292). Next work the bottom bed joint at points 1 and 4. The part between 4 4 is horizontal, and the part between 4 and 1 is inclined. These surfaces may be worked square from the face, or to the lines marked on the inside vertical surface. Next mark on this inclined surface the *developed joint mould No.* 1, which gives the outline for the inside cylindrical surface at that part. The inside cylindrical surface can now be worked, and tested with a *reverse templet* of the curve or a straight-edge held in a horizontal position (Fig. 293).

Now work the top horizontal clean surface on the outside at J, including the portion of the outside vaulting surface between H and J (Fig. 291). The outline of the curve on the inclined joint is obtained from the *developed joint mould*.

Working Stone No. 2 for Interpenetration of Dome (Figs. 525-531). —Moulds required are: *face mould*, *bed mould*, *inside developed face mould*, *developed joint mould*, *developed soffit mould*, and *reverse templets* for the curves. Select a piece of stone to clear the width and height of the *face mould* and the depth of the *bed mould*.

Assuming that the top and bottom bed surfaces are sawn to the height of the *face mould*, and also one face sawn square to these, apply the *face mould* on the face of the stone, keeping it flush with the bed surfaces. Then mark the *bed*

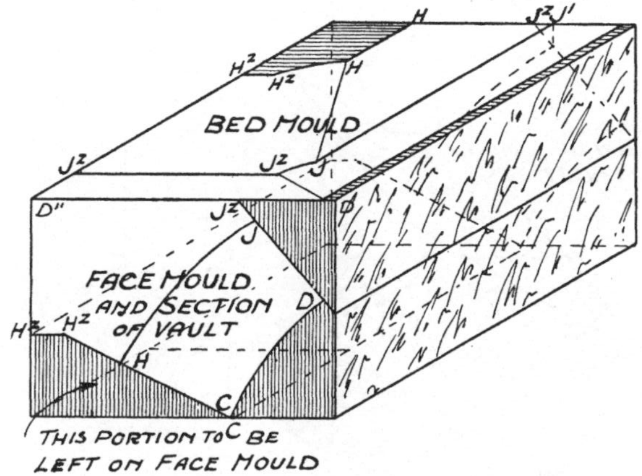

FIG. 286.

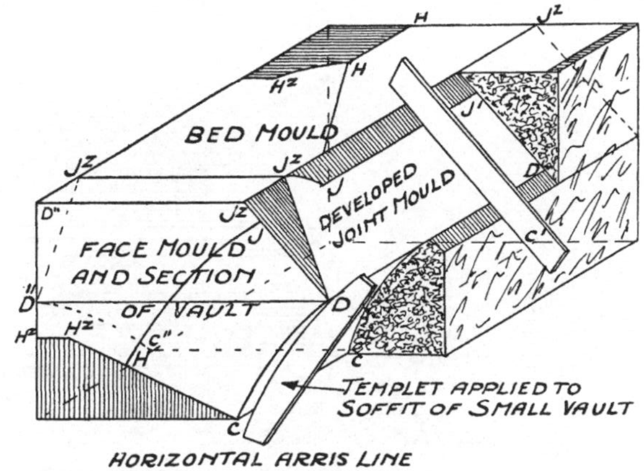

FIG. 287.

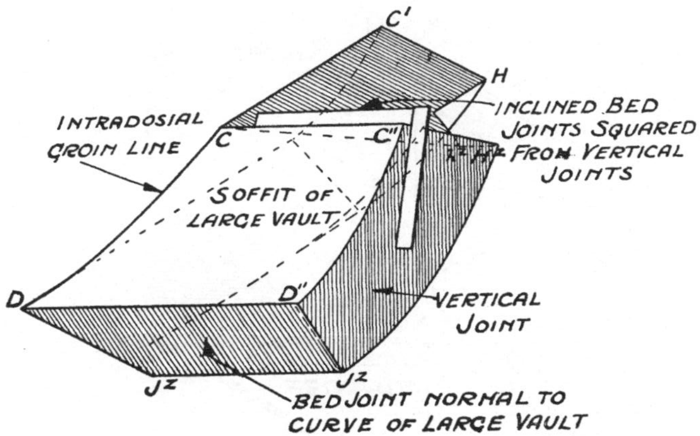

FIG. 288.

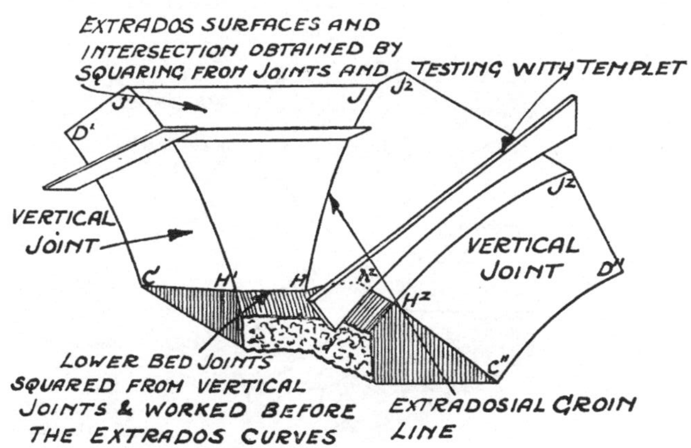

FIG. 289.

WORKING GROIN STONE FOR INTERSECTING VAULTS.

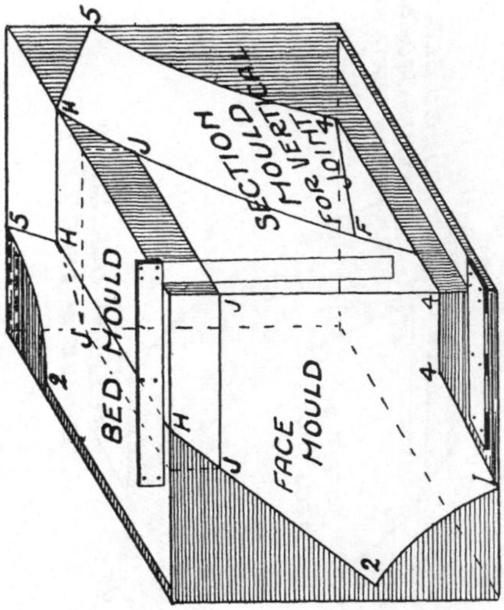

FIG. 290.

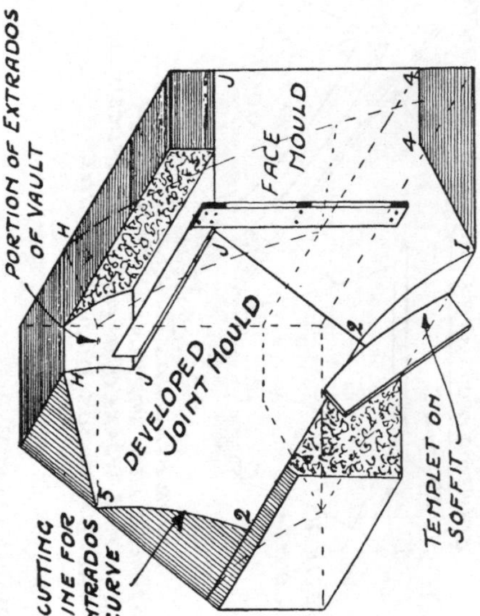

FIG. 291.

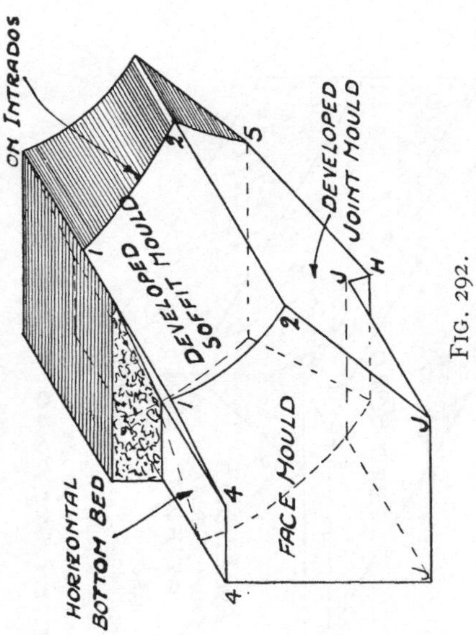

FIG. 292.

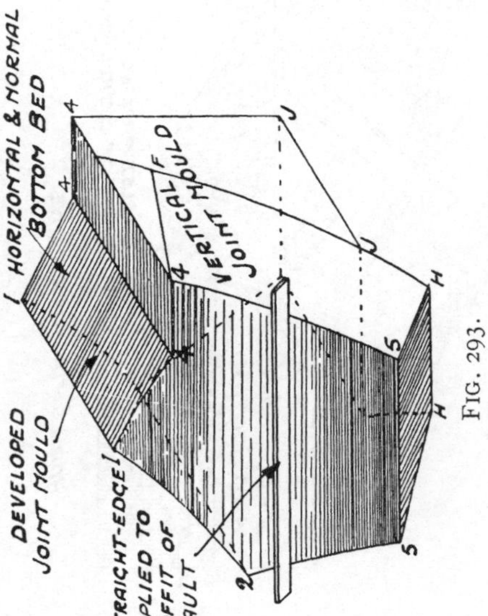

FIG. 293.

WORKING A GROIN STONE FOR WELCH GROIN.

mould on the bed surfaces, adjusting it to the lines on the face mould (Fig. 294). Now work the vertical normal joint P R, and apply the true section mould of the dome, adjusting it to the lines on the beds transferred from the bed moulds.

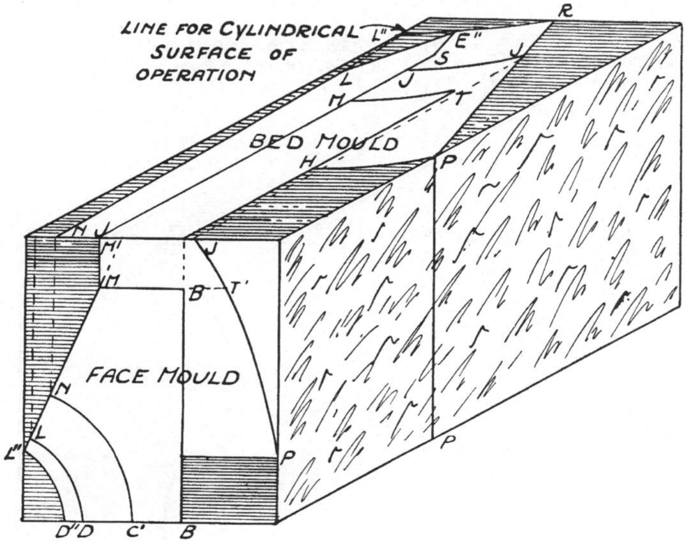

FIG. 294.

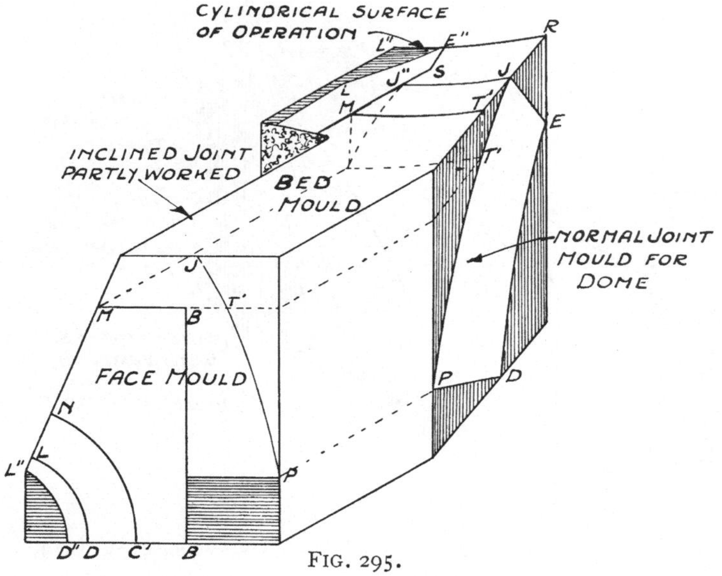

FIG. 295.

WORKING STONE FOR INTERPENETRATION OF DOME.

To work the radiating arch joint and the conical soffit of the arch, it is best to work a vertical and cylindrical surface on the inside of the stone, to the curve marked R E″ L″ (Fig. 295), and apply to this surface the *developed inside face mould*. Work the radiating joint and apply the *developed inclined joint mould*. The conical soffit may now be worked to the lines on the outside face and the inside cylindrical surface, and tested by applying a straight-edge, as suggested in Fig. 296. Squeeze into this surface

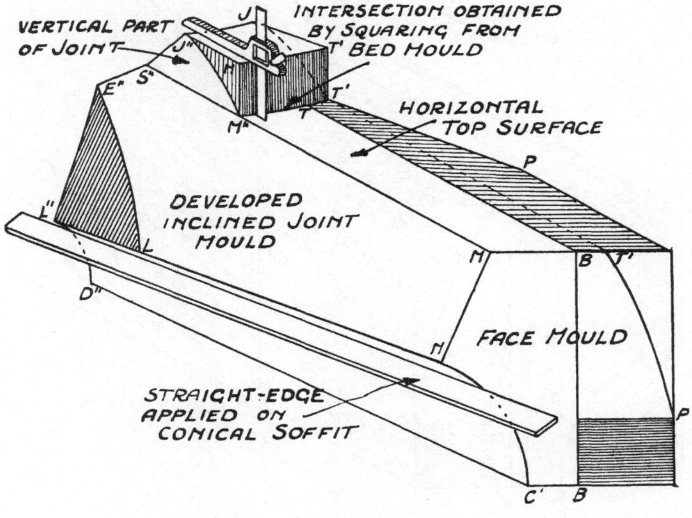

FIG. 296.

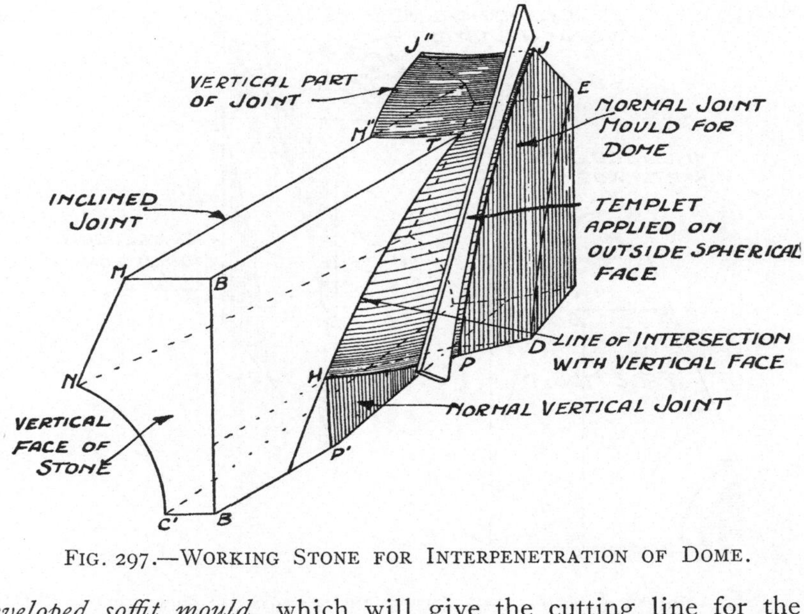

FIG. 297.—WORKING STONE FOR INTERPENETRATION OF DOME.

the *developed soffit mould*, which will give the cutting line for the inside spherical surface, also a portion of the *groin line*. Trammel the horizontal arris line at E″ E along the cylindrical surface, and work the conical bed surface of the dome between J E and J″ E″. Next work the horizontal top surface at M B, obtaining the intersection of the outside domical surface with the horizontal surface, by squaring down from the bed surface at M T′. Now work the external spherical surface to the lines on the stone, testing it with a reverse templet of the curve. This spherical surface dies into the vertical side face B B and also the horizontal surface at M″ T (Fig. 297). Now work the bottom conical bed at P H, D D. The arris line P H at the outside spherical surface is horizontal, so it can be trammelled from the bottom bed surface and worked to a straight-edge adjusted to both arris lines. This lower conical bed surface dies against a vertical normal joint.

The inside spherical surface can now be worked to the lines on the stone and tested with a *reverse templet* of the inside spherical curve (Fig. 298).

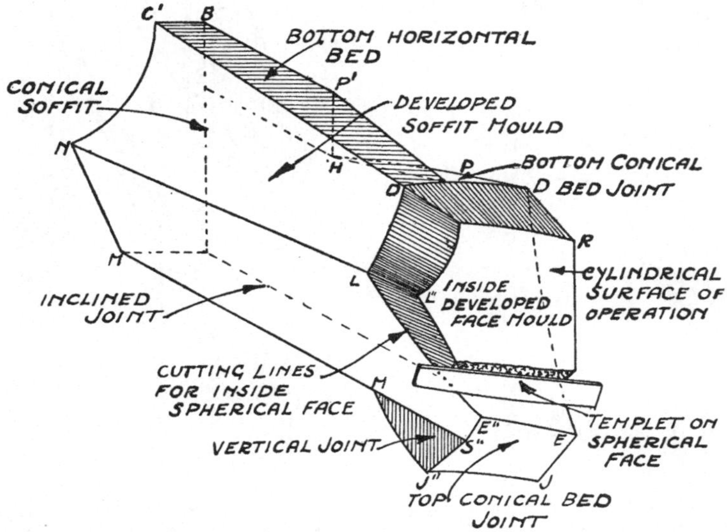

FIG. 298.—WORKING STONE FOR INTERPENETRATION OF DOME.

Working Stone for Wing Wall of Entrance Stair (Figs. 550-563).—The moulds required are: *cover mould*, *plane mould*, and *developed face moulds* for concave and convex surfaces. If the stone is to be moulded, a *joint mould* or *section* would be required. Commence by selecting a slab of stone to the thickness and length of the *cover mould* and sufficiently wide to clear the total width of the *plane mould*. Assuming that the stone is sawn to the size of the moulds, apply the *cover mould*, and transfer the position of the *boning line* on to the stone. This line determines the pitch of the stone. From the points where the *boning line* meets the top and bottom surfaces, as at A and B, draw lines, square from the front surface, across each bed. The *developed plane mould* may now be applied, this being adjusted, so that the *boning line* on the mould coincides with the square line on the stone. It is most convenient to keep the *plane mould* flush with the front surface at points A and B. The mould must be applied *lines-up* on the one bed and *lines-down* on the other.

Now work the convex and concave surfaces in a series of straight drafts parallel to the *vertical boning line*, as at points 1, 2, 3, 4, 5, etc. When these

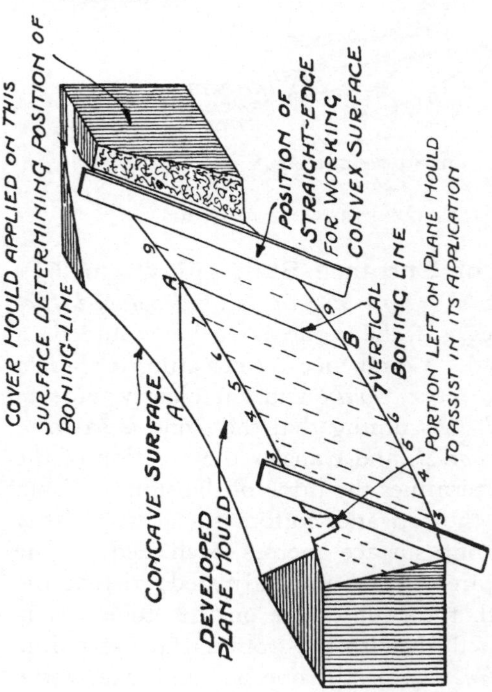

FIG. 299.

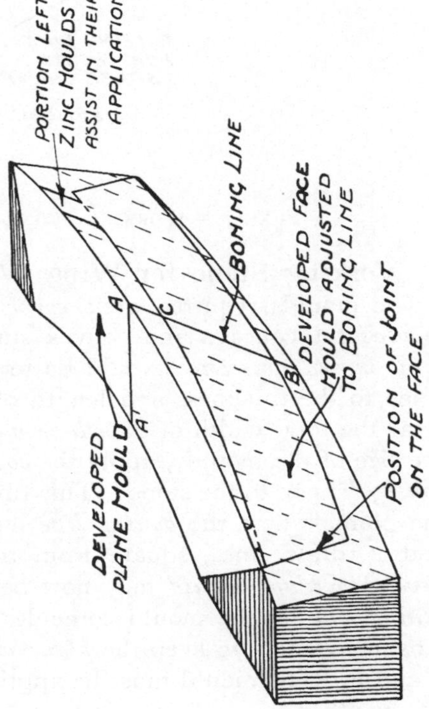

FIG. 300.

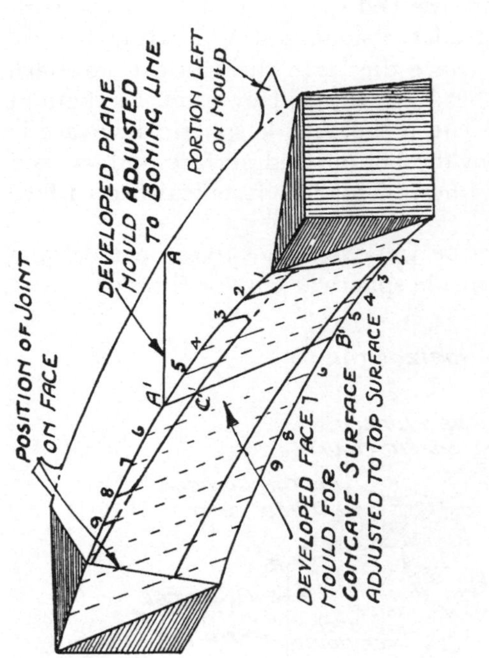

FIG. 301.

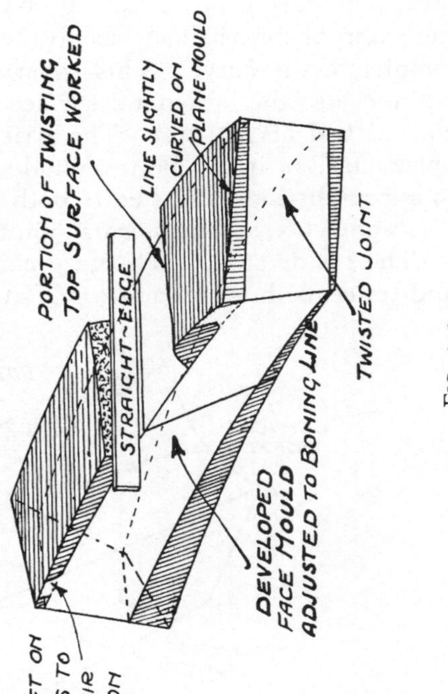

FIG. 302.

surfaces are worked, the *vertical boning line* must be marked down each surface, as shown in Fig. 299. The *developed face mould* should now be applied to these surfaces, and adjusted to the *vertical boning lines*. Great care must be exercised in adjusting these moulds, for their position determines the amount of twist in the top and bottom surfaces of the stone. The exact position of the moulds may be determined by keeping the distance A C on the *boning line* of the convex surface (Fig. 300) equal to the distance A′ C′ on the concave surface (Fig. 301), or by adjusting these moulds to the top or bottom surfaces of the slab by means of the projecting pieces of zinc left on the developed face moulds, as explained in the setting-out of the geometrical stair plinth (Figs. 552-559). These *face moulds* also give the position and lines for working the twisted joints. The top and bottom twisted surfaces may now be worked. In working these surfaces, a straight-edge may be applied in a position across the surface, normal to the curve of the wall in plan, because all normal lines are horizontal, and therefore straight (Fig. 302).

When these surfaces are worked, connect the joint lines, determined by the inside and outside *face moulds*. The joints may now be worked to these lines.

If a moulded section is required, apply the *section mould* to the joints and work the moulding in series of checks, trammelled from the top and bottom surfaces, reverses being cut and applied where desired.

CHAPTER VI

BUILDING STONES

Classification of Rocks—Igneous Rocks—Composition of Granites—List of Granites—Aqueous Rocks—Classification of Sandstones—Limestones—Magnesium Limestones—Magnesium Sandstones—Metamorphic Rocks—Durability of Building Stones—Choosing a Building Stone—Current Bedding Planes or Natural Bed—Preservation of Stone.

ALTHOUGH the craftsman is not usually called upon to select a stone for a proposed building, it is necessary that he should become acquainted with the various stones upon which he is to operate, and so be in a position to give advice in the selection of stones. A great deal of useful information may be gained by actual contact with the various stones in the shop, and keen observation on the part of the craftsman, as to the behaviour of the stones which have been exposed for any length of time, in various buildings.

If this information is to be definitely applied, it is necessary that it should be augmented by a careful study of the *geology* of building stones, so that the student may be in a position to state why certain stones have not good weathering qualities, whilst other stones appear to withstand the attacks of the atmosphere quite well.

It is not the author's intention to deal fully with this important subject in this work, but simply to supply the student, in these notes, with an incentive to seek fuller information from the various books which deal exclusively with the *geology of building stones* and the causes of their decay.

Building Stones.—These may be divided conveniently into three types of rocks :—

1. Igneous Rocks.
2. Aqueous Rocks.
3. Metamorphic Rocks.

In dealing with the classification of these rocks, it is necessary to understand something of their chemical composition and physical structure.

Igneous Rocks.—Granites come under the heading of "Igneous Rocks." Whilst in a practical way we include various rocks under the title of granite, because of their similarity in structure and methods of working, it is essential to realise the difference between *true* granite and other rocks so called.

Granite is a rock made up of granular crystalline particles, which are united without the aid of cementing material, having been formed into a compact mass by crystallisation from a fluid, under extreme pressure,

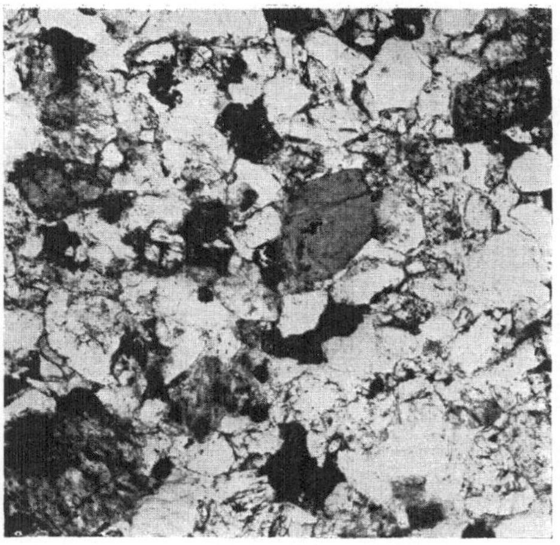

Darley Dale.

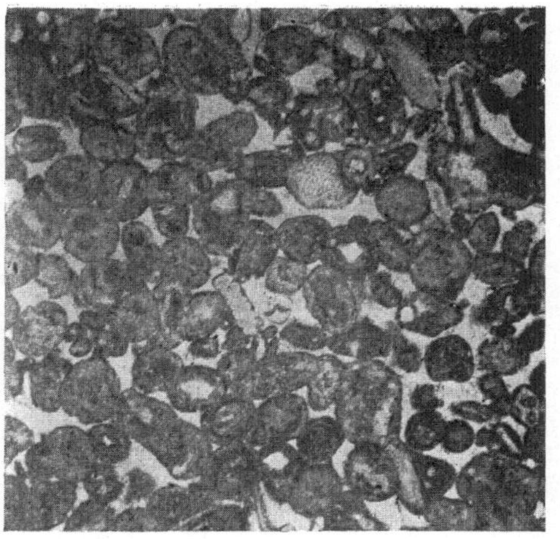

Portland (Whitbed).

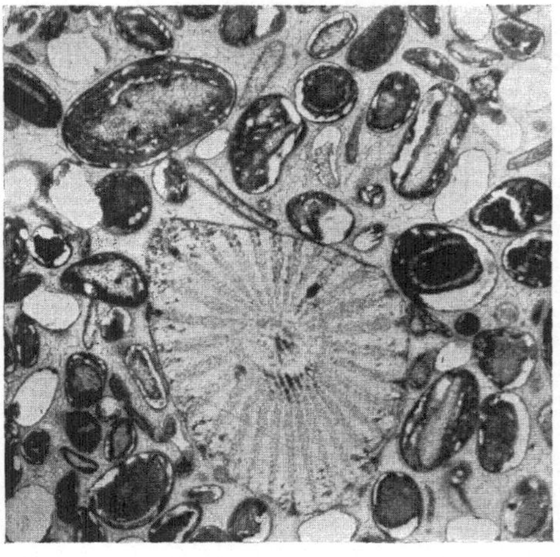

Box Ground.

Hopton Wood.

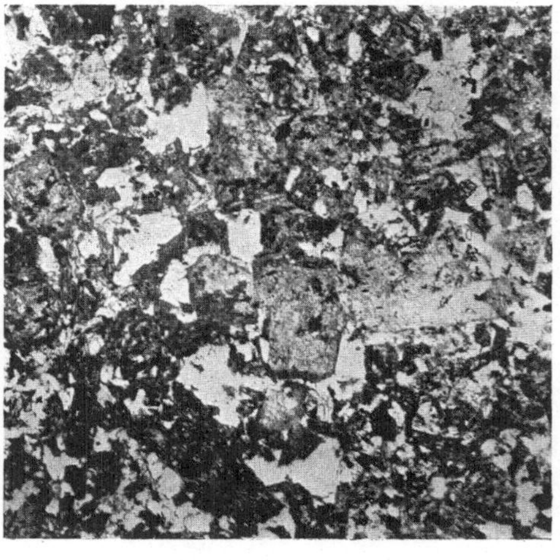

Biotite Hornblende Granite, Mount Sorrel.

Muscovite Biotite Granite, Rubislaw.

FIGS. 303, 304.

below the earth's surface, and having cooled slowly. The essential mineral constituents are *quartz*, *felspar*, and *mica*.

Syenites.—This title is usually given to *igneous rocks* which have *hornblende* among their constituents, although this classification is not strictly correct for *true syenites*. There is also a decrease in the percentage of silica, with a corresponding increase in the proportion of *iron oxide* and *plagioclase felspar*.

Trachyte is a rock of *volcanic* origin equivalent to *syenite*.

Basalt is the *volcanic* equivalent to the basic rocks, the essential constituents being *lime-soda*, *felspar*, and *pyroxene*.

THE CONSTITUENTS OF GRANITE

Quartz may be regarded as chemically unalterable by atmospheric influence, although disruption of the surface of the rock may result from extreme changes of temperature.

Felspar.—This mineral usually occupies about one-half of the bulk of the rock, and is called "*orthoclase*" when it contains *potash*, *alumina*, and *silica*. Sometimes other species are present called "*oligoclase*" (containing more or less *soda*). *Felspars* are the most easily distinguished minerals, and of variable composition: the *potash* felspars decompose less readily than those containing *soda* and *lime*. It is because of this that the composition of the felspars is an important factor in the determination of the durability of a granite.

Mica.—The micas are hydrated silicates of aluminium and potassium, sodium and lithium. They may be of the white variety known as "*muscovite*," or the dark ferro-magnesium mica called "*biotite*." The latter is a source of weakness tending to decompose, whilst the white potash mica stands remarkably well.

Hornblende.—This mineral confers a high degree of toughness upon the rock, but there is an element of uncertainty as to its durability. The best granite should contain small mica crystals, evenly distributed. The uniformity is occasionally spoiled by patches of different grains, or by dark bunches of various minerals.

The hardness and crystalline texture of granites enable them to take a high polish. They are specially suitable for all positions where hard wear is expected—*e.g.*, *plinths*, *piers*, *columns*, and *door-jambs*. *Polished* granite is rightly esteemed among building materials for its combination of strength and beauty. There is no other material so durable, decorative, and so easily cleaned. The variable proportions of the coloured minerals, such as "*pink felspar*," "*green hornblende*," and "*black biotite*," occurring in granite, produce a large variety of tints.

LIST OF GRANITES

Scotch Granites.—CREETOWN (*Biotite*).—Colour, light grey; medium grained; quarried at Creetown, Scotland.

Physical Properties—

Weight per cubic foot	169 lbs.
Crushing resistance per square foot	630 tons
Specific gravity	2.73

DYCE GRANITE (*Biotite*).—Colour, grey; fine texture; quarried at Dyce, near Aberdeen.

Physical Properties—

Weight per cubic foot	165 lbs.
Crushing resistance per square foot	1,105 tons
Specific gravity	2.58

DANCING CAIRNS GRANITE (*Muscovite biotite*).—Colour, light bluish grey; medium grained; quarried near Aberdeen.

Physical Properties—

Weight per cubic foot	170 lbs.
Crushing resistance per square foot	900 tons
Specific gravity	2.74

KEMNAY GRANITE (*Muscovite biotite*).—Colour, light silvery grey; medium grained; quarried near Aberdeen.

Physical Properties—

Weight per cubic foot	160 lbs.
Crushing resistance per square foot	1,212 tons
Specific gravity	2.58

PETERHEAD GRANITE (*Biotite*).—Colour, red; coarse grained; quarried near Peterhead, Aberdeenshire.

Physical Properties—

Weight per cubic foot	158.5 lbs.
Crushing resistance per square foot	1,470 tons
Specific gravity	2.54

PETERHEAD GRANITE (*Muscovite biotite*).—Colour, bluish grey; fine grained.

Weight per cubic foot	171 lbs.

ROSS OF MULL (*Biotite*).—Colour, pink to warm red; rather coarse in texture; quarried in Argyllshire.

Physical Properties—

Weight per cubic foot	176 lbs.
Specific gravity	2.80

RUBISLAW GRANITE (*Muscovite biotite*) (Figs. 303 and 304).—Colour, bluish grey; fine grained; quarried near Aberdeen.

Physical Properties—

Weight per cubic foot	163 lbs.
Crushing resistance per square foot	1,289 tons
Specific gravity	2.61

Sclattie Granite (*Muscovite biotite*).—Colour, light bluish grey; fairly close grained; quarried near Aberdeen.

Physical Properties—

Weight per cubic foot	161 lbs.
Crushing resistance per square foot	850 tons
Specific gravity	2.58

English Granites.—There are several granite quarries in various parts of England, the chief ones being in Cornwall and Devonshire, whilst those of Leicestershire, North Wales, Malvern Hills, produce so-called syenites, or hornblendic granites.

De Lank Granite (*Muscovite biotite*).—Colour, light greenish grey; rather coarse grained; quarried near Bodmin, Cornwall.

Physical Properties—

Weight per cubic foot	165 lbs.
Crushing resistance per square foot	1,171 tons
Specific gravity	2.65

Penryn Granite (*Muscovite biotite*).—Colour, light grey; coarse grained; quarried at Penryn, Cornwall.

Physical Properties—

Weight per cubic foot	165 lbs.
Crushing resistance per square foot	1,250 tons

Shap (*Biotite*).—There are two different shades, known as "*light shap*" and "*dark shap*." The light shap variety has a grey ground with large crystals of pink felspar, whilst the dark variety is dark brown with large pink felspar crystals. These crystals are sometimes 1 in. in length. Coarse grained; quarried at Shap, in Westmorland.

Physical Properties—

Weight per cubic foot	160 lbs.
Crushing resistance per square foot	1,200 tons
Specific gravity	2.57

There are several varieties of foreign granites used in this country, some of which resemble *Labradorites*, and are used extensively for decorative purposes because of the beautiful colourings exhibited on the polished surfaces, due to the felspar crystals (*soda-orthoclase*). The following is a list of a few well-known foreign granites:—

Standard Grey, Norwegian.
Rose Swede, America.
Bonaccord Red, Sweden.
Balmoral Red, Sweden.
Emerald Pearl, Norwegian.
Royal Blue, Norwegian.
Bonaccord Black, Sweden.
Deeside Grey, Sweden.

Aqueous Rocks.—All the sedimentary rocks come under this heading. They include *sandstones* and *limestones*, which are formed in deposits by the agency of water and winds, and are known as *stratified rocks*.

The durability of most of these stones depends upon the kind of cementing material by which the grains are held together. This applies chiefly to sandstones. Their colour is chiefly due to the presence of *carbon* or the various *compounds of iron*. Traces of carbon produce the *blue* and *black* tints. Carbonates and sulphides of iron produce *bluish-grey* colours. Anhydrous sesquioxide of iron produces red and brown tints, whilst the hydrous sesquioxide produces *yellow* hues, and the silicates the *green* tints. The colouring of these stones is very important from an æsthetic point of view, both with regard to their suitability of tint and their liability to change after exposure, these factors determining the architectural value of the stone.

Sandstones.—A sandstone consists chiefly of round or angular grains of quartz, with occasionally other minerals, such as *felspar* and *muscovite mica*. These grains may be cemented into a solid mass by *silica*, *carbonate of lime*, *oxide of iron*, and *alumina*. As these rocks owe their origin to the older crystalline rocks, their chief constituent is naturally least likely to be affected by chemical changes.

The quality of sandstones depends upon two things :—

(1) The nature of the grains ; (2) the quality of the cementing material. The size of the grains determines the texture of the stone, which varies from a fine-grained, compact stone to a coarse grit.

The cementing medium is, however, the chief factor in determining the durability and hardness of the stone. It varies in composition and quantity.

Siliceous Cements produce the most durable sandstone. These cements are usually in the form of silica combining the original grains. Such sandstones are hard and difficult to work.

Calcareous Cement is usually in the form of *crystalline calcite* which has been deposited from solution. *Kentish rag* is an example of a stone having a large proportion of *calcite* as a cementing material, thus producing a *siliceous limestone*.

Although a calcareous cement would produce an easily worked sandstone, it would not tend to increase its durability, for the cement would be gradually attacked by *carbonated water*, thus liberating the original quartz grains.

Ferruginous Cement may consist either of *red oxide* or the *brown hydrated oxide of iron*.

Argillaceous Cements.—These seldom produce a durable stone, owing to the fact that they are very absorptive and liable to the effects of frost. The absorptive power of sandstones varies with the degree of coarseness of the stone. Naturally, as the porosity decreases, there is an increase of hardness, with difficulty in working.

Soft sandstones are also liable to mechanical disintegration by wind-blown sand, near the surface of the ground.

The varieties of sandstones are classified according to the peculiarities of their structure or origin as follows :—

Micaceous Sandstone is one containing muscovite mica, often deposited in the bedding planes.

Felspathic Sandstones are those in which felspar partly fills the spaces between the quartz grains.

Ferruginous Sandstones are those in which the cementing material is coloured by *oxide of iron*.

Calcareous Sandstones are those in which the grains are massed together with a calcareous cement.

Quartzite is an indurated sandstone with a siliceous cement.

Grit Stone is a sandstone with sharp angular grains, generally rather coarse.

Tilestone is a thinly bedded sandstone.

LIST OF SANDSTONES

Bramley Fall.—Colour, light brown; mottled and coarse grained; quarried in Yorkshire; suitable for heavy engineering work. The cementing material is silica.

Chemical Composition—

SiO_2, silica	96.58
Al_2O_3, aluminium oxide	1.10
Fe_2O_3, iron oxide	0.34
MgO and CaO, magnesium and calcium oxides	.49
H_2O, water and loss	1.49

Physical Properties—

Weight per cubic foot	162.3 lbs.
Crushing resistance per square foot	552.2 tons
Specific gravity	2.63

Bristol Blue Pennant.—Colour, dark blue grey; fine grained; suitable for most constructional work; quarried at Fishponds, near Bristol. Cementing material consists of silica supplemented by calcium carbonate and oxides.

Chemical Composition—

SiO_2, silica	73.93
Al_2O_3 and FeO, aluminium and iron oxides	13.87
$CaCO_3$, calcium carbonate	3.64
K_2O and Na_2O, potassium and sodium oxides	6.71
H_2O, water and loss	1.85

Physical Properties—

Weight per cubic foot	172 lbs.
Crushing resistance per square foot	1,001 tons
Specific gravity	2.76

Corsehill Stone.—Colour, pink to dark red; fairly fine grain; slightly micaceous; quarried in Dumfriesshire. The cementing material chiefly silica; suitable for dressings, etc.

Chemical Composition—

SiO_2, silica	95.33
Al_2O_3, aluminium oxide	0.59
Fe_2O_3, ferric oxide	1.28
$CaCO_3$, calcium carbonate	1.49
$MgCO_3$, magnesium carbonate	1.31

Physical Properties—

Weight per cubic foot	141 lbs.
Crushing resistance per square foot	635 tons
Specific gravity	2.26

Darley Dale.—Colour, light buff to grey; close grained; slightly micaceous; suitable for external work, and takes a good arris.

Chemical Composition—

SiO_2, silica	96.4
Al_2O_3 and FeO, aluminium and iron oxides	1.3
$CaCO_3$, calcium carbonate	0.36
H_2O, water and loss	1.94

Physical Properties—

Weight per cubic foot	153 lbs.
Crushing resistance per square foot	455 tons
Specific gravity	2.46

Forest of Dean Stone.—There are two varieties, *blue* and *grey*. They are fine grained and slightly calcareous stones; suitable for monumental work and general stone dressings. Quarried in the Forest of Dean, Gloucestershire.

Chemical Composition of the Grey Bed—

SiO_2, silica	80.16
Al_2O_3, aluminium oxide	14.40
Fe_2O_3, ferric oxide	1.65
MgO, magnesium oxide	0.24
CaO, calcium oxide	2.55
$CaSO_4$, calcium sulphate	1.00

Physical Properties—

Weight per cubic foot	149 lbs.
Crushing resistance per square foot	569 tons
Specific gravity	2.39

Hailes Stone.—There are three varieties, *blue*, *pink*, and *white*, all occurring in the same quarry; medium grained and similar in chemical composition to Craigleith stone, which is now unobtainable. It is suitable for all kinds of building work. Quarried at Slateford, near Edinburgh.

Chemical Composition of the White Beds—

SiO_2, silica	96.52
Al_2O_3, aluminium oxide	2.68
Fe_2O_3, iron oxide (ferric)	0.04
FeO, iron oxide (ferrous)	0.19
CaO, calcium oxide	0.4
K_2O and Na_2O, potassium and sodium oxides	0.17

Physical Properties—

Weight per cubic foot	145 lbs.
Crushing resistance per square foot	523.5 tons
Specific gravity	2.33

Hollington Stones.—There are three varieties, *red*, *white*, and *mottled* (red and drab); even grained to coarse grain in the white beds, and micaceous. The cementing material is silica and subsidiary oxides; suitable for general building work. Quarried at Rocester, Staffordshire.

Chemical Composition of the White Bed—

SiO_2, silica	86.64
Al_2O_3, aluminium oxide	8.78
Fe_2O_3, ferric oxide	1.02
CaO, calcium oxide	0.72
MgO, magnesium oxide	0.44
K_2O and Na_2O, potassium and sodium oxides	0.44
H_2O, water and loss	1.96

Physical Properties—

Weight per cubic foot	133 lbs.
Crushing resistance per square foot	260 tons
Specific gravity	2.13

Howley Park Stone.—Colour, light brown; ferruginous and fine grained; suitable for copings, sills, and thresholds, etc. Quarried at Morley, Yorkshire.

Physical Properties—

Weight per cubic foot	140.3 lbs.
Crushing resistance per square foot	466.7 tons
Specific gravity	2.25

Park Spring Stone.—Colour, yellowish drab; micaceous and medium grained. Quarried near Leeds, Yorkshire.

Physical Properties—

Weight per cubic foot	151 lbs.
Crushing resistance per square foot	487 tons
Specific gravity	2.44

Robin Hood Stone.—Colour, bluish grey; fine grained and micaceous; suitable for copings, sills, etc.

Physical Properties—

Weight per cubic foot	145 lbs.
Crushing resistance per square foot	574 tons
Specific gravity	2.33

CALCAREOUS STONES

Calcareous Stones.—The greater part of this group of building stones owe their origin to chemical or organic agencies, although mechanical action has played its part in their production. The general characteristics of limestones is the presence of a large proportion of carbonate of lime, formed chiefly by the accumulation of shells, or calcareous skeletons of marine or fresh-water organisms, which were deposited as sediment in the waters of seas or lakes. Their composition varies considerably with the condition under which they were formed.

The character of a limestone depends (1) upon the nature of the original grains, whether organic remains or oolitic aggregations; (2) upon the extent of the recrystallisation of the matrix.

The *oolitic limestones* are of marine origin, composed chiefly of carbonate of lime and other substances, such as carbonate of magnesia, silica, alumina, and iron.

The *oolites* resemble the roe of a fish, resulting from the accumulation of carbonate of lime, round a small nucleus of fragmentary shells, or grains of mud and sand. The grains are spherical or oval in shape, and may be easily discerned with the naked eye.

The cementing material which holds the *oolitic* grains together may be more or less *siliceous*, or *argillaceous*, but it is chiefly of the same material as the grains—that is, carbonate of lime.

The hardness of these stones differs with the character of the cementing material. The *oolitic* limestones are the most serviceable for general building purposes, both from the *æsthetic* point of view and for durability if rightly chosen and well selected.

The *porosity* of most limestones causes them to be full of quarry moisture (called *quarry sap*), thus rendering them soft and easy to work whilst in this condition. As this moisture evaporates on the surface of the stone, by exposure to the atmosphere, they become harder. When in their fresh quarried state they are liable to injury by frost.

The suitability of limestones for various architectural uses depends upon their structure, colour, and capability for taking a polish.

The *natural bedding planes* are not so distinct in limestones as in sandstones. There are a few exceptions to this rule, as, for instance, Ham Hill Stone. Although the planes are not visible, it is important that their direction should be carefully considered when choosing stones for certain positions in a building.

LIST OF LIMESTONES

Ancaster Stone (*Oolitic*).—There are two distinct beds—"*Weather Bed*" and "*Free Bed*." The *Weather Bed* is often called "*Ancaster Rag*." Brown to grey in colour, coarse grained, partly crystalline, it has good weathering qualities, will take a polish, and is used extensively for chimney-pieces and internal decorative purposes, etc.

The *Free Bed* is cream in colour, fine grained, and, if selected, is suited for general building work. Quarried near Grantham, Lincolnshire.

Chemical Composition of the Weather Bed Stone—

$CaCO_3$, calcium carbonate	95.99
SiO_2, silica	0.11
Al_2O_3, aluminium oxide	0.32
Fe_2O_3, iron oxide	1.48
$MgCO_3$, magnesium carbonate	1.90
H_2O, water and loss	0.20

Physical Properties—

Weight per cubic foot	156 lbs.
Crushing resistance per square foot	552.6 tons
Specific gravity	2.50

Bath Stones (*Great Oolite*).—These include a series of stones quarried or mined in the vicinity of Bath. All the Bath stones are of marine origin. Their colour is from cream to buff according to the amount of oxide of iron present.

The stones of the Bath series are easily worked, and if well selected are suitable for general building purposes.

The following stones belong to the Bath series:—

St Aldhelm, Box Ground (Figs. 303 and 304).—The bedding planes are visible; coarse grained; suitable for general building purposes. Quarried near Bath.

Physical Properties—

Weight per cubic foot	129 lbs.
Crushing resistance per square foot	107 tons
Specific gravity	2.09

Corsham Down.—Moderately fine grained, and fairly regular in texture; suitable for ashlar work, etc. Quarried near Bath.

Physical Properties—

Weight per cubic foot	129 lbs.
Crushing resistance per square foot	128 tons
Specific gravity	2.08

Hartham Park.—Colour, cream to buff; fine to coarse grain; the bedding planes are more or less distinct; suitable for exterior work. Quarried near Bath.

Weight per cubic foot	123.5 lbs.

Monk's Park.—Colour, cream, becoming much lighter in colour when dry; fine grained, fairly compact, and moderately hard; a good, serviceable stone for external and internal dressings, if well selected. Quarried near Bath.

Chemical Composition—

$CaCO_3$, calcium carbonate	95.56
$MgCO_3$, magnesium carbonate	0.40
SiO_2, silica	1.20
Al_2O_3 and Fe_2O_3, aluminium and iron oxides	1.52
H_2O, water and loss	1.32

Physical Properties—

Weight per cubic foot	137 lbs.
Crushing resistance per square foot	223.5 tons
Specific gravity	2.19

Beer Stone (*Cretaceous*).—Colour, dead white, from the chalk series; close grained, and made up of minute fragments of shells; suitable for internal work only. Quarried near Seaton, Devonshire. Chemical composition is practically pure carbonate of lime ($CaCO_3$).

Physical Properties—

Weight per cubic foot	131.7 lbs.
Crushing resistance per square foot	151.2 tons
Specific gravity	2.12

Chilmark Stone (*Oolitic*).—Colour, buff to brown, and fairly fine grain. There are three beds—*Trough Bed*, *Green Bed*, and *Pinney Bed*. All the beds are suitable for general dressings, but should be used with caution externally in town atmospheres. Quarried at Tisbury, Wilts.

Chemical Composition—

$CaCO_3$, calcium carbonate	78.80
$MgCO_3$, magnesium carbonate	3.90
SiO_2, silica	10.34
Al_2O_3 and Fe_2O_3, aluminium and iron oxides	2.51
H_2O, water and loss	4.45

Physical Properties—

Weight per cubic foot	154 lbs.
Crushing resistance per square foot	136.6 tons
Specific gravity	2.48

Clipsham Stone (*Oolitic*).—Colour, buff to cream; medium grained, but shelly. If selected with caution, the stone is suitable for general dressings. Quarried near Oakham, Rutlandshire.

Chemical Composition (A. R. Warnes)—

$CaCO_3$, calcium carbonate	97.56
MgO, magnesium oxide	0.54
Al_2O_3 and Fe_2O_3, aluminium and iron oxides	0.83
Insoluble residue (chiefly SiO_2)	0.84
H_2O, water and loss	0.23

Physical Properties—

Weight per cubic foot	150 lbs.
Crushing resistance per square foot	291.6 tons
Specific gravity	2.42

Doulting Stone (*Oolitic*).—Colour, light brown, or buff, to cream. There are two beds—*Bramble-ditch* and *Chelynch Bed*; the first-named is a fine-grained stone, whilst the latter is coarse. Suitable for general external dressings. Quarried at Shepton Mallet, Somerset.

Chemical Composition—

$CaCO_3$, calcium carbonate	95.83
Al_2O_3, aluminium oxide	0.91
Fe_2O_3, iron oxide	0.78
SiO_2, silica	1.96
H_2O, water and loss	0.52

Physical Properties—

Weight per cubic foot	150 lbs.
Crushing resistance per square foot	211.6 tons
Specific gravity	2.41

Ham Hill Stone (*Oolitic*).—There are three beds—*Yellow*, *Grey*, and *Brown*. Composed of fragmentary shells and a high percentage of iron. The *current bedding* is very distinct and regular in the grey and brown beds, but not so defined in the yellow bed. It is a coarse-grained stone, and most suitable for panels as a contrast in colour to the surrounding stones. It is not suitable for external dressings. Quarried at Norton, Somerset.

Chemical Composition of Yellow Bed, approximate (A. R. Warnes)—

$CaCO_3$, calcium carbonate	79.12
$MgCO_3$, magnesium carbonate	4.91
SiO_2, silica	4.56
Fe_2O_3 and Al_2O_3, iron and aluminium oxides	7.96
H_2O, water and loss	3.45

Physical Properties—

Weight per cubic foot	135.6 lbs.
Crushing resistance per square foot	207 tons (Kirkaldy)
Specific gravity	2.19

Ketton Stone (*Oolitic*).—Colour, cream to yellowish brown; even texture. The *Oolitic grains* are very pronounced and well developed, with a scarcity of cementing material. It is fairly soft to work, but hardens on exposure. Suitable for external work. Quarried at Stamford, Lincolnshire.

Chemical Composition—

$CaCO_3$, calcium carbonate	92.17
$MgCO_3$, magnesium carbonate	4.10
Fe_2O_3, iron oxide	0.90
H_2O, water and loss	2.83

Physical Properties—

Weight per cubic foot	156.7 lbs.
Crushing resistance per square foot	101.7 tons
Specific gravity	2.50

Nailsworth Stone (*Oolitic*).—Colour, light cream; fine grained; suitable for interior work. Quarried at Nailsworth, Gloucestershire.

Weight per cubic foot	142 lbs.

Portland Stone (Figs. 303 and 304) (*Oolitic*).—Quarried in the isle of Portland, Dorset. Generally speaking, there are three beds of this stone—*Roach bed*, *Whitbed*, and *Basebed*. The *Roach bed* is full of cavities and moulds of marine organisms, only suitable for *rock-faced work* or marine construction.

The *Whitbed* is cream to light brown in colour; hard and fairly fine grains; suitable for all external work.

Basebed.—Colour, whitish cream; fine, even-grained stone; suitable for enrichments and carvings and external work if well selected.

Chemical Composition of Whitbed—

$CaCO_3$, calcium carbonate	95.80
$MgCO_3$, magnesium carbonate	1.20
Al_2O_3 and Fe_2O_3, aluminium and iron oxides	0.30
SiO_2, silica	1.30
H_2O, water and loss	1.40

Physical Properties—

Weight per cubic foot	132.3 lbs.
Crushing resistance per square foot	204.7 tons
Specific gravity	2.12

Weldon Stone (*Oolitic*).—Colour, buff to light brown; coarse grained; very shelly and moderately hard; quarried at Kettering, Northamptonshire.

Physical Properties—

Weight per cubic foot	150 lbs.
Specific gravity	2.42

MAGNESIUM OR DOLOMITIC LIMESTONE

This is a term given to any limestone which has as a base an appreciable amount of magnesia, combined as a carbonate. In texture these stones are compact and of a semi-crystalline to crystalline character. The following is a list of a few of the stones which come under this heading:—

Anston Stone.—Colour, light cream. It is a fine-grain stone, but liable to great variations in texture. This stone may be used for external dressings if well selected. Quarried near Sheffield, Yorkshire.

Chemical Composition—

$CaCO_3$, calcium carbonate	54.88
$MgCO_3$, magnesium carbonate	43.08
Al_2O_3 and Fe_2O_3, aluminium and iron oxides	0.73
SiO_2, silica	0.56
H_2O, water and loss	0.75

Physical Properties—

Weight per cubic foot	134 lbs.
Crushing resistance per square foot	302 tons
Specific gravity	2.12

Bolsover Moor Stone.—Colour, warm yellowish brown. It is of very fine grain and semi-crystalline in texture. Suitable for external dressings in some situations, but its use is very limited because of the small supplies obtainable. Quarried in Derbyshire.

Physical Properties—

Weight per cubic foot	152 lbs.
Crushing resistance per square foot	484 tons
Specific gravity	2.44

Huddlestone Stone.—Colour, light cream. It is of fairly fine grain and crystalline in texture. Quarried in Yorkshire.

Physical Properties—

Weight per cubic foot	138 lbs.
Crushing resistance per square foot	278 tons
Specific gravity	2.21

Mansfield Woodhouse Stone.—Colour, warm yellow; very fine grain; compact and crystalline in texture. This stone is rather poor as regards its weathering properties. Quarried in Nottinghamshire.

Physical Properties—

Weight per cubic foot	145 lbs.
Crushing resistance per square foot	577 tons
Specific gravity	2.34

Roche Abbey.—Colour, cream, with numerous brown ferruginous specks. It is of fine grain and fairly crystalline in texture. Quarried in Nottinghamshire.

Physical Properties—

Weight per cubic foot	139 lbs.
Crushing resistance per square foot	250 tons
Specific gravity	2.23

MAGNESIUM SANDSTONES

Mansfield Stone.—There are two principal varieties under this heading, red and white in colour. They are magnesium stones, containing a large percentage of silica. Because of this, they are usually classified as magnesium sandstones. The *red* stone is hard and of even grain, suitable for external dressings, but great care should be exercised in making sure that a similar stone is not used as a substitute. It is quarried at Mansfield, Nottinghamshire.

Chemical Composition of the Red Stone—

SiO_2, silica	50.52
$CaCO_3$, calcium carbonate	26.80
Al_2O_3 and Fe_2O_3, aluminium and iron oxides	3.72
$MgCO_3$, magnesium carbonate	17.91
H_2O, water and loss	1.05

Physical Properties—

Weight per cubic foot	139 lbs.
Crushing resistance per square foot	601 tons
Specific gravity	2.26

METAMORPHIC ROCKS

Under this heading there is a large variety of marbles and slates. Although slates are a natural rock, they are not intended to come within the scope of this book.

Marbles are limestones or dolomites that have been *metamorphosed*, that is, transformed from non-crystalline to crystalline rocks; or those that have been recrystallised in structure.

The term "marble" is commonly, though erroneously, given to any fairly hard stone that will take a polish and can be suitably used for decorative purposes, and includes *Travertines* or *Onyx*, *Serpentines*, *Breccias*, and sometimes *Alabaster*. Such stones as *Lapis-lazuli*, *Jasper*, *Verdite*, and *Porphyry* are often included under this heading.

The chemical composition of true marble is calcium carbonate ($CaCO_3$), although, generally speaking, the composition of marble varies considerably. All marbles contain a greater or lesser amount of foreign substances, such as carbonate of magnesium ($MgCO_3$), silica (SiO_2), clay (Al_2O_3), carbonaceous matter, oxides of iron (Fe_2O_3), etc.

The veinings, markings, and the various colours of marbles are due to the admixture of one, or the combination of more than one, of these minerals with the carbonate of lime. The beauty of *polished* marbles depends upon the nature of the colouring matter, and its disposition in streaks and veins through the stone. Thus, when the composition of the stone is practically pure carbonate of lime, the colour is white, as, for instance, statuary marbles. The black and grey varieties owe their colouring to the intrusion of *carbon*, *magnetite*, or highly *ferriferous silicates*; the green marbles to *ferro-magnesium silicates*, or *copper carbonates*. The yellow, brown, and red colourings are due to *iron oxides*.

Onyx Marble or Travertine.—The so-called marbles under this heading are of aqueous origin, being deposited from water, carrying carbonate of lime in solution. There are two kinds: the former is formed in caves, and is a cold-water deposit, forming *stalactites* and *stalagmites*, whilst the latter is a hot-spring deposit formed in beds. The colourings are due to metallic oxides present in the deposit. These colours vary from pure white to a rose tint, red, and golden yellow. There is also the *ribboned* or *veined onyx*. The onyx from Algeria is of the *hot-spring variety*, whilst those from Egypt are of the *cave variety*.

Alabaster is a variety which comes under the general heading of marble. It resembles onyx in its translucency, and is termed *Oriental Alabaster*. The so-called alabaster produced in this country is not true alabaster, which is composed of *carbonate* of lime, but is *sulphate* of lime, and is a variety of the mineral gypsum, which is formed by hydrated sulphate of lime.

Serpentine is an alteration product of other rocks rich in magnesium minerals, especially *Olivine*, which merges into *Serpentine* by simple hydration, combined with a loss of silica and lime. The variations in colour are chiefly due to the presence of iron in various stages of oxidisation, whilst its varied appearance is due to the admixture of other minerals.

DURABILITY OF BUILDING STONES

The conditions affecting the durability of a building stone are varied, and the precise cause of decay in any particular case is very often difficult to determine. In order that the student may be in a position to discriminate between a good or bad weathering stone, it is necessary that he should have a knowledge of the agencies causing the decay. There are two chief agencies which cause either decay or destruction, each assisting the other—(1) mechanical, (2) chemical. The chief mechanical agencies are: dust-borne winds, changes in temperature, frost, friction, and rain. The surfaces of stonework are liable to disintegrate through the agency of sand-borne winds, especially in some coast towns. This has a distinct abrasive effect on the surfaces of the stone. The expansion and contraction of the particles due to changes of temperature have a remarkable power of disintegrating the surfaces of the stones. Close-grained and compact stones are very subject to this action.

To avoid the possibility of eruption of the surfaces of the stone by frost, it is important that the moisture contained in the stone when freshly quarried, known as *quarry-sap*, and the moisture due to the saturation of the stone whilst passing through the various stages of manufacture, should be allowed to dry out before the stone is set in position in the building. *Quarry-sap* is dilute acid, with silica, lime, and other deposits in solution. This moisture evaporates on the surfaces of the stone, leaving a deposit of crystals which fills, or partly fills, the pores of the stone, thus forming a film on the surface. The removal of this film greatly reduces the weather-resisting properties of the stone.

Friction.—This often occurs at the lower parts of a building, causing the surfaces of the stones to be worn away.

Chemical Agencies.—Chemical action is due to the absorption of rain water, charged with impurities, taken up from the air and deposited on the surfaces of the stones. Acids readily act upon carbonate of lime, which is almost entirely soluble in hydrochloric acid. Sulphurous acid in moist air becomes oxidised to sulphuric acid, which changes carbonate of lime into sulphate of lime, which is slightly soluble in water and crystallises out from solution. This chemical action is associated with an alteration of the structure of the surfaces, and is attended by a large expansion on crystallisation, which results in the loosening of the particles, thus rendering them easily displaced by mechanical agencies.

The durability of sandstones depends to a very large extent upon the kind of cementing material by which the grains of quartz are held together. If the cementing material is calcareous—that is, containing a large percentage of carbonate of lime in crystalline form, known as calcite—it is readily attacked by acids, thus loosening the silica grains, which become disintegrated. This action is more noticeable in such siliceous stones than in limestones generally, for whereas the most obvious effect of decay in limestone is the gradual removal of practically the whole of the exposed surface, in the case of sandstones, the removal of a portion of the binding material alone tends to make noticeable the decay which is taking place.

CHOOSING A BUILDING STONE

Knowing the agencies at work causing the decay of building stones, the student should now be in a better position to choose a stone for building purposes. A stone may be suitable in structure under certain conditions, but quite unsuitable with regard to colour, so that building stones are often chosen from the æsthetic point of view rather than with respect to their weather-resisting qualities. To an extent, the chemical analysis of a stone is very useful as a guide to the expected behaviour of the various constituents forming the structure of the stone; but very often it is misleading with regard to the behaviour of the stone as a whole. The practical points to notice in choosing a stone are:—

1. *Situation of Building*.—Some stones are known to withstand the effects of the atmosphere of an inland air, but, when exposed to sea air, to decay rapidly. Generally speaking, stones are attacked more readily in town atmospheres than in country districts.

2. *Aspect.*—Frequently the stonework of a building with a *southern* or *south-western* aspect has decayed rapidly owing to the changes in temperature due to alternating periods of sunshine and rain.

3. *Appearance and Colour.*—Many sandstones are durable, but owing to the angular grains in the structure of the stone, they soon become discoloured in town atmospheres, producing a drab to almost a black appearance, whereas some limestones, after exposure, produce some remarkably pleasing colouring effects. Portland stone is used to such a considerable extent chiefly because of this reason. The style of architecture should be taken into account, and also the purpose for which the stone is to be used.

4. *Possible Supply of Blocks from the Quarry.*—This is an important item in choosing a stone. It is useless to specify the use of a good stone if the supplies required are unobtainable. The condition of the Houses of Parliament to-day is the result of an oversight, in this respect, by those who were responsible for the choice of Bolsover Moor Stone—the stone with which that building was commenced.

5. *Facility for Working.*—The most durable so-called building stone is undoubtedly granite, and for certain positions its use is advisable, but the high comparative cost in producing the finished article, especially the *polished* surfaces, is often prohibitive. It has strength and beauty and can easily be cleaned, which is of great advantage when it is used in positions such as main doorways, plinths, etc.

A sandstone merging into a *quartzite*, although extremely durable, would not justify the installation of the machinery requisite for its conversion into wrought stonework at the rate required.

6. *Surface Finishings.*—In choosing a stone it is necessary to know the surface finishings required, since the structure of some stones lends itself to *rock-face* work, that of others to *polished*, *rubbed*, or *tooled surfaces*, so that this should be kept in view when the stone is chosen.

7. *Architectural Details.*—It would be fatal to choose some very durable stones if delicate details are required. The texture of some stones is eminently suited for this class of work, whilst with others it is not. To-day stones are selected in which the architectural details of the design can be reproduced, but the ideal arrangement would be to design the details suitable for execution in a particular material or stone. Under these conditions many of our more durable stones could be used to advantage.

8. *Destruction of Stonework Caused by the Oxidisation of Iron and Steel.*—It is necessary that a few remarks should be made with reference to damage caused by the insertion of iron or steel into the surfaces of the stonework, or in such positions as will allow of moisture penetrating to the metal. In almost all instances the metal used in stonework for structural purposes is in confined spaces, such as metal connections to steel structures, dowels, or cramps. When iron or steel is used for these purposes, unless special precautions are taken to prevent moisture from coming into contact with the metal, layers of rust are eventually formed on the surface of the metal. This is caused by oxidisation, set up by the acids and salts, which are in solution in atmospheric moisture, coming into contact with the metal and thereby causing expansion to take place.

The serious fractures in the stonework of pinnacles, etc., which is stated to be due to decay, is often entirely due to the use of *iron vanes*, or *dowels*. The author has particularly noticed this to be the chief cause of the destruction of numerous pinnacles of one of our most important and historical buildings. As to the effect of oxidisation on the actual stonework used in facing steel-framed structures, we have as yet no reliable evidence, as this method of construction has not been in use a sufficient length of time to ascertain this with any certainty; but it is the author's opinion that great damage to the stonework will eventually take place in instances where the steelwork is within a few inches of the face of the stone. Most limestones used for building purposes are porous, and unless great care is exercised in effectively creating a thick coating of neat cement around the surface of the steel, atmospheric moisture is bound to set up oxidisation sooner or later.

The use of painted iron or steel in the joints of stones, as a means of connecting stonework to the steelwork, is a practice not to be recommended, and is only resorted to for cheapness. Admitting that the metal is encased in cement after grouting, the risk of failure by its use is too great to warrant the saving.

It often happens that the advantage of a good selection of a stone for a proposed building is nullified by overlooking one of the main causes of failure with regard to building stones—namely, the failure to ensure a careful selection of the individual blocks from the quarry.

Blocks of stone from the same quarry vary considerably, and samples rarely exhibit a true description of the texture of the stone throughout the quarry. Natural faults, and variation in texture and structure, continually occur. Some blocks are good, whilst others should be rejected. The selection of the individual blocks can only be done successfully by some one who is thoroughly acquainted with the stone and the quarry.

Current Bedding Planes or Natural Bed.—All stratified rocks, which include sandstones and limestones, were deposited in layers from time to time. This building up of layers was often accompanied by a variation of material. The joints between these layers are called the *natural bed* of the stone. In most sandstones the planes are easily discernible, but in some limestones it requires an intimate knowledge of a stone before a definite statement in regard to the direction of the planes can be made.

It is very important that the ends of the *laminæ* should be exposed when the stones are placed in the building. The *bedding planes* in arch-stones should be placed as near as possible normal to the curve of the arch. It is often suggested that cornices and overhanging courses should be placed with the bedding planes vertical, to prevent the falling off of the *drip*; but with the exception of distinctly laminated stones, it is preferable to select the stones for these courses, and place them in the wall of the building, with their planes horizontal.

For the guidance of the mason, the direction of the *bedding planes* is usually indicated on the blocks before they leave the quarry.

In Bath and similar stones a *kerf* is drawn with an axe through one of the surfaces of the block perpendicular to the *planes of bedding*.

THE PRESERVATION OF STONE

The subject of stone preservatives is a very complex one, and one that calls for an extensive knowledge of chemistry and physics. As yet there is no known preservative which entirely protects the stone surfaces from the action of acids.

No definite rules can be laid down for the universal preservation of stone, for in almost every example the cause and effect of the decay will vary. The reason why some of the well-known stone preservatives do not realise the expectations raised by a description of their merits, so glowingly extolled in the trade circulars, is because the treatment of the stone is usually postponed until buildings become unsightly or dangerous through decay, and the trouble has become so deep seated that it is almost futile to apply any preservative whatever, irrespective of what its merits may be. Preservatives may be grouped under three headings—oils, wax, or solution giving rise to the formation of insoluble salts.

Boiled Oil has been used extensively, but its use is accompanied by discoloration of the face of the stone.

Paraffin Wax is an effective preservative, especially if driven into the stone by heat, but this presents a difficulty not easily overcome in practice.

Those preservatives which come under the remaining heading are solutions which are intended to act upon the carbonate of lime, making insoluble compounds, such as *silicate of soda* and *fluosilicic acid* and its salts. These silicon derivatives certainly have some value if administered correctly and at the right time. If the surfaces of the stones show distinct signs of decay, there is no alternative to their removal.

In order that the preservatives may be effective, it is essential that the surface of the stone should be treated immediately after cleaning down and when the stone is in a dry state. The binding of the particles of the stone by the application of the above treatment can be regarded as a temporary remedy only, so that it becomes necessary that the application of the preservative should be periodically performed. Unfortunately, once a preservative is applied, the necessity for its periodical renewal is not usually entertained until attention is drawn, by the dangerous condition of certain portions of the stonework, to the continued decay of the stone.

BOOKS ON THE SUBJECT

BLAGROVE, G. H., " Marble Decoration."
ELSDEN, J. M., " Applied Geology."
HOWE, J. ALLEN, " Geology of Building Stones."
LAURIE, Prof. A. P., " Stone Decay and Preservation."
MUMBY, ALAN E., " Chemistry and Physics of Building Materials."
RENWICK, W. G., " Marble and Marble Working."
" Report of the Stone Preservation Committee Department of Scientific and Industrial Research."
WARNES, A. R., " Building Stones."
WATSON, J., " Building Stones."

SECTION II

THE GEOMETRY OF MASONRY

CHAPTER I

PLANE GEOMETRY AND SETTING OUT

Properties of a Circle—Tangents to a Circle—Circles in Contact—Loci—Application to Tracery—Tracery Windows—Construction of Mouldings—The Ellipse—Tangents and Normals to the Ellipse—Semi-elliptical Arch—Semicircular Arch—Segmental Arch—Substitute for Semi-elliptical Curves—Tudor Arch—Rampant Arch—Setting Out Arches of Large Span—Enlarging and Reducing Mouldings by Radial or Polar Projection—Spiral Curves—Setting Out Ionic Volute—Entasis of Columns—Setting Out Ionic Capital.

THE geometrical problems in this section are not intended to form a complete work on trade geometry, but are given to enable the masonry student to appreciate that a careful study of geometrical principles is the only means whereby a clear understanding of masonry problems can be obtained.

It is advisable for the student to obtain a good book on geometry, and to study carefully the principles explained therein, in conjunction with the following examples, before he attempts to work out the masonry problems given in this book. If this is done, there should be no difficulty in following the illustrations and text, which are all explained in a simple, practical manner. When these are thoroughly mastered, the student should be able to solve any masonry problem, especially if the information gained is used in combination with such practical experience as is only to be obtained in the setting-out shop.

Properties of the Circle.—The angle contained in a semicircle is a right angle (Fig. 305). The angles contained in the same segment of a circle are equal (Fig. 306). As an application—the span and rise of an arch are given. Draw the curve without using the centre.

A straight line which bisects a chord at right angles passes through the centre of the circle (Fig. 307). The span and rise of an arch being given, determine the centre from which to draw the curve, passing through the three given points A, B, C (Fig. 308).

A Tangent to a Circle (Fig. 309) is a straight line which touches the circumference of the circle at a point P (point of contact) and is at right angles to the radius or normal C P, drawn from the point of contact to the centre of the circle.

A normal to a circle is a straight line drawn from the point of contact at right angles to the tangent, at that point.

All normals to a circle pass through the centre of the circle.

To draw a tangent to a circle from a point outside the circle by applying the above principles, as in Fig. 310:—

C is the centre of the given circle, P the point outside. Join P C, and on it erect a semicircle, cutting the given circle in R. Then, because the angle contained in a semicircle is a right angle, if C R and R P be joined, R C would be the normal and R P the tangent, and R the point of contact.

To draw a tangent to two given circles:—

A and B are the centres of the given circles (Fig. 311). Join A and B, cutting the circles in C and D. Make D E equal to B C. Draw an arc with centre A, radius A E. Erect a semicircle on A B, cutting the arc in F. Join A F, cutting the circle, centre A, in H. Draw B K parallel to A F, then H and K are the points of contact through which to draw the tangent.

Circles in Contact.—When two circles touch one another, the straight line which joins their centres, or that line produced, passes through the point of contact (Fig. 312). Therefore, if the centre A were known, and any point, such as P, were to be the point of contact, the line A P produced would contain the centres of all circles which would touch the given circle, centre A, at the point P. The above principles are important in the drawing of tracery and compound curves. The determining of the point of contact for the setting out of tracery is very important.

Loci.—In order that the student may appreciate the correct method of determining the centres for the various curves in tracery, it is necessary that he should understand something concerning a *Locus* (plural *Loci*).

A locus is the path traced by a moving point. The path of the point may be a straight line, a curve, or a combination of both, its shape depending upon the conditions which govern the motion of the tracing point.

A *circle* is a simple example of a *locus*, because it represents a point moving in a plane at a constant distance from a fixed point (the centre).

An *ellipse* is a *locus* of a point, which moves so that the sum of its distances from two points known as the *Foci* of the ellipse is constant.

Given a circle with centre A (Fig. 313) and a tangent to the circle, describe a circle to touch the tangent at point P, and the given circle tangentially. Draw a perpendicular to the tangent as at D, and produce the normal A C from centre A, beyond the circle. Mark off on the normal produced from point C any number of equal divisions, as at 1 2 3 4 5 6. Now mark on the perpendicular line from D a corresponding number of equal divisions of the same magnitude, as at 1′ 2′ 3′ 4′ 5′ 6′. With A as centre, draw parallel arcs from points 1 2 3 4 5 6 to meet lines drawn parallel to the tangent from points 1′ 2′ 3′ 4′ 5′ 6′. A fair curve drawn through the points of intersection will give the *locus*, or *path*, containing the centres of circles that will be tangential to the given circle and the given tangent. To determine the point required on this line, from point P draw a perpendicular line, cutting the locus in point E, E being the centre required. Join A E, cutting the two circles in F, then F is the point of contact between the two circles.

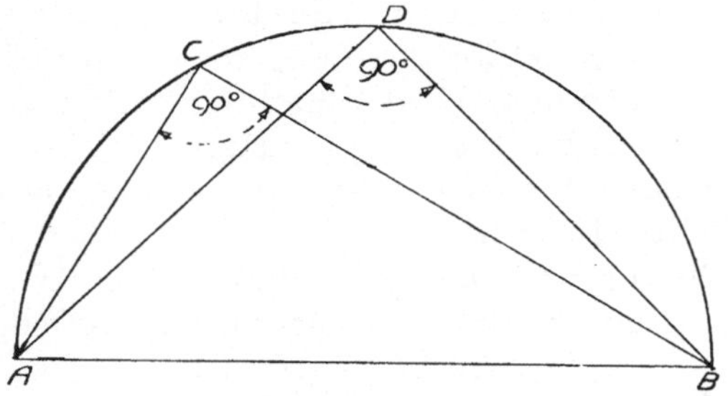

FIG 305

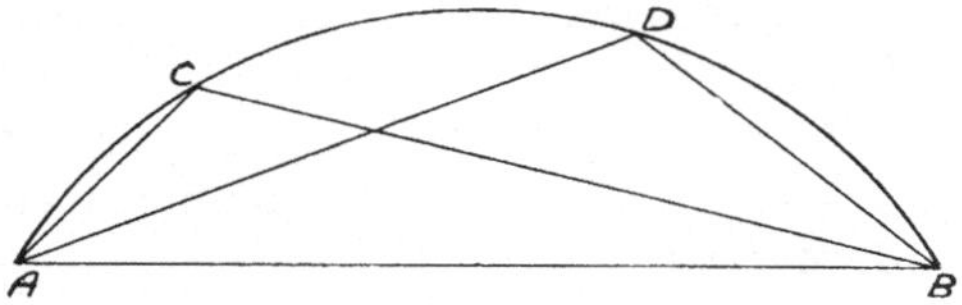

FIG. 306

ANGLES CONTAINED IN A CIRCLE

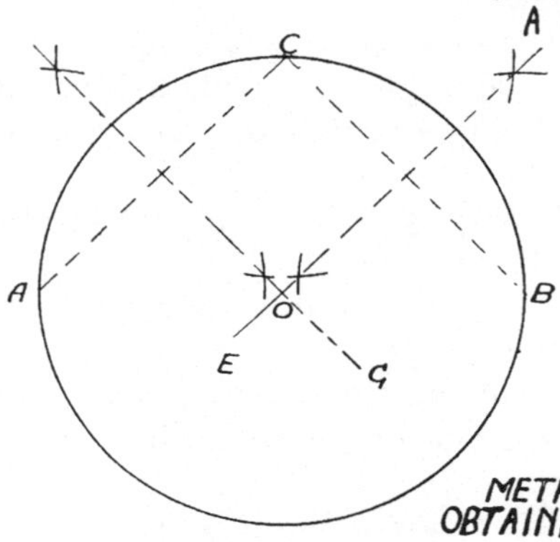

FIG 307

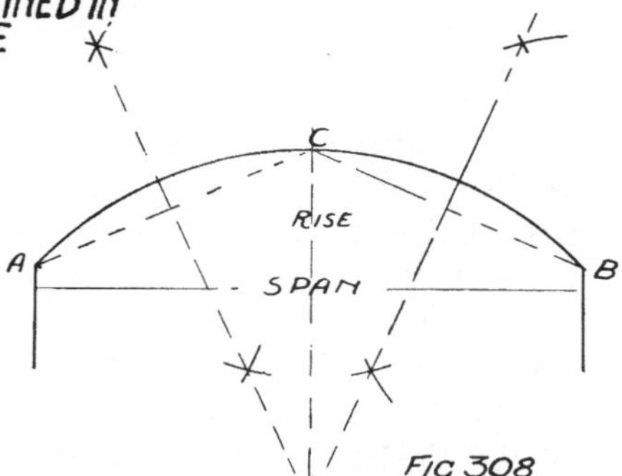

FIG 308

METHOD FOR OBTAINING CENTRE

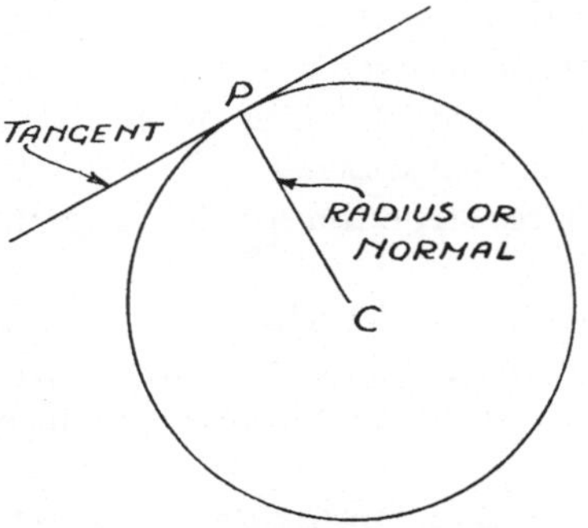

FIG 309

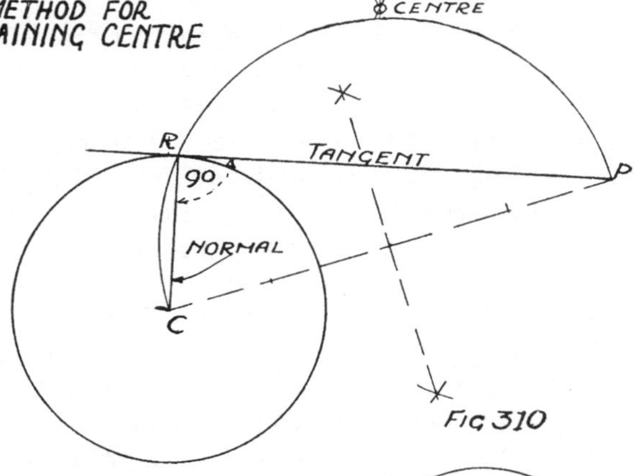

FIG 310

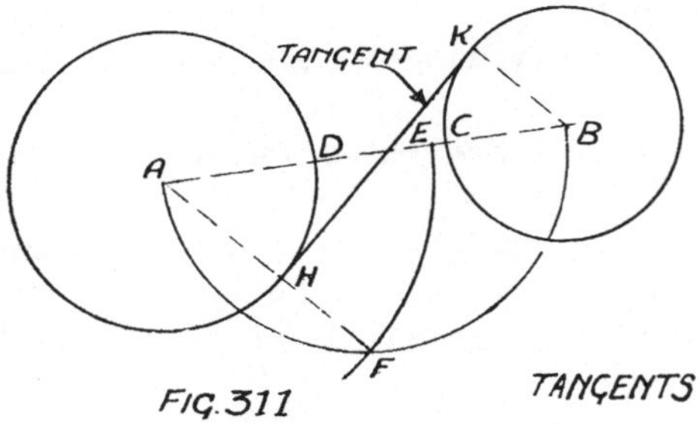

FIG. 311

TANGENTS & NORMALS

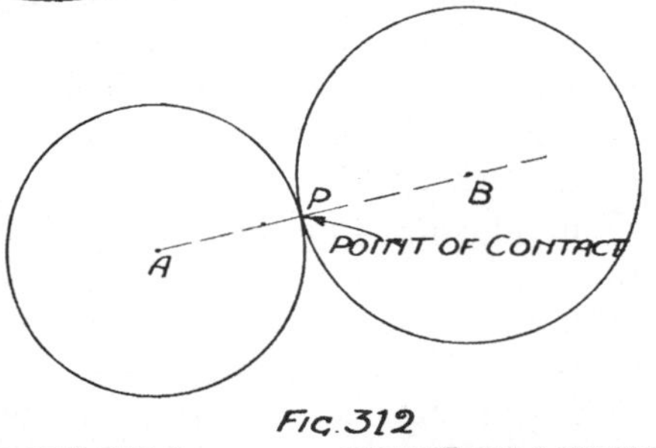

FIG. 312

POINT OF CONTACT

The centre line diagram of a tracery window is given. Set this out to the scale of $1\frac{1}{2}$ in. = 1 ft., showing clearly the method for obtaining the centres of the inscribed circles by means of *loci*. The method of obtaining loci, and the points of contact, is clearly shown in the diagram (Fig. 314).

Having proceeded thus far, we are now able to consider the setting out of Trefoils and Quatrefoils, and also Foliated Circles.

There are certain rules governing the correct drawing of these, which the following examples will demonstrate, although short-cut methods are often adopted in practice.

In a given sector of a circle A B C inscribe a circle tangential to the lines bounding the sector (Fig. 315).

Bisect the angle A C B by the line C D, and at D draw a tangent to meet A C and B C produced in E and F. Bisect the angle D F C by a line cutting C D in G. This is the required centre, G D being the radius, D H K the points of contact.

Inscribe six equal circles in a given circle (Fig. 316). Divide the circle into twice as many sectors as there are circles to be inscribed.

First decide on the position one circle is to occupy, and by using the method explained in the previous example, obtain the centre C.

With centre O, radius O C, draw a circle cutting the other diameters in the points D E F, etc. These are the centres required for the remaining inscribed circles.

To determine the point of contact, from C draw a line C H perpendicular to O B, meeting O B in H. H is the point of contact. The lines in this construction form the basis lines upon which the detail is built.

In setting out various forms of tracery, it is often necessary to draw reverse curves. In such cases, first determine the point of contraflexure, or the point where the curve changes from concave to convex, or vice versa.

Fig. 317 is a diagram showing the geometrical construction for drawing the reverse curve.

Draw the axes A B and O C. Join A C, and on A B describe a semicircle cutting A C in E. Bisect C E. The point D, where this bisector cuts a line drawn perpendicular to O C from C, is the centre required for the reverse curve, E being the point of contraflexure.

Fig. 318 shows the centre line diagram for the head of a *Reticulated Tracery Window*. Set out the window to a convenient scale, add the thickness of the *Tracery Bars*, and place in the *Cuspings*.

Examples showing the setting out of tracery windows are given in Figs. 319, 320, 321, and 322.

The above examples on compound curves are based on the following theorem:—

When two circles touch one another, the straight line which joins their centres, or that line produced, passes through the point of contact.

The geometrical construction of mouldings is based on the principles already outlined.

Examples showing the geometrical construction for a few of the classic mouldings are given in Fig. 323—cyma recta comprising quadrants; Fig.

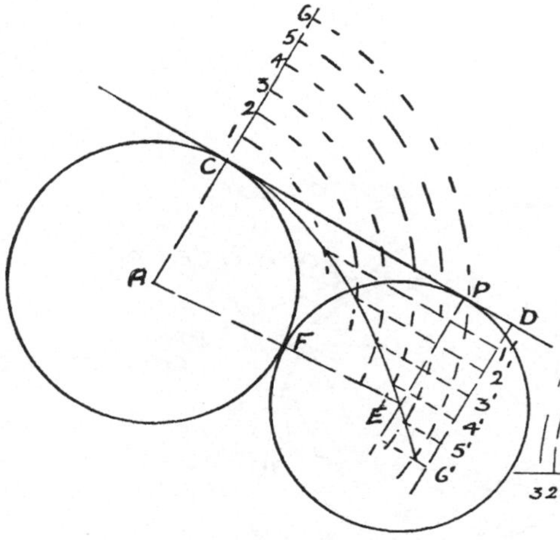

FIG. 313.—LOCI.

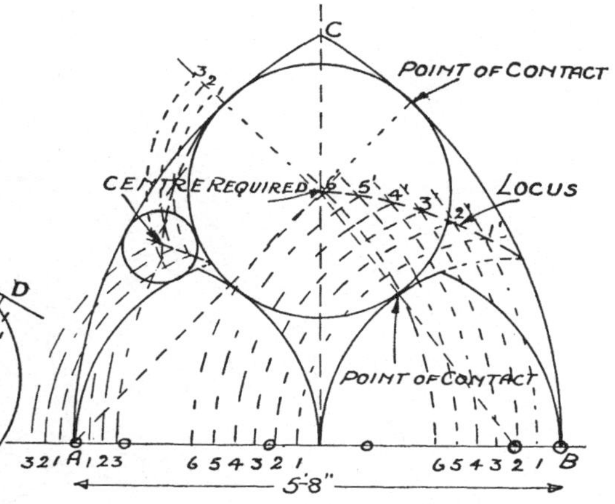

FIG. 314.—LOCI.

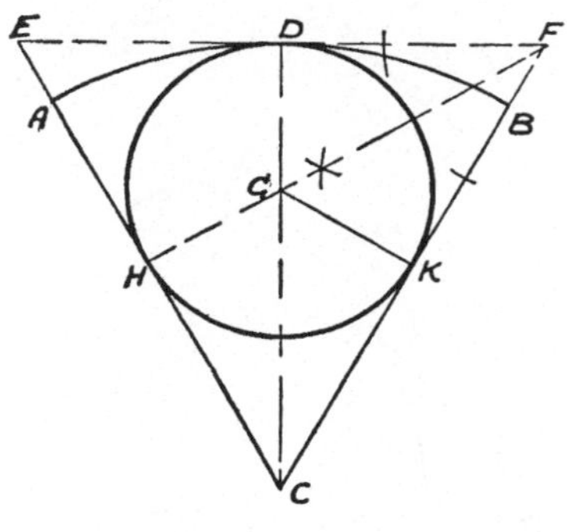

FIG. 315.

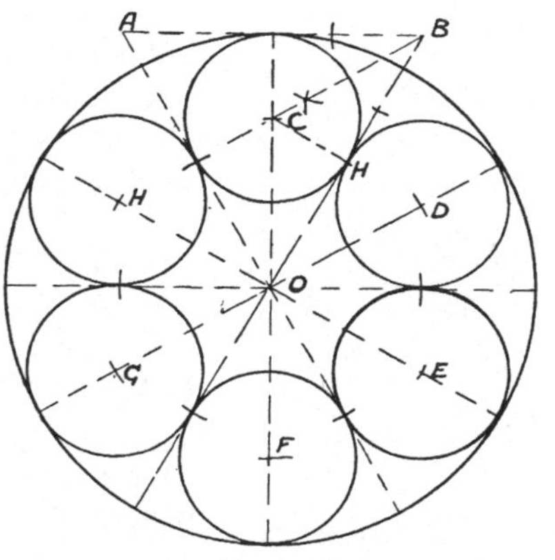

FIG. 316.

CIRCLES (INSCRIBED).

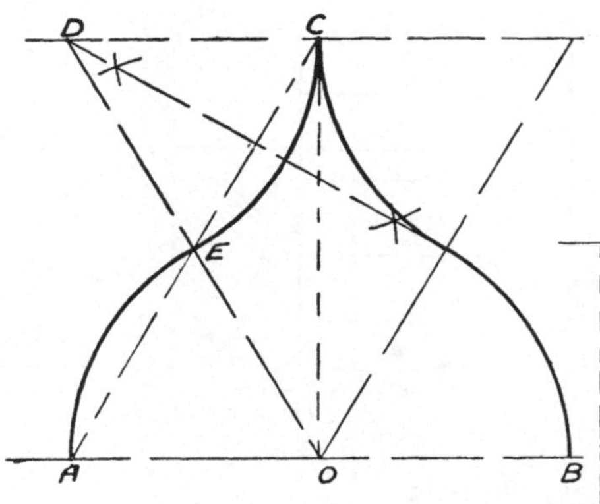

FIG. 317.—POINT OF CONTRAFLEXURE.

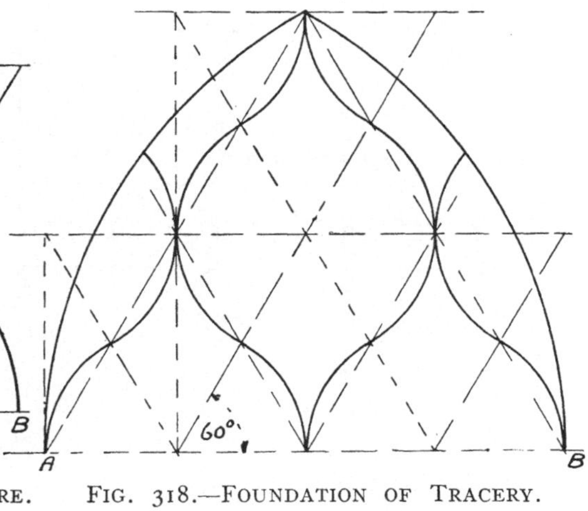

FIG. 318.—FOUNDATION OF TRACERY.

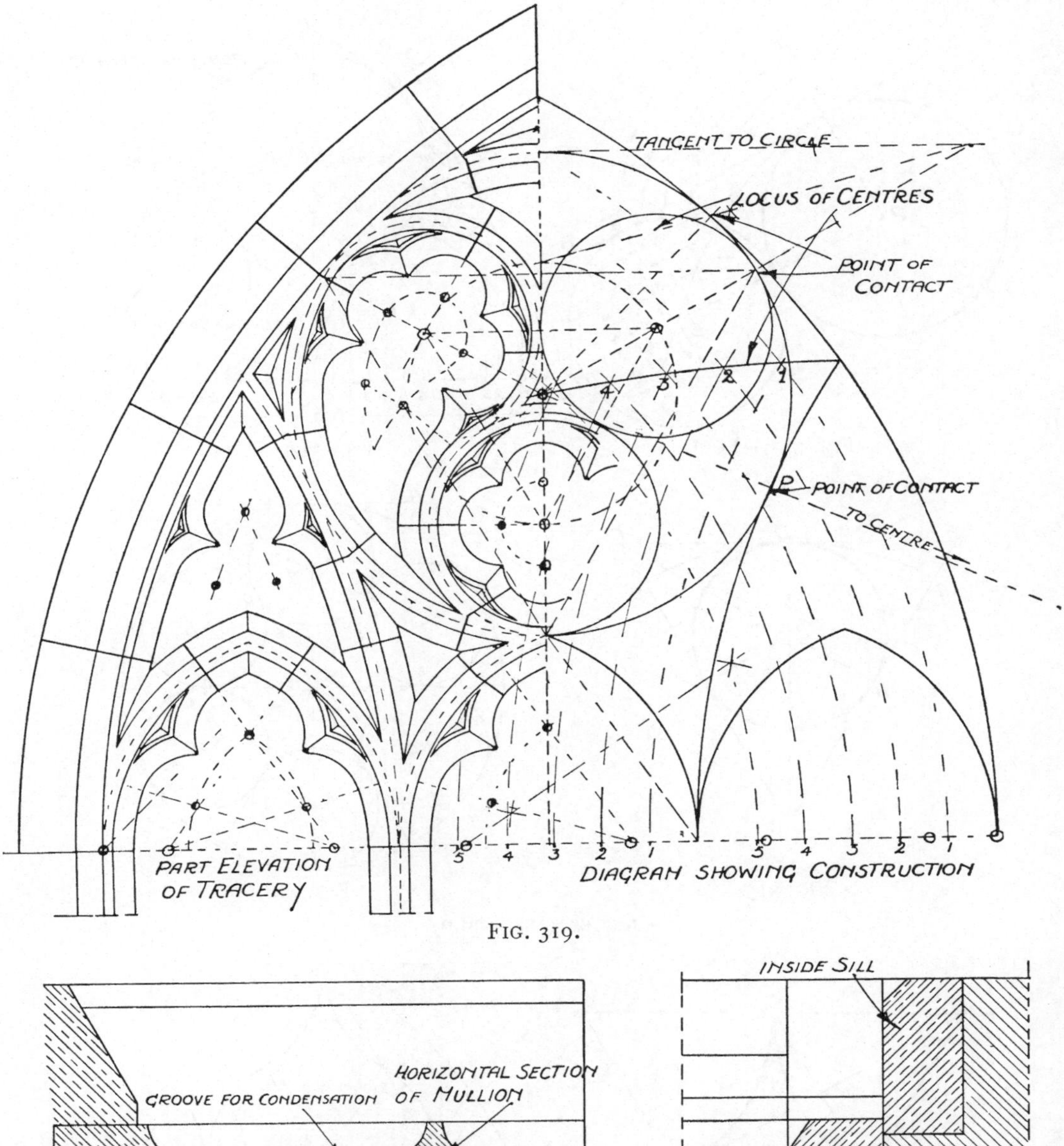

FIG. 319.

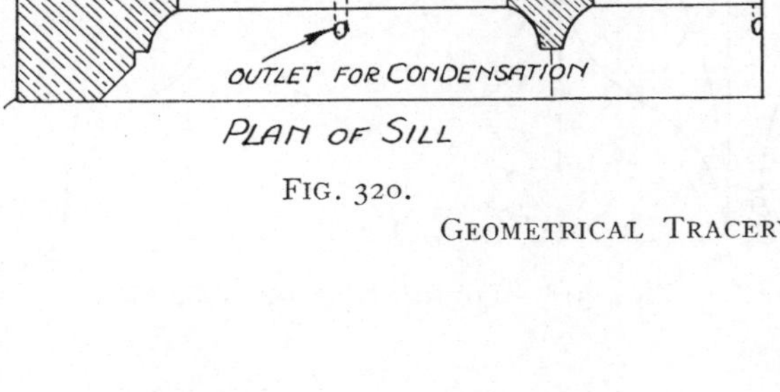

FIG. 320.

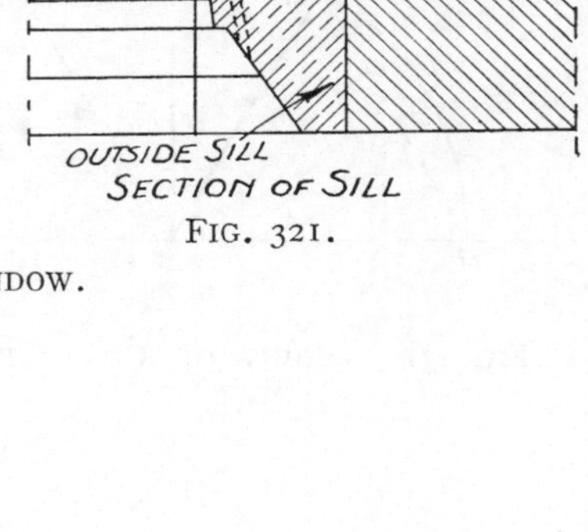

FIG. 321.

GEOMETRICAL TRACERY WINDOW.

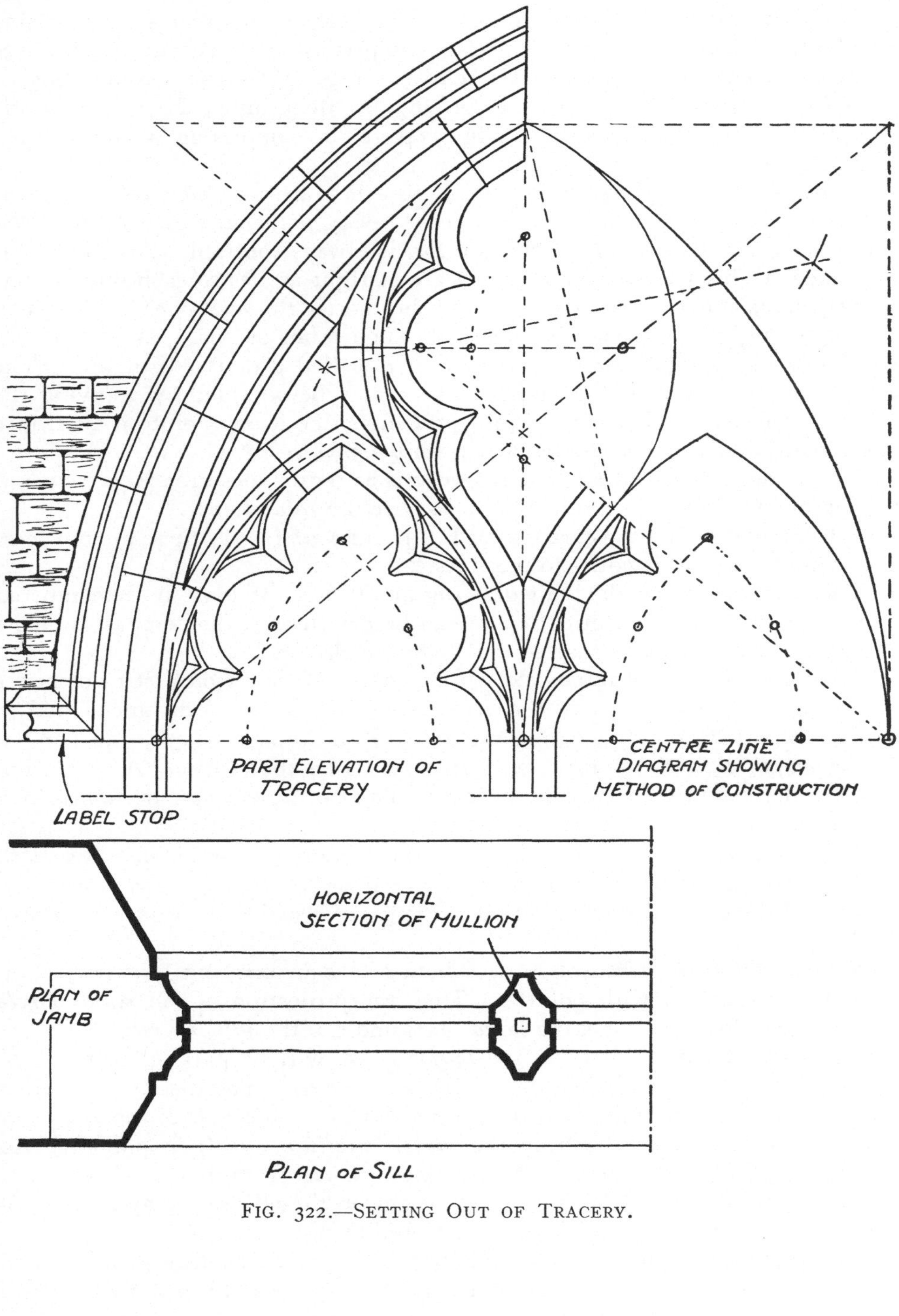

FIG. 322.—SETTING OUT OF TRACERY.

324—cyma recta comprising segments; Fig. 325—cyma recta comprising parabolic curves; Fig. 326—cyma reversa comprising quadrants; Fig. 327—cyma reversa comprising segments; Fig. 328—cyma reversa comprising parabolic curves; Fig. 329—cavetto comprising a quadrant; Fig. 330—cavetto comprising an elliptic curve; Fig. 331—torus comprising a semicircle; Fig. 332—ovolo comprising a quadrant; Fig. 333—scotia comprising an elliptical curve; Fig. 334—scotia comprising a combination of quadrants.

Considered as a plane curve, the ellipse has already been referred to as the *locus* of a point which traces a curve, every point on which is such that the sum of the distances from the two *foci* is always constant. An ellipse has two unequal diameters or axes, which are at right angles to each other. The longer one is called the *major axis*, and the shorter the *minor axis* (Fig. 335).

Any straight line perpendicular to the axis is an ordinate, as at H K. If continued to meet the curve at point J, it is called a double ordinate. The ellipse has two *foci*, as at F and F′, which may be obtained by applying the following theorem: The sum of the *focal distances* at any point on an ellipse is constant, and equal to the length of the *major axis*.

With C as centre, and half the major axis as radius, describe an arc cutting the major axis in points F and F′. These are the *focal points*.

Select any point on the curve, as at P. Join P F and P F′; these are the *focal lines*, and their sum is equal to the lines C F and C F′.

There are several methods of drawing an ellipse. It must be remembered that no portion of an elliptical curve can be drawn with the compasses. The method usually adopted is called the *Trammel Method*.

Draw the major and minor axes (Fig. 335). Mark off on a strip of paper the distance G H equal to half the *major axis*, and R from G distance equal to half the *minor axis*. By keeping the point H on the *minor axis*, and point R on the *major axis*, the point G will give points on the ellipse. A succession of these points, through which the curve may be drawn, can be found in a similar manner.

The ellipse may also be considered as a projection of a circle, as in Fig. 336. The circle described on the *major axis* of an ellipse is called the *major auxiliary circle*, and the circle described on the *minor axis* is called the *minor auxiliary circle*.

Suppose the *major circle* be turned about A B as an axis, until the projection of point E coincides with point C. Then the projection of the *major circle* will be the ellipse, with A B the *major axis* and C D the *minor axis*.

The *path* of any point G on the *major circle* will be parallel to the *path* of the point E, and perpendicular to the *major axis*. The distance travelled by the point G is determined by drawing a radius, or normal, to the *auxiliary circles*, through point G, cutting the *minor auxiliary circle* in point G′ and drawing a line perpendicular to the *minor axis* from point G′, to meet the line from G in point G″, which is a point on the ellipse. Any number of points on the curve may be obtained in a similar manner.

Tangents and Normals to an Ellipse (Fig. 337).—To draw a tangent and a normal to an ellipse from any given point P on the ellipse, draw the *focal*

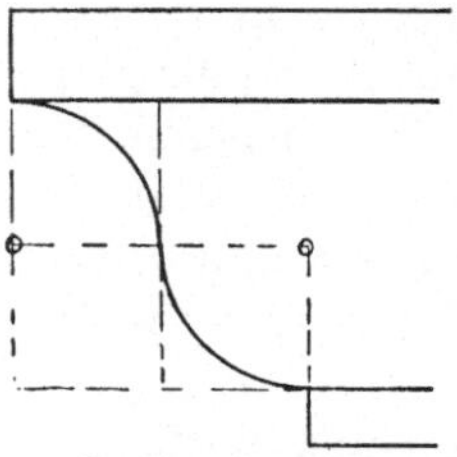

CYMA-RECTA (QUADRANTS)

FIG. 323.

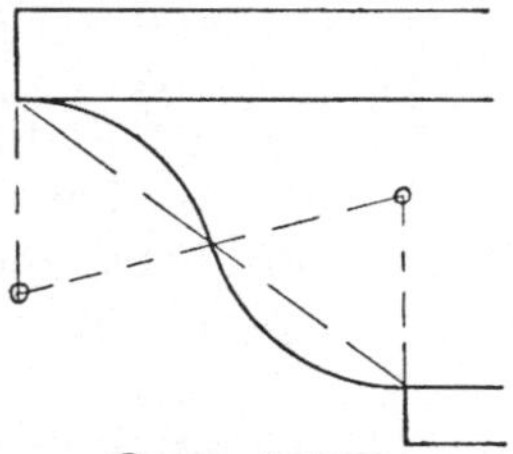

CYMA-RECTA (SEGMENTS)

FIG. 324.

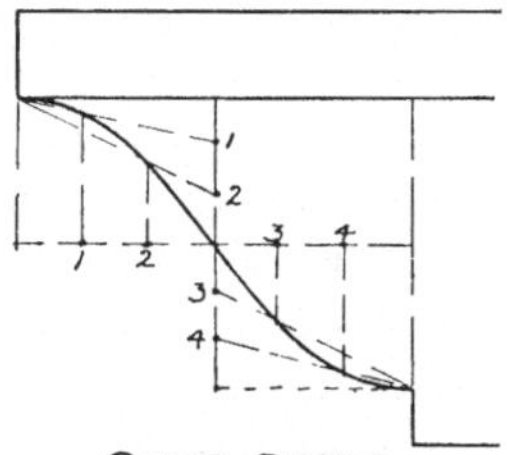

CYMA-RECTA (PARABOLIC)

FIG. 325.

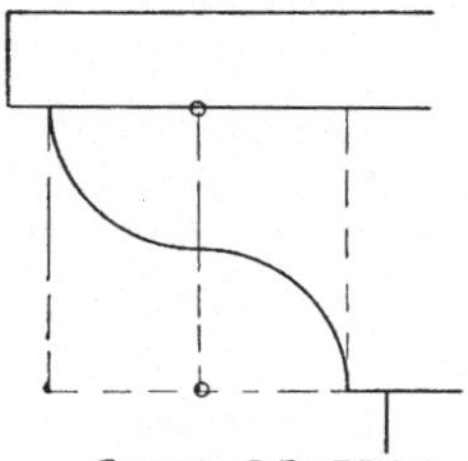

CYMA-REVERSA (QUADRANTS)

FIG. 326.

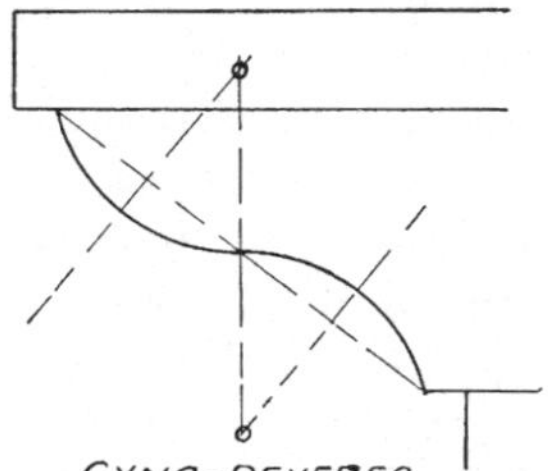

CYMA-REVERSA (SEGMENTS)

FIG. 327.

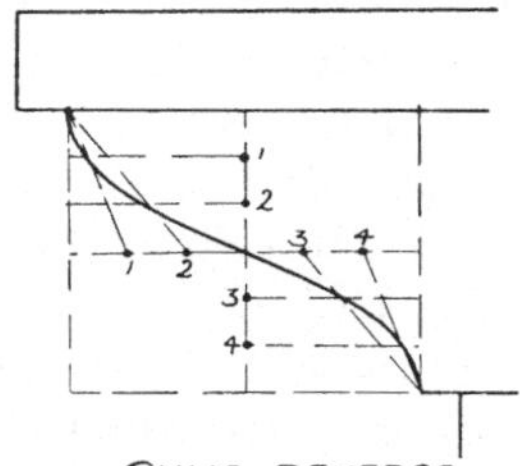

CYMA-REVERSA (PARABOLIC)

FIG. 328.

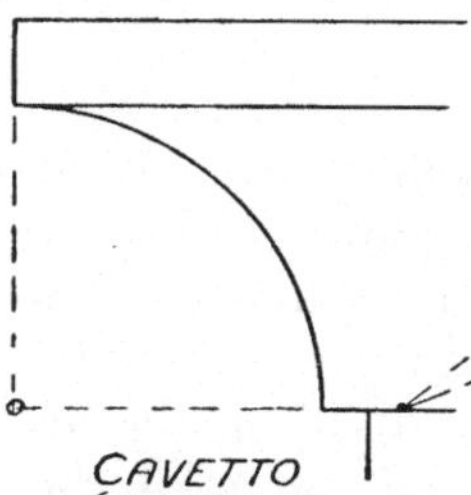

CAVETTO (QUADRANT)

FIG. 329.

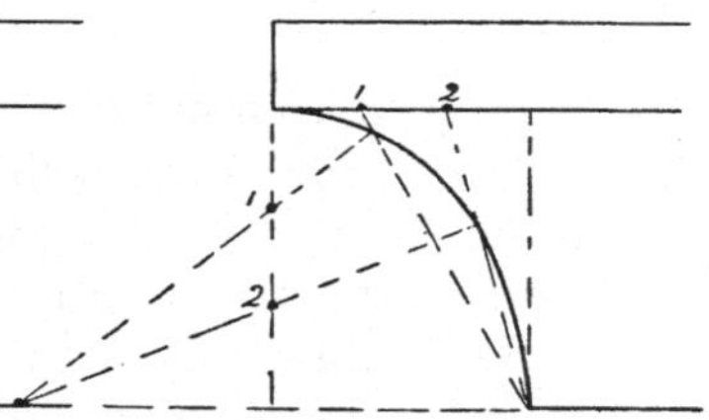

CAVETTO (ELLIPTICAL)

FIG. 330.

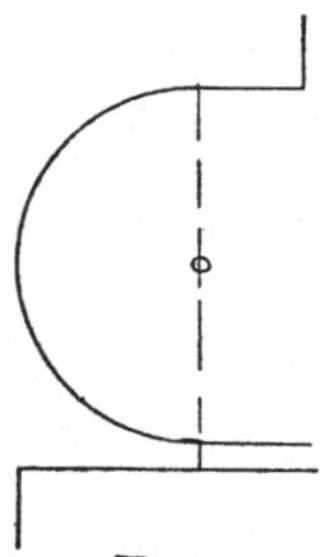

TORUS

FIG. 331.

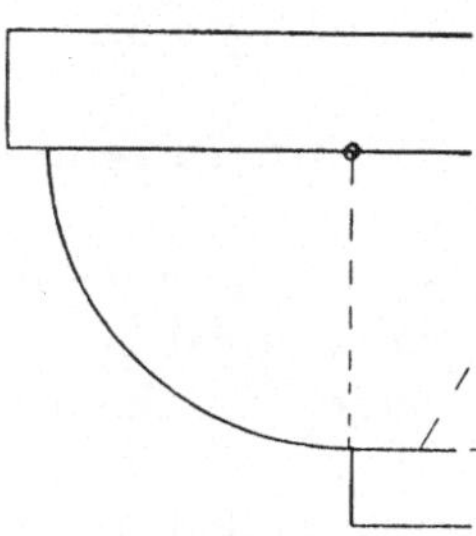

OVOLO (QUADRANT)

FIG. 332.

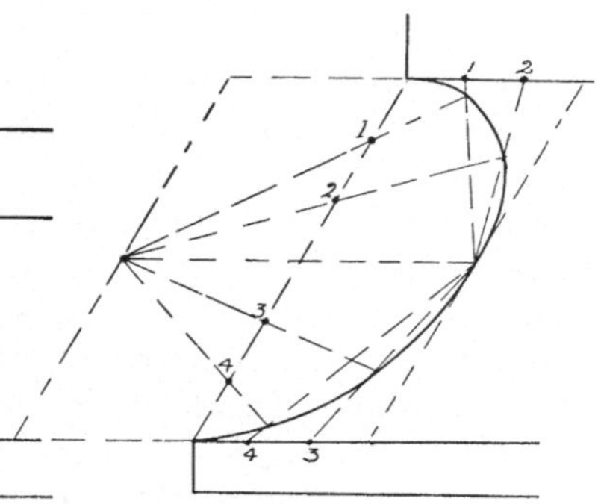

SCOTIA (ELLIPTICAL)

FIG. 333.

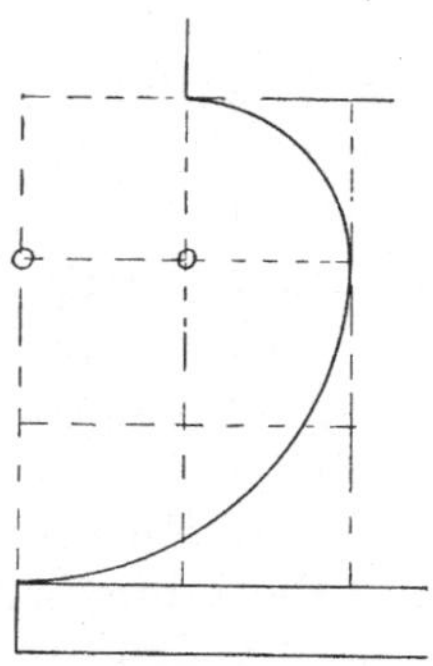

SCOTIA (QUADRANTS)

FIG. 334.

GEOMETRIC CONSTRUCTION OF MOULDINGS.

lines P F and P F′, and produce them to D and E. Bisect the angle E P F′, then the bisector is a tangent to the curve, P being the point of contact. To obtain the normal, bisect the angle F P F′ or D P E, then the bisector P H is the normal required at point P on the curve. The above principles have been dealt with at length, because it is essential that the student should understand their application when considering the problems connected with the setting out of arches and vaults.

The following is a very important principle connected with the ellipse, and its application is clearly demonstrated in the setting out of the intersecting vaults of unequal span (Figs. 532-539).

To obtain the normal joints for the semi-elliptical vault as a projection from the small semicircular vault:—

Theorem.—A tangent to an ellipse, and the corresponding tangent to the *major auxiliary circle*, intersect the *major axis* at the same point, also the tangent to the ellipse, and the tangent to the *minor auxiliary circle*, intersect the *minor axis* at the same point, and the two *auxiliary tangents* are parallel.

Draw the ellipse as the projection of the *major* and *minor auxiliary circles* as previously described (Fig. 338). Select any point P on the ellipse as the point of contact for the tangent, and at R on the major circle draw an auxiliary tangent R T to the *major circle*, meeting the *major axis* produced in T. Join T P, meeting the *minor axis* produced in T′. Join T′ R′: this is the *auxiliary tangent* to the *minor circle*, R′ being the point of contact. The *auxiliary tangents* are parallel.

The principles just described apply to arch curves.

Semi-elliptical Arch (Fig. 339).—Set out a semi-elliptical stone arch, the face of the arch on one-half to be 18 in. deep and parallel to the arch line, the other half being bonded to intersect the surrounding ashlar courses. Show the bonding clearly, and the method of obtaining the normal joint lines.

In the drawing (Fig. 339) is shown a method for dividing the arch curve into the number of voussoirs required. The extrados curve of the arch is trammelled from the intrados curve.

Note.—A line drawn parallel to an ellipse is not an ellipse. A test of the drawing will verify this statement.

Semicircular Stone Arch.—Set out a semicircular arch containing seven stones, bonded to suit the horizontal ashlar courses of the wall.

This is shown in Fig. 340.

Draw the springing line A B and centre line C D. With C as centre, C A radius, describe the curve line of the arch.

Place in the horizontal bed lines of the ashlar courses, and draw the normal bed-joints of the arch to the centre C. Next draw the vertical joint lines bonding the voussoirs with the ashlar. An alternate method for the arrangement of the joints is shown in the drawing.

The plan should now be drawn and the details required added to the plan and elevation. *Beam compasses* are used for striking the arch curves shown.

Segmental Stone Arch (Fig. 341).—The span and rise of a segmental arch are given. Set out the arch and obtain the necessary moulds for working the stones comprising the arch.

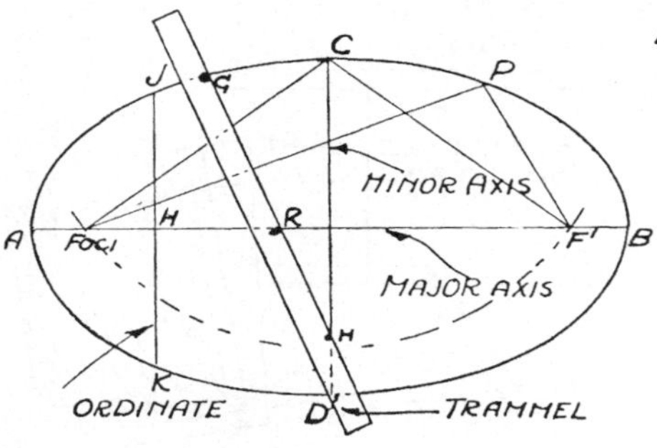

FIG. 335.—THE ELLIPSE (TRAMMEL METHOD).

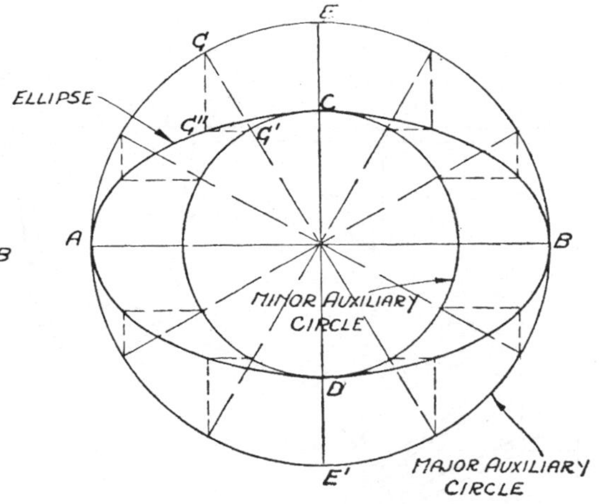

FIG. 336.—THE ELLIPSE (PROJECTION OF AUXILIARY CIRCLES).

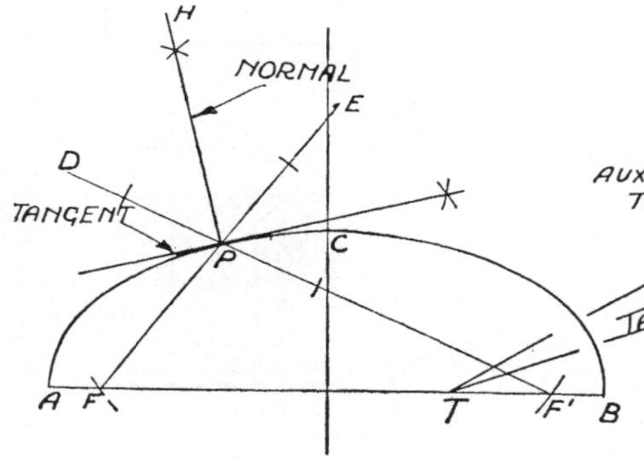

FIG. 337.—TANGENTS AND NORMALS TO AN ELLIPSE.

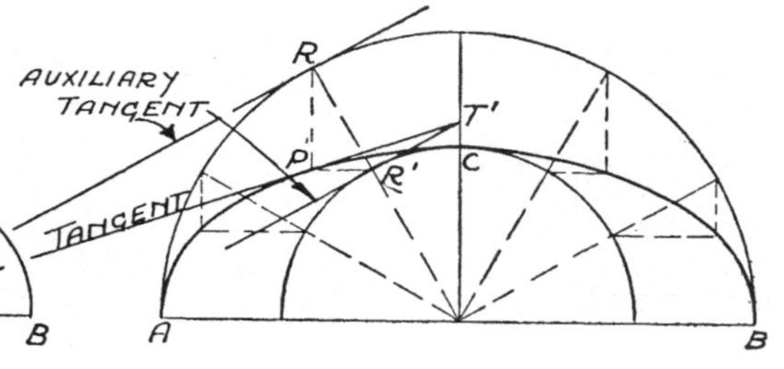

FIG. 338.—TANGENTS TO THE ELLIPSE.

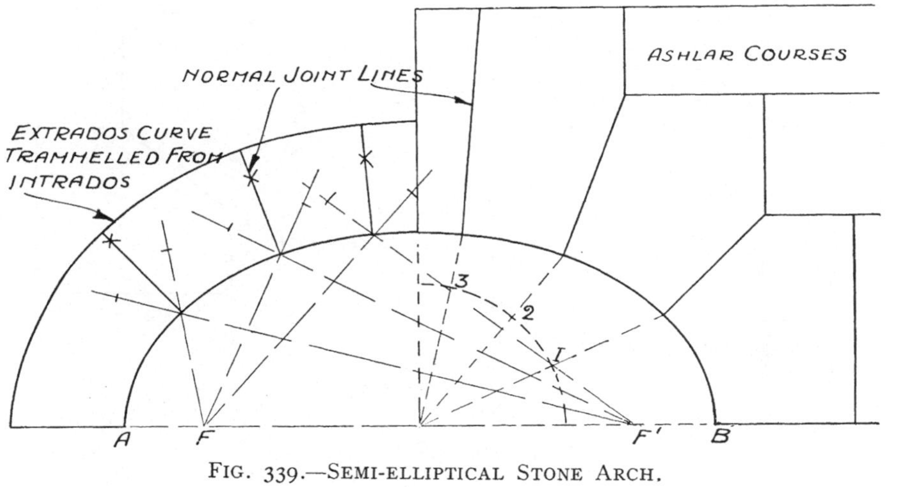

FIG. 339.—SEMI-ELLIPTICAL STONE ARCH.

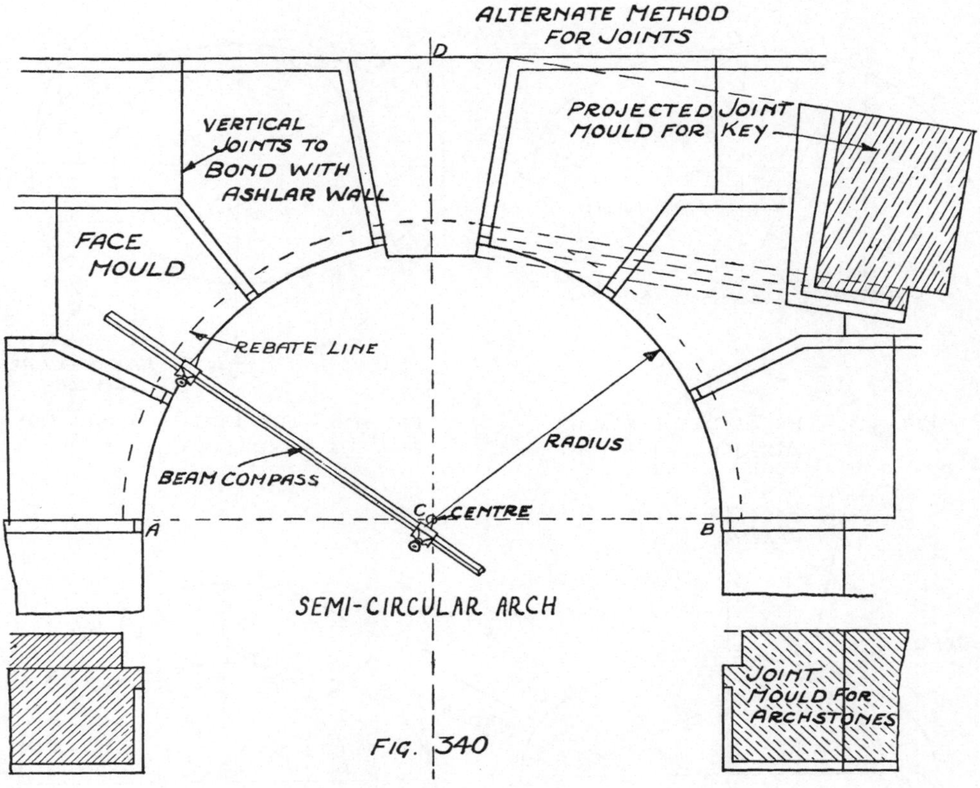

FIG. 340

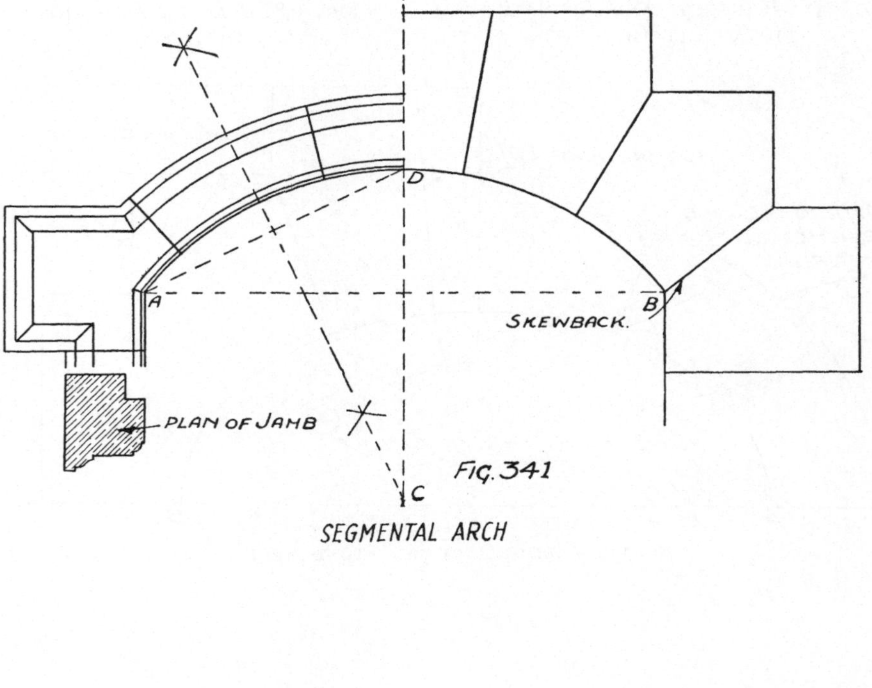

FIG. 341

SEGMENTAL ARCH

First draw the springing line A B and centre line C D. Join A D and B D. The bisectors of these lines meet the centre line D C in the point C, which is the centre from which to strike the curve.

Two methods of bonding are shown, giving the lines required for the face moulds. Draw the plan to determine the bed mould and section for bed-joints of arch stones.

Substitutes for Semi-elliptical Arch Curves or Approximate Semi-elliptical Arches.—Two methods of drawing the arch curve are given, Fig. 342 showing the setting out for a three-centred arch.

A B and C D represent the *major* and *minor axes* of the ellipse.

Draw part of the *major auxiliary circle*, as at B E. Join C B. With C as centre and C E radius, draw an arc cutting C B in F.

Bisect F B, the bisector cutting the *minor axis* produced in G. Then G is the centre for the top portion of the curve, and H is the centre for the lower curve. K is the point of contact. The normal joint lines for the arch-stones converge to the centre of the curve, which is cut by the joint plane.

Five-centred Arch (Fig. 343).—Draw the major and minor axes A B C D. From C draw a line parallel to O B, and from B draw a line parallel to C D, meeting the line drawn from C in point E. Divide B E into three equal parts and O B into a similar number of equal parts, as at 1 2 and 1′ 2′. Join C 1 and C 2 and D 1′ and D 2′, producing these to cut the lines C 1 and C 2 in points 3 and 4.

These are points on the ellipse. Draw a bisector to a chord between points C 4, meeting the centre line produced in C′. C′ is the centre for striking the top curve, and point 4 is the point of contact.

Draw a bisector between points 4 and 3, meeting the first bisector in point C^2. With C^2 as centre, radius C^2 4, draw an arc passing through point 3. Continue this arc until it cuts a horizontal line drawn from C^2 in point F. From F draw a line through B, cutting the arc in point K. Draw a line from K to C^2, cutting the springing line at C^3. With C^3 as centre and C^3 K radius, the arc K B may now be drawn.

Tudor or Four-centred Arch.—The span only being given in Fig. 344, draw the axial lines and divide the span into four equal parts, as at 1 O 2. With 1 and 2 as centres, draw two tangential circles.

Now construct a square below the springing line from points 1 and 2, thereby obtaining points C and D. Draw the diagonal D 1 and produce it to cut the circle already drawn at E. Then with D as centre and radius D E, draw the arc E F. Complete the other side of arch in a similar manner.

Tudor Arch, the Span and Rise being given (Fig. 345).—Draw the axial lines A B and C O. In this example only half is drawn. From B erect a perpendicular B D, making B D equal to two-thirds the rise O C. Join D C, and from C draw a line perpendicular to D C. With B as centre, radius B D, draw an arc cutting the springing line in F. With the same radius from point C, mark off point G on the perpendicular line. Join F G and bisect it, producing the bisector to meet the perpendicular line from C in H. Join H F. This line produced determines the point of contact between the two arcs. With F as centre, F B radius, draw an arc cutting F H produced in J. With H as centre

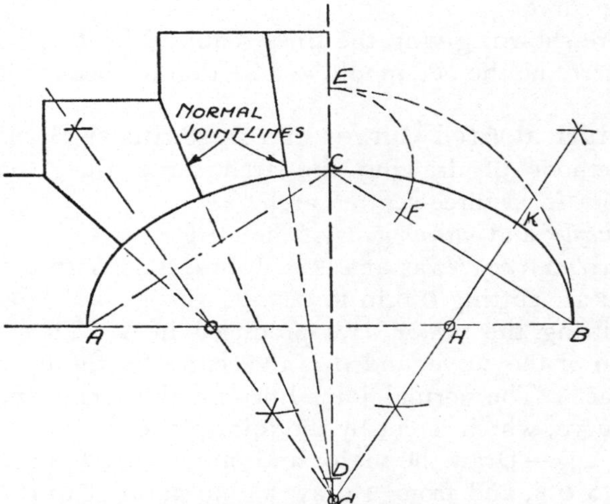

APPROXIMATE SEMI-ELLIPTICAL ARCH

FIG. 342

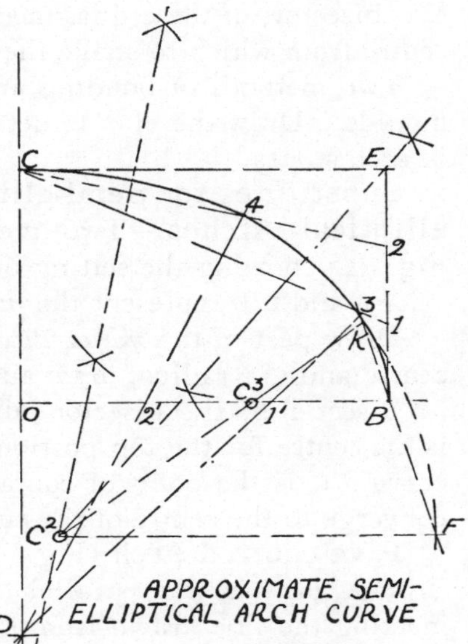

APPROXIMATE SEMI-ELLIPTICAL ARCH CURVE

FIG. 343

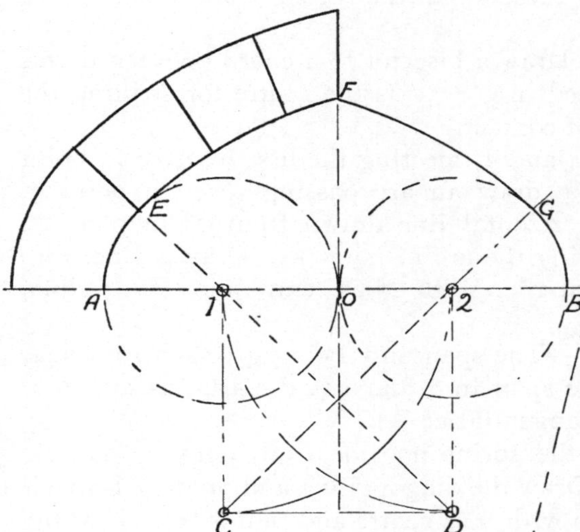

TUDOR OR FOUR CENTRED ARCH

FIG 344

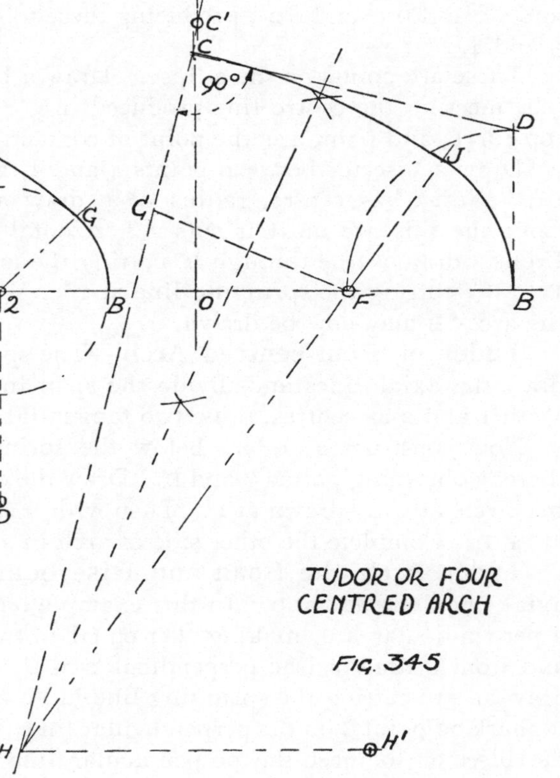

TUDOR OR FOUR CENTRED ARCH

FIG. 345

and H J radius, draw the arc J C. To complete the arch, draw a horizontal line from H to H′, measuring the same distance from the centre line to obtain the centre for drawing the other side of the arch.

Rampant Arch (Fig. 346).--Let A B represent the springing points and pitch of the stairs, C being the crown of the arch. Draw a horizontal line from A, to cut a vertical line drawn from C in point E. With E as centre, radius E A, draw a semicircle to cut the line A E produced in F. From B draw a line through F, cutting the semicircle in point H. From H draw a line through centre E, producing it to meet a horizontal line drawn from B in K. With K as centre, radius K H, draw the curve between H and B.

Setting Out Arches of Large Span and Walls Circular on Plan.—The radius may be found by the following formula, and may be applied when the *span* and *rise* only are given. Add together the squares of the rise and half the span, and divide by double the rise :—

$$A = \text{Rise.}$$
$$B = \text{Half the span.}$$
$$\text{Radius} = \frac{A^2+B^2}{2A}.$$

Let

$$A = 2 \text{ ft.}$$
$$B = 10 \text{ ft.}$$
$$R = \text{Radius.}$$

Then

$$\frac{A^2+B^2}{2A} = \frac{2^2+10^2}{2\times 2} = \frac{104}{4} = 26 \text{ ft. radius.}$$

To calculate the rise when the span and radius are known (Fig. 347): Subtract the square of half the span from the square of the radius. Take the square root of the result and subtract it from the radius :—

$$A = \text{Rise.}$$
$$B = \tfrac{1}{2} \text{ Span} = 10 \text{ ft.}$$
$$R = \text{Radius} = 13 \text{ ft.}$$
$$A = R - \sqrt{R^2 - B^2}.$$
$$A = 13 - \sqrt{13^2 - 10^2}.$$
$$A = 13 - \sqrt{169 - 100}.$$
$$A = 13 - \sqrt{69}.$$
$$A = 13 - 8.3 = 4.7 \text{ ft. rise.}$$

When a curve with a large radius is required, it often becomes necessary to plot portions of the curve. This may be accomplished, when the radius and chord are known, by determining the rise as already stated, and obtaining the length of numerous ordinates. These lengths may be ascertained by the following formula :—

Any ordinate Y, distance X, from the centre of chord $= \sqrt{R^2 - X^2} - (R - A)$. Take an ordinate Y, 4 ft. from the centre of chord :—

$$Y = \sqrt{13^2 - 4^2} - (13 - 4.7).$$
$$Y = \sqrt{169 - 16} - (8.3).$$
$$Y = 12.36 - 8.3. \quad \therefore \; Y = 4.06 \text{ ft.}$$

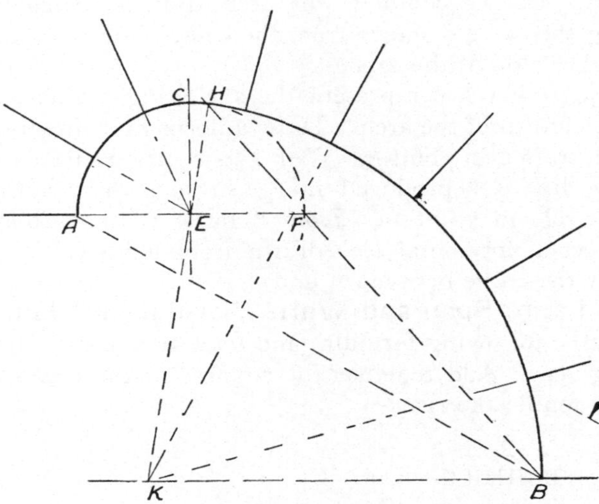

FIG. 346.—RAMPANT ARCH.

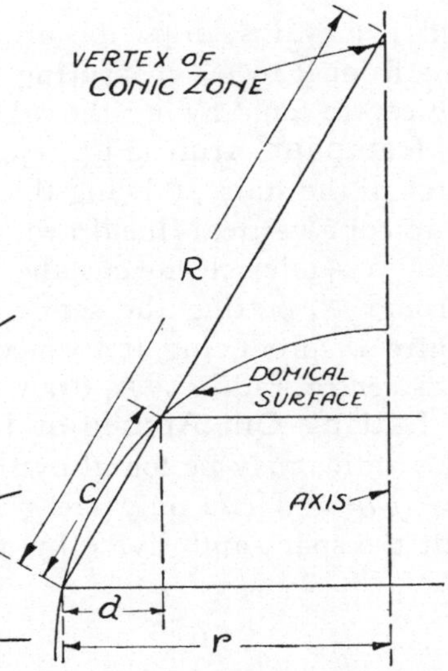

FIG. 348.—DIAGRAM ILLUSTRATING METHOD FOR CALCULATING THE DEVELOPMENT CURVES FOR CONICAL ZONES.

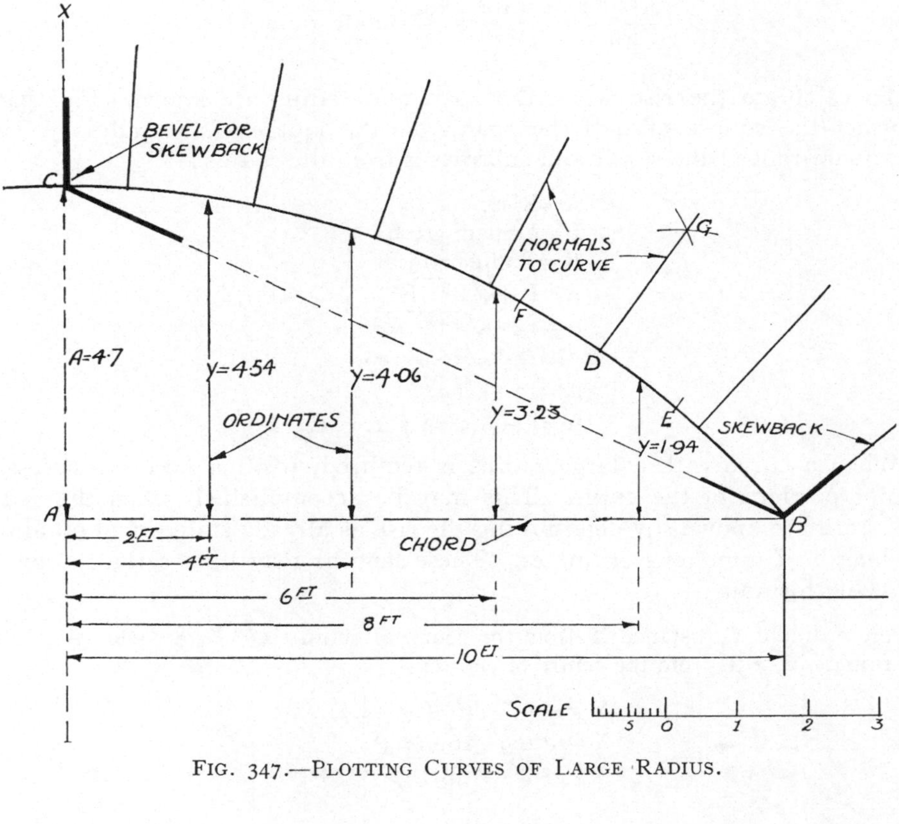

FIG. 347.—PLOTTING CURVES OF LARGE RADIUS.

The application of the above is given in Fig. 347, showing the plotting of the curve for a segmental arch.

The following may be applied to the approximate development of a domical surface, when divided into *conical zones*.

To calculate the radius for drawing the development, when the vertex of the cone is inaccessible (Fig. 348) :—

Let C=the slant height of the zone=5 ft.
d=the amount the zone leans towards the axis=3 ft.
r=half the diameter of the base of the cone=18 ft.

Then by similar triangles :—

$$\frac{R}{r}=\frac{C}{d}. \quad \therefore R=\frac{C\times r}{d}. \quad \therefore R=\frac{5\times 18}{3}=30 \text{ ft.}$$

To plot the curve by the formula applied in Fig. 347: Let us assume a chord of 20 ft. First we must obtain the rise :—

Let A=Rise.
$B=\frac{1}{2}$ Chord=10 ft.
R=Radius=30 ft.

Then

$$A=R-\sqrt{R^2-B^2}$$
$$=30-\sqrt{30^2-10^2}$$
$$=30-\sqrt{900-100}$$
$$=30-\sqrt{800}$$
$$=30-28.16. \quad \therefore A=1.84 \text{ ft.}$$

Any ordinate Y, distance X, from centre of chord$=\sqrt{R^2-X^2}-(R-A)$. Take an ordinate Y, 4 ft. from centre :—

Then

$$Y=\sqrt{30^2-4^2}-(30-1.84).$$
$$Y=\sqrt{900-16}-28.16.$$
$$Y=29.73-28.16.$$
$$\therefore Y=1.57 \text{ ft.}$$

Any other points in the curve may now be plotted in a similar manner.

To obtain the skewbacks for the arch in Fig. 347, the striking centre being inaccessible: Draw a line from the crown of the curve at C to B, produce the centre line A C beyond C to X, then a templet cut to the angle B C X, and held on the line C B, with point C at B, will give the correct position of the skewback.

To draw the other normal joint lines, determine the positions at which the joints intersect the arch curve, as at D, then with any convenient radius draw arcs, cutting the arch line in E and F. With E and F as centres and any convenient radius, greater than E D or F D, draw arcs intersecting in G. Join G D, and this is the position for the normal joint line.

Enlarging Mouldings by Polar or Radial Projection.—The pole to which the lines radiate may be taken within or without the figure, but it should be chosen so that all the radial lines may be drawn without overlapping.

Enlarge the given cyma-recta moulding to twice the size (Fig. 349). Draw the given figure, and select a pole P in any convenient position. Draw radial lines from the point P to the points A B C D, etc. Measure the radial line P A, and from A mark off this distance on P A produced, determining A′. From A′ draw line A′ D′ parallel to A D, cutting the line P D produced in D′. From D′ draw a line D′ C′ parallel to D C, cutting the line P C produced in C′. Draw A′ B′ to cut P B produced in B′. Then join C′ B′. The rectangle containing the figure is then complete.

Produce P E to E′ and P J to J′. A line drawn from J′ parallel to J K to meet P K produced, determines point K′. Produce P G to G′ and P F to cut a line drawn from G′ to K′. Join F′ E′, thus obtaining the point of contraflexure H′. The curves may now be drawn.

To reduce a given section, reverse the process described.

Spiral Curves.—A spiral curve is the *locus* of a point which moves in a plane surface round a fixed point in such a way as to approach or recede from the fixed point.

To construct an Archimedean spiral (Fig. 350): Draw a circle and divide it into any number of equal parts (eight in this example). Divide the radius O P into a corresponding number of equal parts, numbering them as shown. With O as centre, O 7 radius, draw an arc cutting the first radius in 7′, and so on, until the whole of the divisions are complete. By drawing a fair curve through these divisions, the spiral is obtained. This completes one revolution. If two revolutions are desired, divide the radius O P into twice the number of parts to that of the circle, and proceed as before explained.

Setting Out Ionic Volute (Gibb's Rule) (Fig. 351).—Divide the height into eight equal parts and the width into seven similar divisions. Bisect the division 3 4 in height, in point F, and draw a horizontal line from F to meet a vertical line drawn from point 3 in G. The point of intersection gives the centre for striking the eye of the volute.

A large scale detail of the drawing of the eye is given in Fig. 352.

From the points 3 and 4 of the vertical divisions, draw horizontal lines. This gives the diameter for the eye. Form a square within the circle, and draw the diagonals. Divide each of the cross lines into six equal parts, numbering them as shown. From these points draw lines as indicated, cutting the outline of the volute in points 1′ 2′ 3′, etc.

With 1 as centre, 1 ① radius, draw an arc to meet line drawn from point 2 in ②; with 2 as centre, draw an arc ② ③ to meet the line from point 3 in ③. Repeat this process until the volute is complete.

Entasis of Columns.—The *entasis* of a column is the delicate swelling which is worked on the shaft in order to counteract an appearance of hollowness in outline.

The *entasis* may commence from the bottom of the shaft, or about one-third of the height of the shaft from the base line.

There are various methods for producing the curve required; the method to adopt is usually suggested by the architect. The curve generally used is known as the *Conchoid of Nicomedes*.

Fig. 353 shows the setting out of this curve. Draw the base line A B, and produce it to F. Draw the centre line of the shaft perpendicular to the base line and to the correct height: also draw the top bed line C D. Determine the

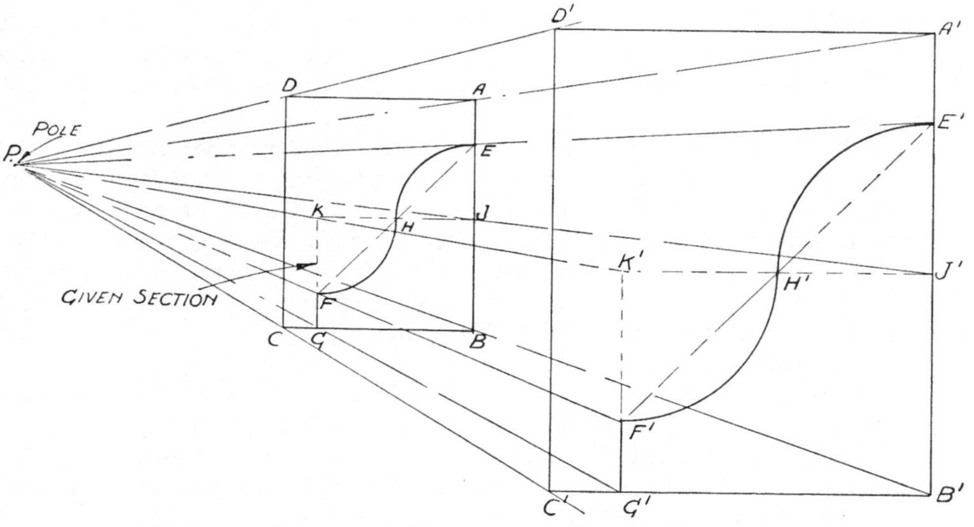

FIG. 349.—ENLARGING MOULDINGS BY RADIAL PROJECTION.

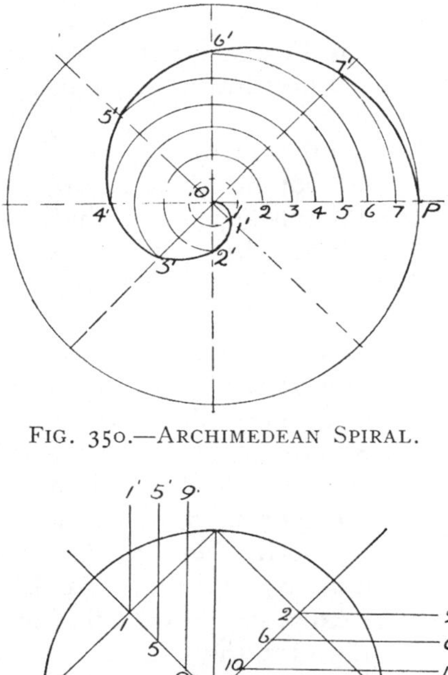

FIG. 350.—ARCHIMEDEAN SPIRAL.

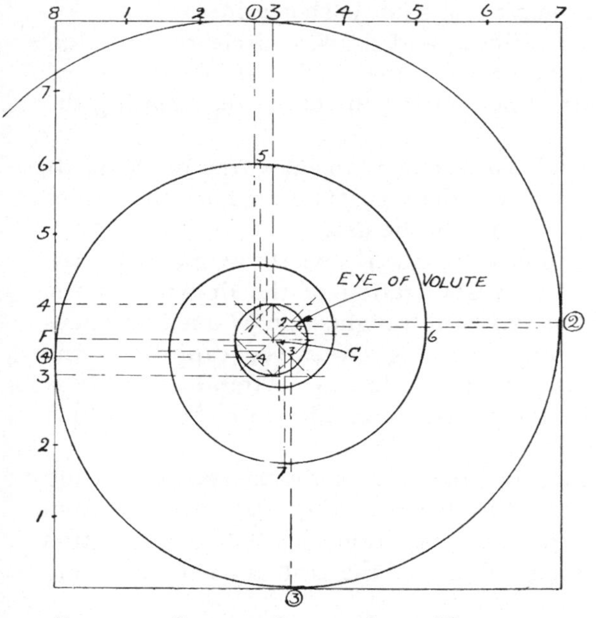

FIG. 351.—SETTING OUT OF IONIC VOLUTE.

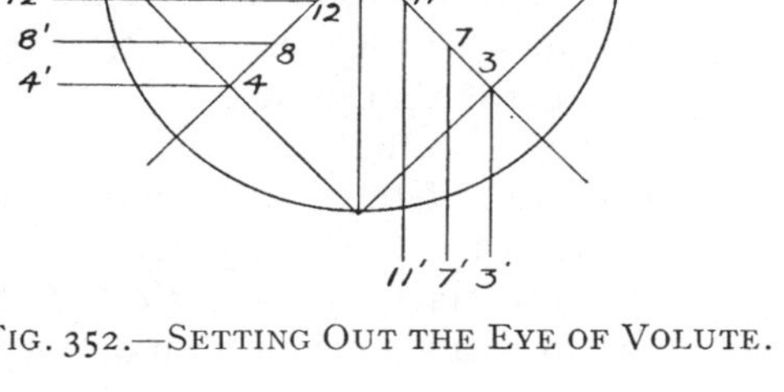

FIG. 352.—SETTING OUT THE EYE OF VOLUTE.

diameter of the shaft at the base A B, and also the top diameter C D. With O A as radius (half bottom diameter) and the compasses at C, draw an arc cutting the centre line in 8. Join C 8, and produce this line to meet the base line produced in F.

Next divide the centre line O 8 into any convenient number of parts, as at 1 2 3, etc., and draw lines from these points converging to F on the base line. Produce these lines past the centre line as shown, and mark off on these lines a distance equal to half the lower diameter O A from the centre line. A fair curve drawn through these points will produce the curve required. To obtain the curve on the opposite side of the shaft in elevation, draw horizontal lines cutting the centre line as shown. Make the distance from the centre line to the curve equal on each side.

For setting out the *entasis* of large columns by the above method, a large surface area is required to obtain point F, but actually this point is not required, for if any line is drawn parallel to the centre line O 8, cutting the triangle O 8 F in O′ 8′, then the line O′ 8′ will be divided into a similar number of parts, as the line O 8.

Draw a line O′ 8′ parallel to O 8 at any convenient distance from O 8, divide O 8 into any number of equal parts, produce C 8 to cut the parallel line in 8′, and divide O′ 8′ into a similar number of parts, as at 1′ 2′ 3′, etc. Then lines drawn through 1 1′, 2 2′, 3 3′, etc., will, if produced, meet at point F.

Another method of drawing the *entasis* to a shaft is shown in Fig. 354.

Set out the shaft as previously described, and draw a circle representing the bottom bed of the shaft. From point 4′ on the extremity of the top bed line draw a line parallel to the centre line, cutting the circle representing the base of the shaft in point 4.

Divide the arc 4 O′ into four equal parts, and the centre line of the shaft into a similar number of equal parts. From these points draw horizontal lines to meet vertical projectors from points 1 2 3 on the arc.

A fair curve drawn through these intersections will give the *entasis* required.

Setting Out Ionic Capital.—As an application of the drawing of the volute just described, and to assist the masonry draughtsman who is sometimes called upon to draft a full-size boasting drawing for carved capitals of the Ionic, Corinthian, and Composite Orders, the following example is given as a guide, but the author would advise the perusal of architectural books, which deal exhaustively with this subject.

Fig. 355 shows a method of setting out the plan of the capital for cutting the bed mould, whilst the part elevation is shown in Fig. 356 projected from the plan. Moulds cannot be cut from this elevation, so that its projection can only be used as a picture for the guidance of the craftsman in executing the details required.

To draw the plan, divide the lower diameter of the shaft into twelve parts. Take ten of these parts for the upper diameter of shaft, and draw the circles representing these in plan.

Now mark on each side of the centre line in plan nine similar divisions, and form a square in plan on the ninth division. This square gives the extreme

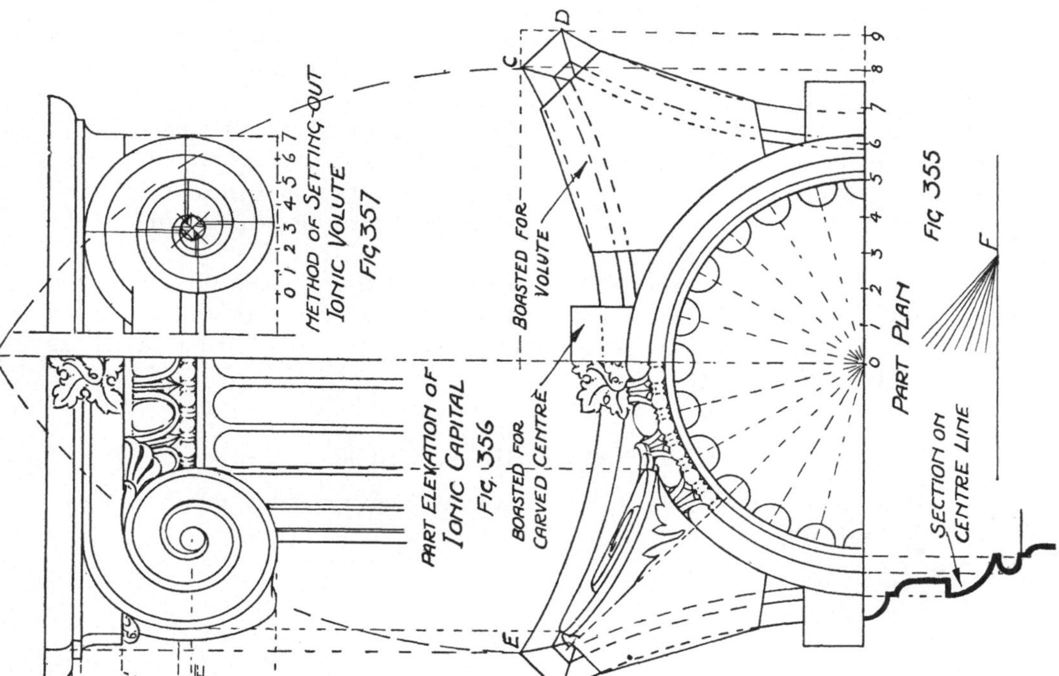

SETTING OUT OF IONIC CAPITAL.

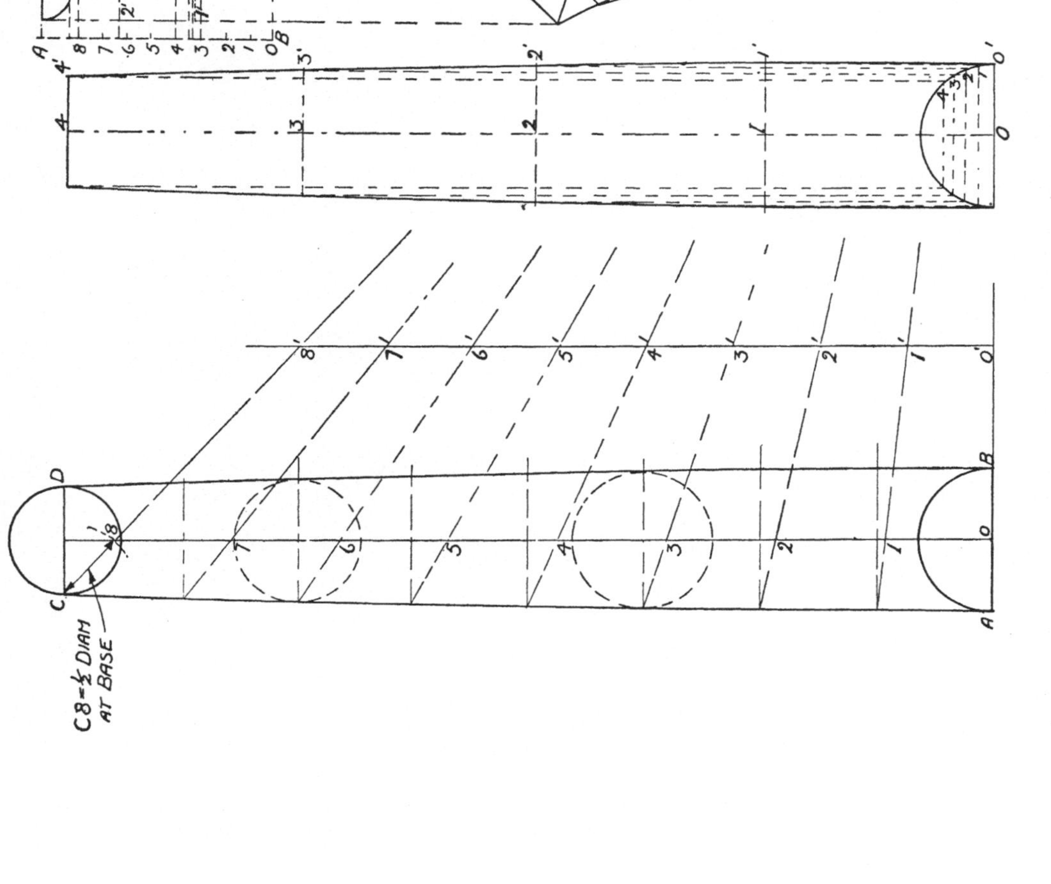

METHODS FOR DRAWING THE ENTASIS OF COLUMN SHAFTS.

projection of the abacus. For the splayed part of the abacus at C and D allow one division on each side of the square.

To form the curve of the abacus in plan, with C and E as centres, radius C E, draw arcs meeting in point F; then F is centre for drawing the plan outline of the abacus, the radius of which is sixteen parts.

To draw the elevation, divide the height A B into three equal parts, as at 1′ 2′, the upper division to form the abacus. For its subdivision, divide the abacus into two equal parts, as at point 8. Now divide B 8 into eight equal parts. Bisect the division 3 and 4. This point determines the height for the centre of the eye of the volute.

The volute in elevation and plan must be drawn free-hand and the required details added. A portion of the finished detail is shown in the plan and elevation. The application of the geometry required for the setting out of the volute is given in Fig. 357.

BOOKS ON THE SUBJECT

BATES, E. L., and CHARLESWORTH, F., "Practical Geometry and Graphics."
Low, "Geometry and Graphics."

CHAPTER II

SOLID AND DESCRIPTIVE GEOMETRY AND SETTING OUT

Orthographic Projection—Auxiliary Projection—Traces of a Plane—Angles between Lines and Planes—Intersection of Arch Mould and Splayed Jamb—Sections and Developments of Solids—Sections of a Cylinder—Development of the Surface of a Cylinder—Sections of a Cone—Development of the Surface of a Cone—Sections of a Sphere—Approximate Development of the Surface of a Sphere—Sections for Weathering of Stone Bridge Cut-water—Oblique Plane—Intersections of Planes—Application to Marble Window Lining—Setting Out Marble Lining for Conical Head over Window Opening—Marble Lining for Niche—Intersections of Solids—The Geometry of Intersecting Vaults—Intersection of Cone and Pyramid—The Helix—Axial and Normal Sections of Helical Solids—Approximate Development for Joint Moulds—Setting Out Pediment in Straight Wall—Setting Out Segmental Pediment in Straight Wall—Setting Out Semi-elliptical Skew Arch—Setting Out Semicircular Arch in Cylindrical Wall—Setting Out Flat Arch with Projecting Key and Wing Stones in a Cylindrical Wall—Setting Out Segmental Pediment in a Cylindrical Wall—Setting Out Corbelling at Splayed Angle of Building—Setting Out Hemispherical Hooded Niche—Setting Out Dome—Setting Out Dome Stone by Rectangular Block Method—Setting Out Pendentive Dome—Setting Out Ribbed Pendentive Ceiling—Setting Out Interpenetration of Dome—Setting Out Intersecting Vaults of Unequal Span—Setting Out Welch Groin or Lunette Vault—Ramp and Twist Work—Setting Out Wing-wall for Entrance Stair—Setting Out Plinth or Capping for a Geometrical Stair Balustrade—Setting Out Marble Cut String for Geometrical Stair—Setting Out Rib and Panel Vaulting—Setting Out Helicoidal Skew Arch.

PRACTICAL solid geometry is that branch of geometry which enables us to represent on a plane—such as a sheet of paper or setting-out board—solid objects and sections and developments of solids in order to determine the shapes of figures, the lengths and inclinations of lines, and the relative proportions of the parts of a solid. Accurate scale representation is essential.

Orthographic Projection.—In solid geometry all objects are represented as they would appear when projected on to the planes of projection, by means of parallel projectors, at right angles to the plane upon which the object is to be projected. This projection is called orthographic projection.

Planes of Projection.—These planes are known as the "*Vertical*" and "*Horizontal*" *Planes of projection*. To ensure a clear understanding of projection, reference must first be made to the projection of a point. A point in space is represented by drawing a perpendicular line from it to each of the planes of projection. The point represented on either of these planes is the projection of the point upon it. Points in space are the boundaries of lines, so that by the projection of points lying in curves of any description, the projections of the curves are obtained.

If we suppose a surface to be bounded by a system of lines, then these lines projected on to the planes of projection will project the surface. For the projection of solids, it must be remembered that the surfaces by which these bodies are bounded form at their junction *arrises*, which may be projected as lines. The arrises of solids with plane surfaces are represented by straight lines, terminating in solid angles formed by these surfaces. When considering a combination of solids, as, for instance, a piece of combined masonry, we must imagine that the units comprising the whole are simply *arrises* or *lines*, which form the extreme surfaces of the units. In order to obtain the projection of any line bounding the unit, we must project the arris lines on to one of the planes by drawing perpendiculars to the plane from the extremities of the line.

If the arris is parallel to the plane, then the line which represents its projection will be the same length as the original line.

If it is inclined to both planes, its representation on each plane of projection will be shorter than the original line.

The projection of a curved line on to a plane parallel to the surface in which it lies is similar to the curve.

If the plane of projection is not parallel, then the projection of the curve on to the plane will be foreshortened.

In orthographic projection the image of any object is projected on to the planes of projection by means of parallel rays.

To illustrate this, assume a block of stone to be lying flat on the floor of a room. The outline of the block on the floor is the plan of the block. Assume the block to be raised vertically from the floor, the base being kept horizontal, then the outline of the plan of the block, which will not change, will be the *horizontal projection* of the block on to the Horizontal Plane (H.P.).

Now assume that the block of stone is resting on the floor close against the wall, then its outline on the wall will be the elevation of the block. Slide the block away from the wall, at right angles to the wall; this will not change the outline of the stone on the wall, which is known as the *vertical projection* of the object, on to the Vertical Plane (V.P.). The line of intersection between the floor and the wall, or the line separating the H.P. from the V.P., is called the *ground line*, and is usually referred to as the X Y line.

In order to obtain a complete knowledge of the shape of the stone, it is necessary to determine the outline of the block when viewed from the side or the end.

To obtain this outline, as before described, projectors must be drawn from various points on the stone on to a plane at right angles to the projectors. The outline thus obtained is called the *side elevation* of the stone, and the plane upon which the outline is projected is called the *Side Vertical Plane* (S.V.P.), as in Fig. 358.

For convenience in drawing, the side plane is revolved into the plane of the paper, as shown in Fig. 359.

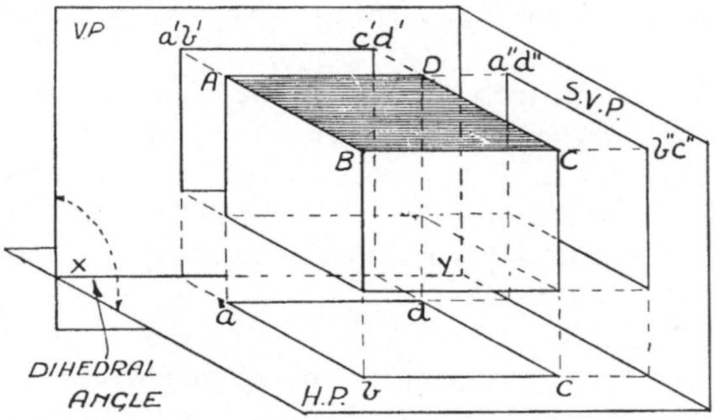

FIG. 358.

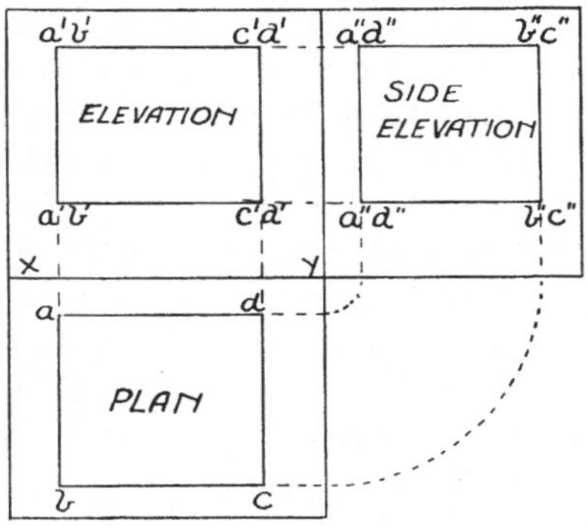

FIG. 359.

ORTHOGRAPHIC PROJECTION.

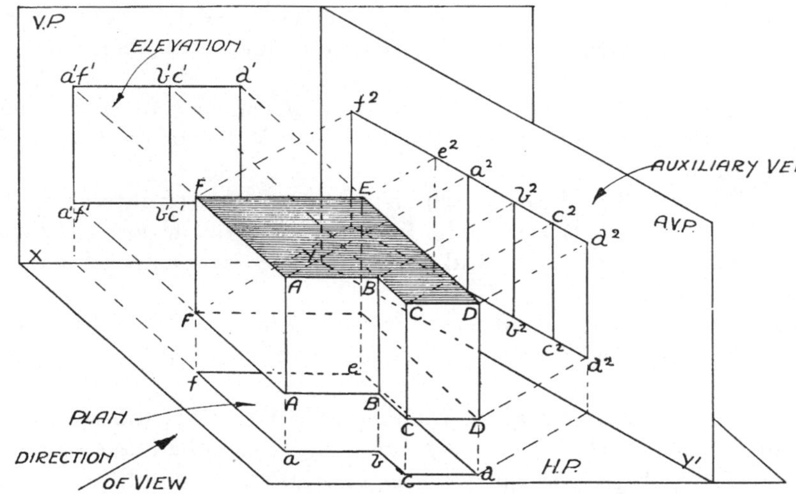

FIG. 360.—AUXILIARY ELEVATION.

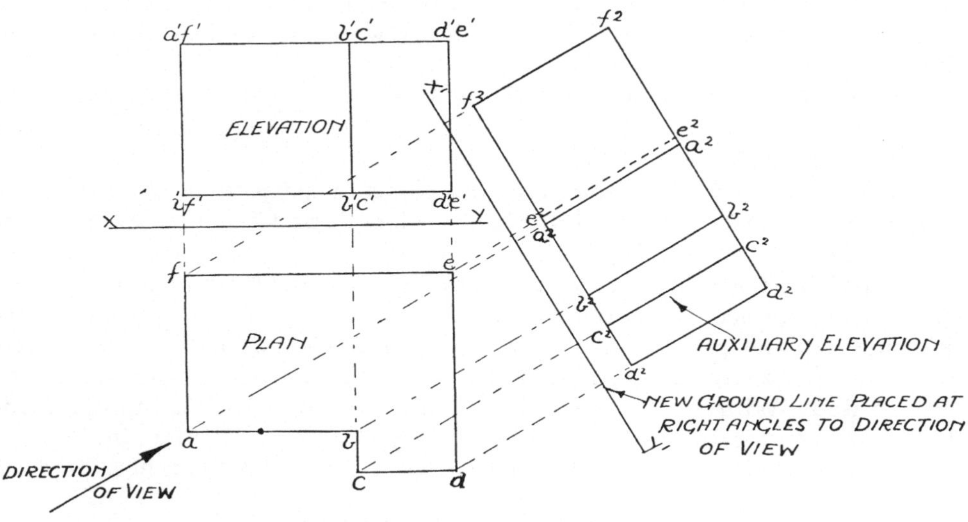

FIG. 361.—AUXILIARY ELEVATIONS.

We have now discussed the geometrical rules governing the drawing of plan elevation and section of any object.

Auxiliary Projection is a view or projection of any line, plane surface, or solid on to a plane other than the *planes of projection* already mentioned. To obtain an *auxiliary elevation* of the stone, the following rule should be noted. Place a plane at right angles to the direction of view. Project all the required points from plan at right angles to the new plane, and measure the heights from elevation, as shown in Figs. 360 and 361.

To obtain an auxiliary plan, place in the ground line of the new plane at 90° to the direction of view. Project all necessary points from elevation at 90° to the new ground line, making the distance of each projector beyond this *new ground line* correspond with its respective distance in front of the V.P. or the *original ground line*, as in Fig. 375.

It is now necessary for the student to understand something about the planes, other than the co-ordinate planes, *i.e.*, the H.P., V.P., and S.V.P., so that he may be able to visualise the various cutting planes and their relation to sections of solids.

Fig. 362 shows a plane inclined to the V.P.; Fig. 363 shows a plane inclined to the H.P.; Fig. 364 shows a plane inclined to both V.P. and H.P., and is called an Oblique Plane.

Traces of a Plane.—The lines in which a plane meets or intersects the co-ordinate planes are called *Traces*. The intersection with the *vertical plane* is called the *Vertical Trace* (V.T.) and that with the *horizontal plane*, the *Horizontal Trace* (H.T.).

The planes may be completely represented by their *Traces*. When the planes are opened out, they give the position of the plane in relation to the *planes of projection*.

Fig. 365 shows the Traces of a plane inclined to V.P.; Fig. 366 shows the Traces of a plane inclined to H.P.; Fig. 367 shows the Traces of a plane inclined to both H.P. and V.P.; Figs. 368 and 369 are pictorial views of oblique planes in addition to those already mentioned; Fig. 370 shows the Traces of the plane inclined to both H.P. and V.P.; Fig. 371 shows the Traces of the oblique plane whose real angle is obtuse. The angle between two planes is called the dihedral angle.

Angles between Lines and Planes.—The method of determining the angles contained between lines and planes is of great importance to masonry students, especially those engaged in marble works.

When a line is parallel to one of the planes of projection, the projection of the line on that plane exhibits its true length and its inclination to the other plane.

By applying the above rule to an oblique line, its true length and inclination may be determined by swinging either the plan or the elevation of the line into a plane parallel to the *planes of projection*, as in Fig. 372. Here $a\ b$ is the horizontal projection, and $a'\ b'$ is the vertical projection of the oblique line. With b as centre and $b\ a$ radius, draw an arc $a\ a''$ to cut a line drawn from b, parallel to X Y. From a'' project a perpendicular line to meet a horizontal line in elevation drawn from a' in A. Join A b', thus obtaining the

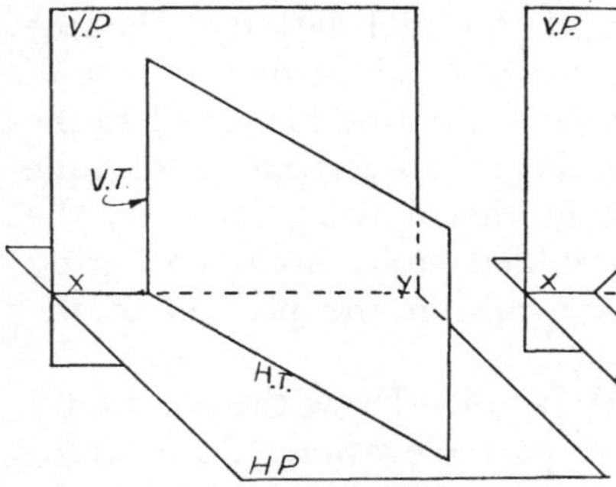

FIG. 362.

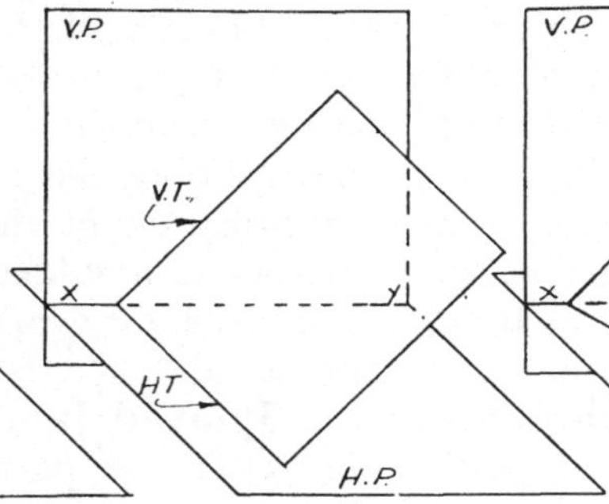

FIG. 363.

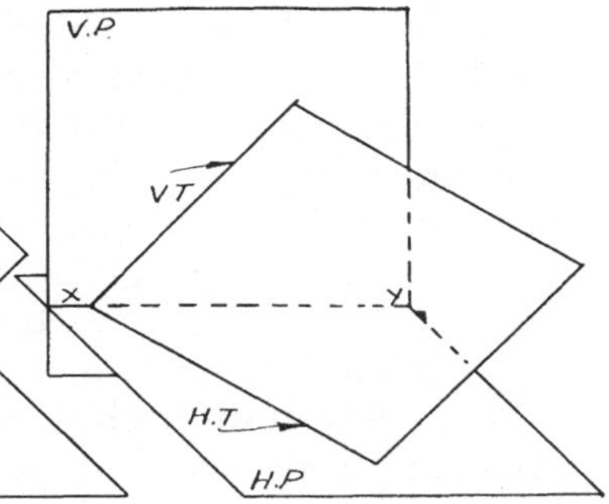

FIG. 364.

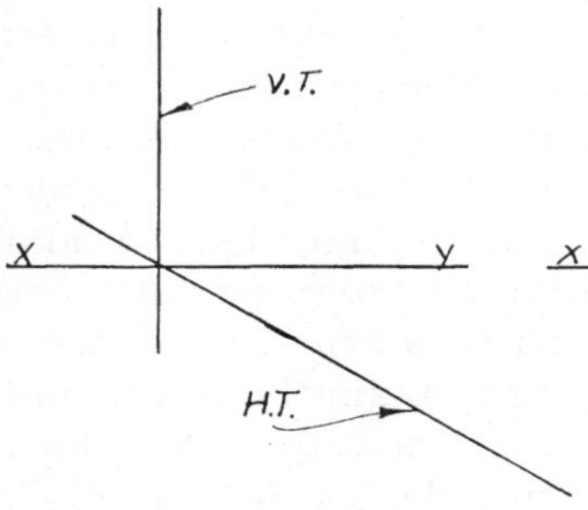

FIG. 365.

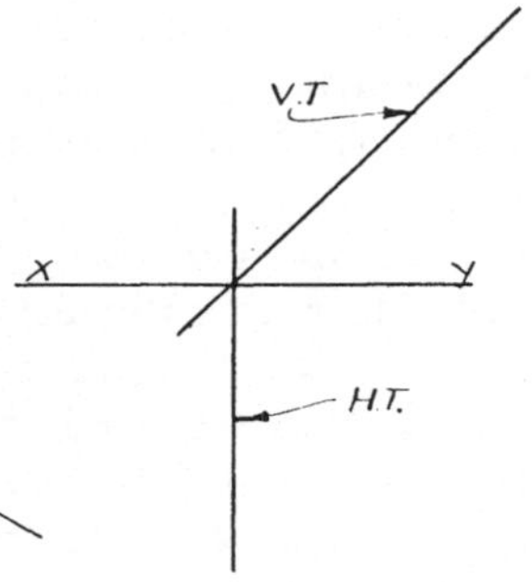

FIG. 366.

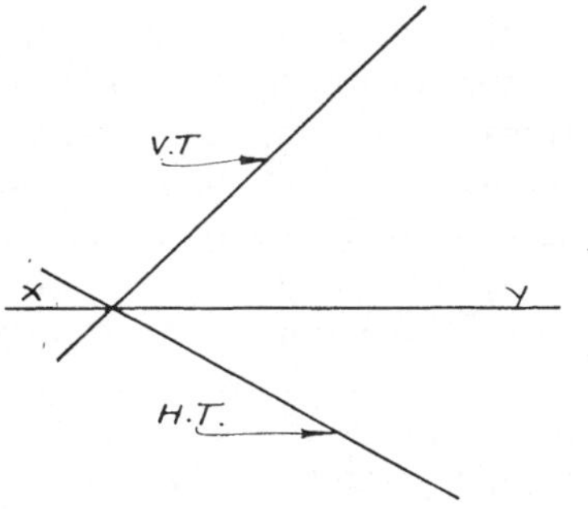

FIG. 367.

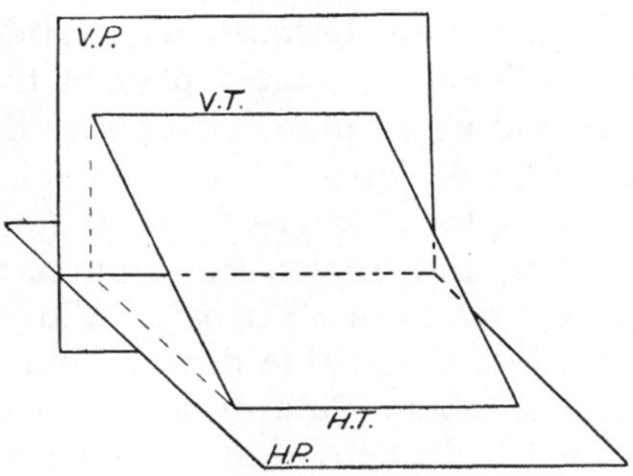

FIG. 368.

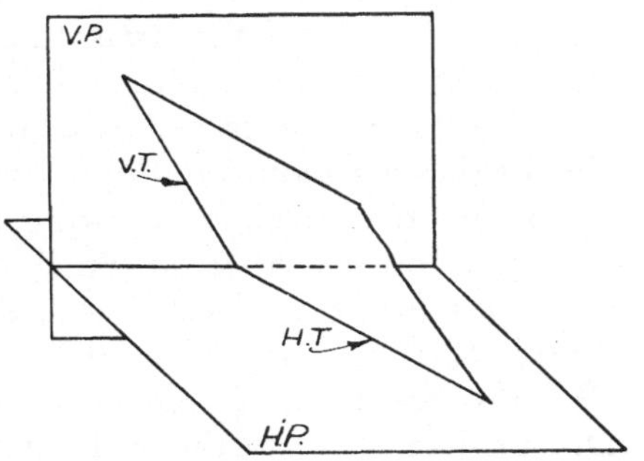

FIG. 369.

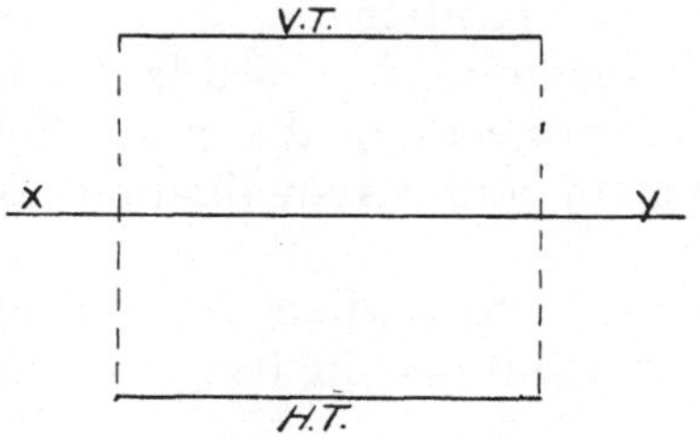

FIG. 370.

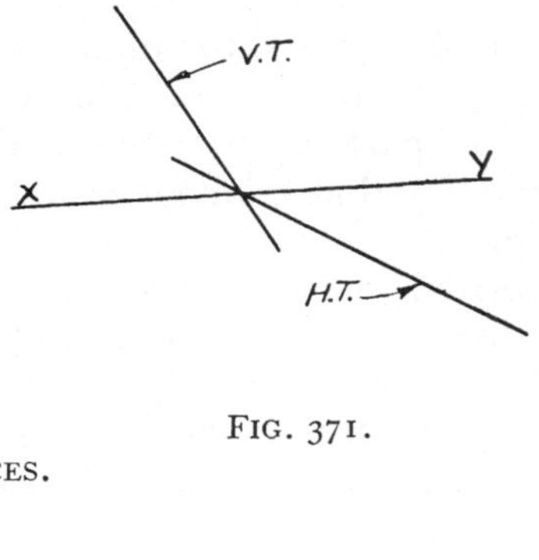

FIG. 371.

PLANES AND TRACES.

true length of the line, while the angle θ (Theta) is the inclination of the line with H.P. The student should notice that the vertical height of a' above b' has not changed. If, instead of rotating the line $a\ b$, the line is viewed in the direction of the arrow, the same results are obtained by converting the oblique line into an inclined line. As an application of the above principles, the following example is given, which shows a moulded stone arch, springing from a splayed jamb. A bevel is required for working the portion of the splayed jamb marked X on the second stone.

Intersection of Arch Mould with Splayed Jamb.—Draw the horizontal projection of the splayed jamb (Fig. 373) and part elevation of the arch, placing in the normal joints for the voussoirs, as in Fig. 374. Draw the arch section and project it into the H.P. by rotating about its axis O. To do this, take any number of points, as at $1'\ 2'\ 3'\ 4'\ 5'$, etc., swing them round to the springing line or X Y, and project them vertically to plan, to intersect lines drawn parallel to X Y, the position of these lines being taken from the section at $1^x\ 2^x\ 3^x$, etc., and transferred to the plan. Now obtain the intersection of the splayed jamb and the arch moulding in elevation. From points 1 2 3 4 5 6 in plan, draw horizontal lines to the splayed jamb line $a\ b$, and from these points erect perpendiculars to intersect their corresponding arcs in elevation drawn from points $1'\ 2'\ 3'$, etc. A curve drawn through these points is the elevation of the intersection required. To obtain the bed mould to apply on the bed joint $c'\ d'$, proceed as follows: From $a'\ b'$ in X Y draw vertical lines to the V.T. of the bed joint $d'\ c'$, meeting the bed joint in $e'\ f'$. Swing these points round into X Y, and project vertically to plan to meet horizontal lines drawn from a and b in points e and f, and join $e\ f$. This line is the rotated plan of the arris line of the intersection between the vertical chamfer plane and the inclined joint plane. Where the rotated plan of this arris line cuts in points g and k, the arch moulding in plan, determines the outline of the bed mould to apply on the inclined bed joint.

To obtain the bevel for working the vertical splayed intersection on stone No. 2, select any point M on the line $a\ b$ in plan, and draw a line from M at right angles to $a\ b$ to meet a line drawn vertically from d' in d''. With M as centre, M d'' radius, draw an arc to meet a horizontal line drawn from M in d^3. Project d^3 up to elevation to meet a horizontal line drawn from d' in D. Project M up to the V.T. of the inclined bed plane at M′. Join D M′ and produce M M′; then the angle D M′ N is the angle required for the working of the vertical chamfer plane, the stock of the shift-stock being held at right angles to $a\ b$, whilst the blade is held in a vertical position.

The geometrical principle upon which the foregoing is based is that when a line is parallel to one of the planes of projection, the projection of the line on that plane exhibits its true length and its inclination to the other plane.

The plan of the oblique line M d'' is rotated into a plane parallel to the V.P., hence its true length and inclination is exhibited in its projection in the V.P., as shown in Fig. 372.

Sections and Developments of Solids.—The surface produced when a plane cuts a solid is called a section. If the true form of a section is required,

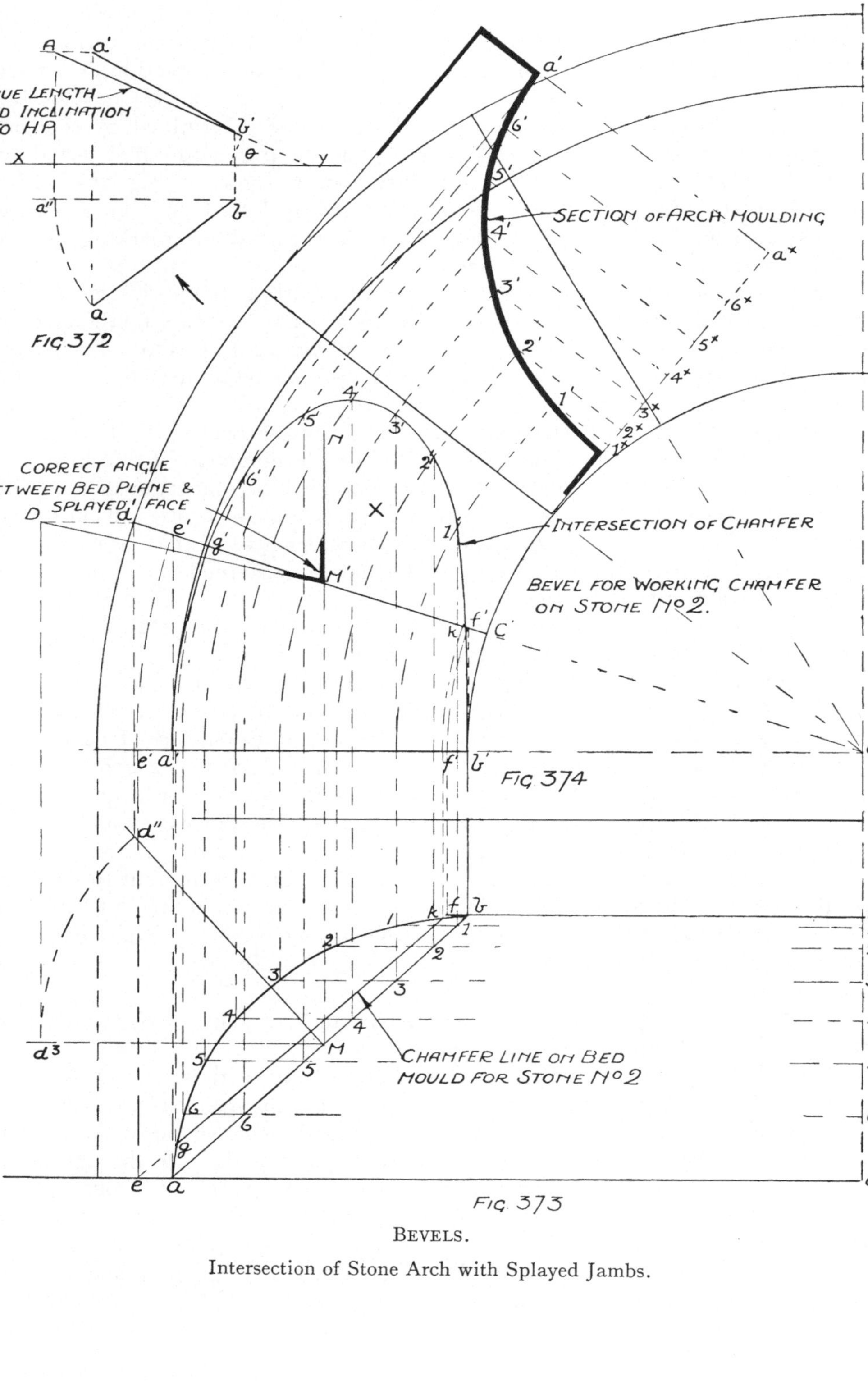

BEVELS.

Intersection of Stone Arch with Splayed Jambs.

it must be either projected on to a plane parallel to the section—that is, by auxiliary projection—or by rabatment into one of the planes of projection.

The true shape of the section of a solid is outlined on the section or cutting plane.

The true form of the figure on the plane can be determined by rotating the plane about one of its traces into the plane of projection intersected by its trace, or by the projection of an auxiliary plan, as shown in Fig. 375, which shows a piece of moulding cut by a plane inclined to H.P. This process can be also applied when a raking section is required for marking on the splayed surfaces of a polygonal stone.

If we can determine the correct position of a cutting plane, irrespective of the complicated outline of the section made by a plane, we may, by the method of rabatment, determine its true shape, as shown in Fig. 376. The above principles enable us to solve a great many problems connected with joint planes in masonry.

Developments.—A surface is said to be developed when it is laid out on a plane, the figure on the plane being called the development of the surface.

Only the surfaces of solids which are bounded by planes, such as *cubes*, *pyramids*, *prisms*, and surfaces of *single curvature*, such as cones and cylinders, can be developed. Surfaces of *double curvature*, such as a *sphere*, cannot be developed, but approximate developments may be obtained by dividing the surface into a number of parts.

The development of the vertical faces of a prism, when standing on its base, produces a rectangle containing all the faces of the prism.

An application of the development of a prism is given in Fig. 448.

The Development of the Surface of a Cylinder.—Fig. 377 shows the method for graphically determining the development of the surface of a cylinder; the application of this principle occurs in various examples contained in this book, as, for instance, the development of the surfaces of a semicircular arch in a cylindrical wall (Figs. 450-452), and also the examples dealing with vaulting and geometrical stairs.

Divide the surface into any number of equal parts, then lay out these parts on the plane of the paper. Any figures contained on the surface will be *laid out*, or *developed*, on the development of the surface of the cylinder, which is a rectangle whose height is equal to the height of the cylinder and whose base is equal in length to the circumference.

Sections of a Cylinder.—If a cylinder is cut by a plane at right angles to its axis, the resulting section is a *circle*.

If cut by a plane inclined to the axis, the resulting section is an *ellipse*.

If cut by a plane containing the axis, the resulting section is a *rectangle*. Fig. 377 shows the construction for determining the true shape of the section made by a cutting plane inclined to the H.P. passing through the cylinder.

Divide the plan into any suitable number of equal divisions, as at A 1 2 3 4 5 B, etc. Project these points up to cut the V.T. of the cutting plane in points A′ 1′ 2′, etc. To obtain an auxiliary plan of the section, draw a new ground line X′ Y′ parallel to V.T., and project the points A′ 1′ 2′ B, etc., perpendicular to V.T., producing them beyond X′ Y′. Measure the distance

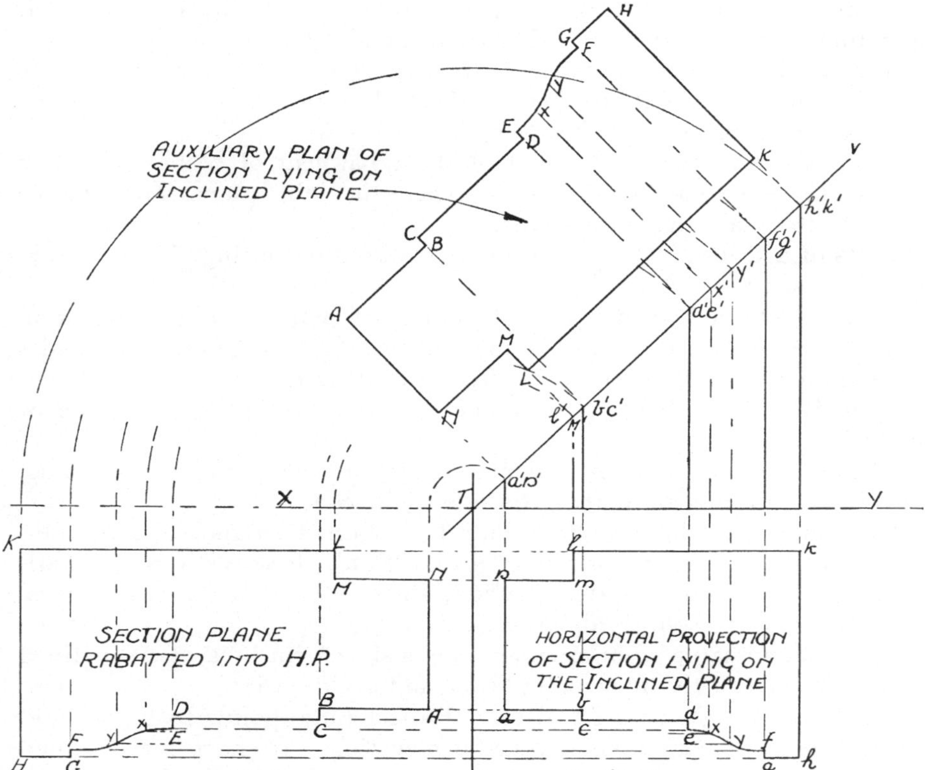

FIG. 376
AUXILIARY PLANS

FIG. 375

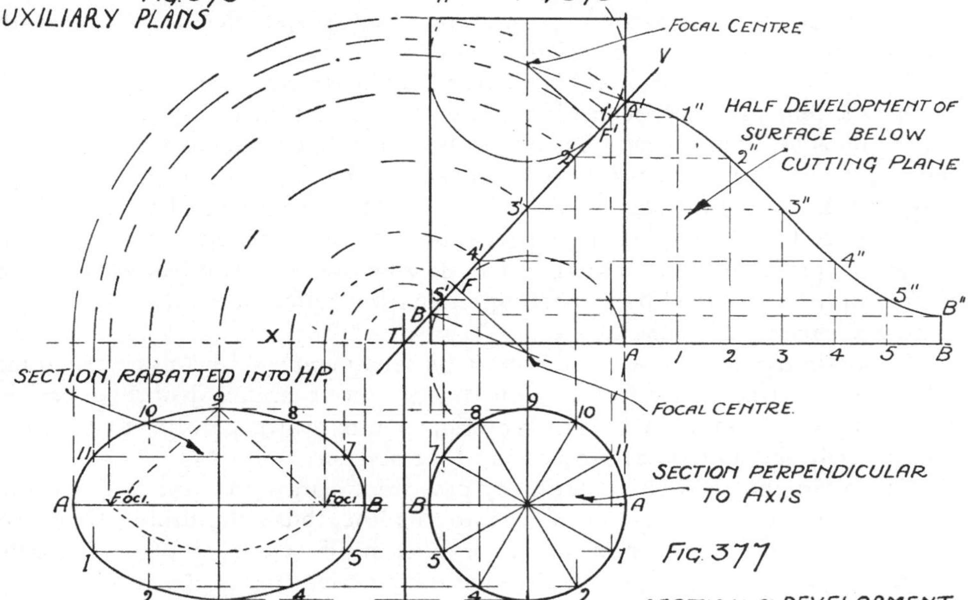

FIG. 377
SECTION & DEVELOPMENT OF SURFACE OF CYLINDER

FIG. 378

of each point in front of X Y in plan, and transfer these distances to their corresponding lines beyond X′ Y′. The resulting section will be an *ellipse*.

The true section may be obtained by rotating the points contained in the cutting plane into the H.P. to meet horizontal projectors drawn from the points A 1 2 3, etc., in plan, as shown in Fig. 378.

Examples showing the application of the rabatment of planes cutting a cylinder are to be seen in Figs. 540-549, where the joint planes are rabatted into the H.P. to obtain their true shape.

Sections of a Cone.—The four sections produced by cutting planes passing through a cone are :—

1. A *circle*—obtained when a plane cuts the cone perpendicular to its axis.
2. An *ellipse*—obtained when a plane intersects all the generators on the same side of the vertex, and not perpendicular to its axis.
3. A *parabola*—formed when the plane cuts the cone parallel to one of its generators.
4. A *hyperbola*—formed when the plane cuts the cone parallel to the axis, or inclined to it at a less angle than the side of the cone.

The true shapes of these sections may be obtained by the rules of solid geometry. Space will not permit us to deal with all these sections in detail, but we will consider the elliptical section, since it is a section which most frequently occurs in masonry problems.

In Figs. 379 and 380 a right circular cone is shown cut by a plane inclined 45° to the H.P. Draw the plan of the cone and divide it into any convenient number of generators, as at O 1 2 3, etc. Project these points to the ground line X Y, and connect them to the vertex N′ in elevation. We have now obtained the plan and elevation of these several generators. Where the V.T. of the section plane intersects these generators, drop perpendiculars to meet the plan of the corresponding generators.

The points through which a fair curve may be drawn, representing the plan of the section, are now determined, with the exception of points 3 3, which occur opposite the axis line. To obtain these points, proceed as follows. From point 3′ in elevation draw a horizontal line to cut the side of the cone in 3′. Project this point into plan at point 3. With N as centre, and N 3 as radius, draw an arc to cut the plan of the generators 3 and 9 in points 3 and 3. The horizontal projection of the section may now be drawn. The true shape of the section may be obtained by projecting an auxiliary plan, as shown in Fig. 381, or by rabatment (Fig. 382).

To obtain the true shape of the section by rabatment : With T as centre, rotate all the intersecting points in V.T. into X Y produced, and project them perpendicular to X Y to meet their corresponding points drawn parallel to X Y from the points in the horizontal projection of the section.

To obtain the section by auxiliary projection, draw the line M K parallel to V.T. Project points A B on V.T. into this line, thus determining the major axis. To obtain the length of the minor axis, divide A B in point C. Through C draw a horizontal line to cut the side of the cone in H′. On C H′, with R as centre, draw an arc, cutting a perpendicular line from C in H. Then C H is the length of the semi-minor axis.

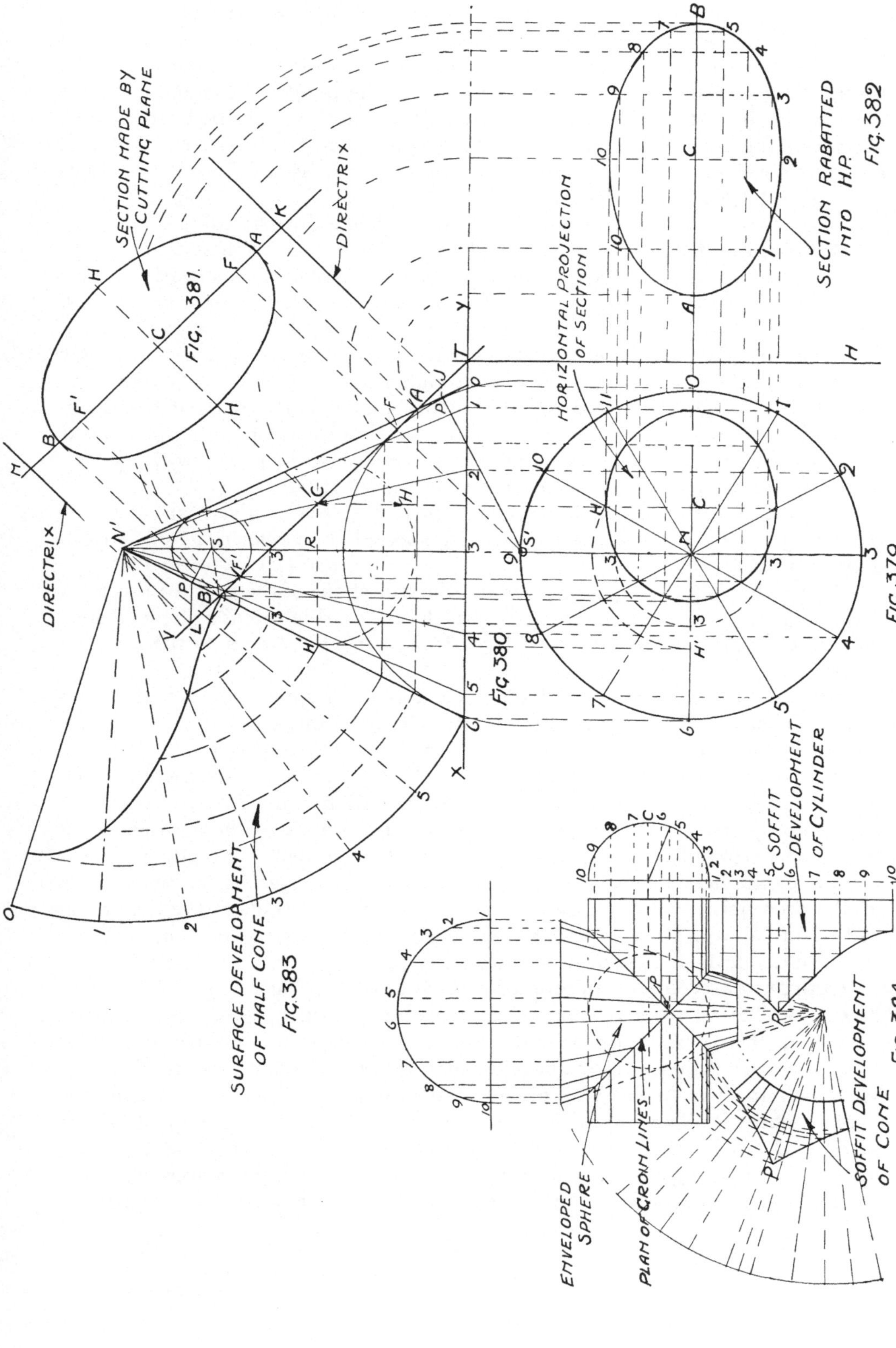

THE CONE, ITS SECTIONS AND SURFACE DEVELOPMENT.

INTERSECTION OF CONIC AND CYLINDRICAL SURFACES.

The ellipse may now be drawn by the Trammel method. To determine the position of the foci and directrices, draw circles inscribed in the cone representing spheres touching the sides of the cone and the section plane. S is the centre of one of these circles, and P and F′ are the points of contact. Project F′ into the line M K. With S′ as centre, draw another inscribed circle, thus obtaining the points of contact F and P. Project F to the line M K. The foci are now obtained. To determine the directrix, from P draw horizontal lines to meet V.T. produced in points L and J.

Project these points to the line M K; these are the directrices.

Development of the Cone.—To develop the surface of the cone, with N′ as centre, N′ 6 radius, draw an arc and mark off distances 6 5 4, etc., corresponding to the plan of the generators at 6 5 4, etc., or, in other words, unfold along this arc the circle representing the plan of the base of the cone. To determine the development of the surface of the portion of the cone below the cutting plane (Fig. 383), draw horizontal lines from the points where the elevation of generators intersect the cutting plane to the side of the cone. With N′ as centre, swing them round to intersect their corresponding generators in development, as shown. A curve drawn through these points will give the development outline of the portion required.

As an application of the above principles, Fig. 384 is given, showing the plan diagram of a *semicircular vault* intersected by a *conical vault* at right angles to it.

Obtain the plan of the *groin lines* and the true shape of the groins in section. Assuming the soffit of the vaults to be lined with marble, obtain the developed moulds to apply on the curved surfaces for cutting the stones to the required shape.

Section of a Sphere.—All plane sections of a sphere are *circles*. If a section be made by any plane, then the circle of the section will have its diameter visible in the edge view of the section, on the cutting plane.

If a cutting plane inclined to the H.P. passes through a sphere, the horizontal projection of the section on the H.P. will be an *ellipse*, because the oblique projection of a circle is an ellipse. An example of the application of this is to be found in Figs. 504-513, where the inclined joint planes of the arches cut the hemispherical dome. The horizontal projection of the portion of the arch joint, which passes across the spherical surface to the horizontal bed joints above, as at 3^x and 4^x, are portions of an elliptical curve whose axes are determined as shown on the setting out.

Approximate Development of the Sphere.—It has already been stated that the surface of a sphere cannot accurately be developed, but an approximate development may be obtained by dividing the sphere into a number of *Conical Zones*, which then requires the development of the surfaces of a series of cones, as shown in Fig. 386. The greater the number of *Zones*, the nearer the approximation. The application of the approximate development of the surface of a sphere often occurs in connection with the development of face moulds for marking the lines forming the panelling or intersections required.

A geometrical application of the approximate development of a sphere is

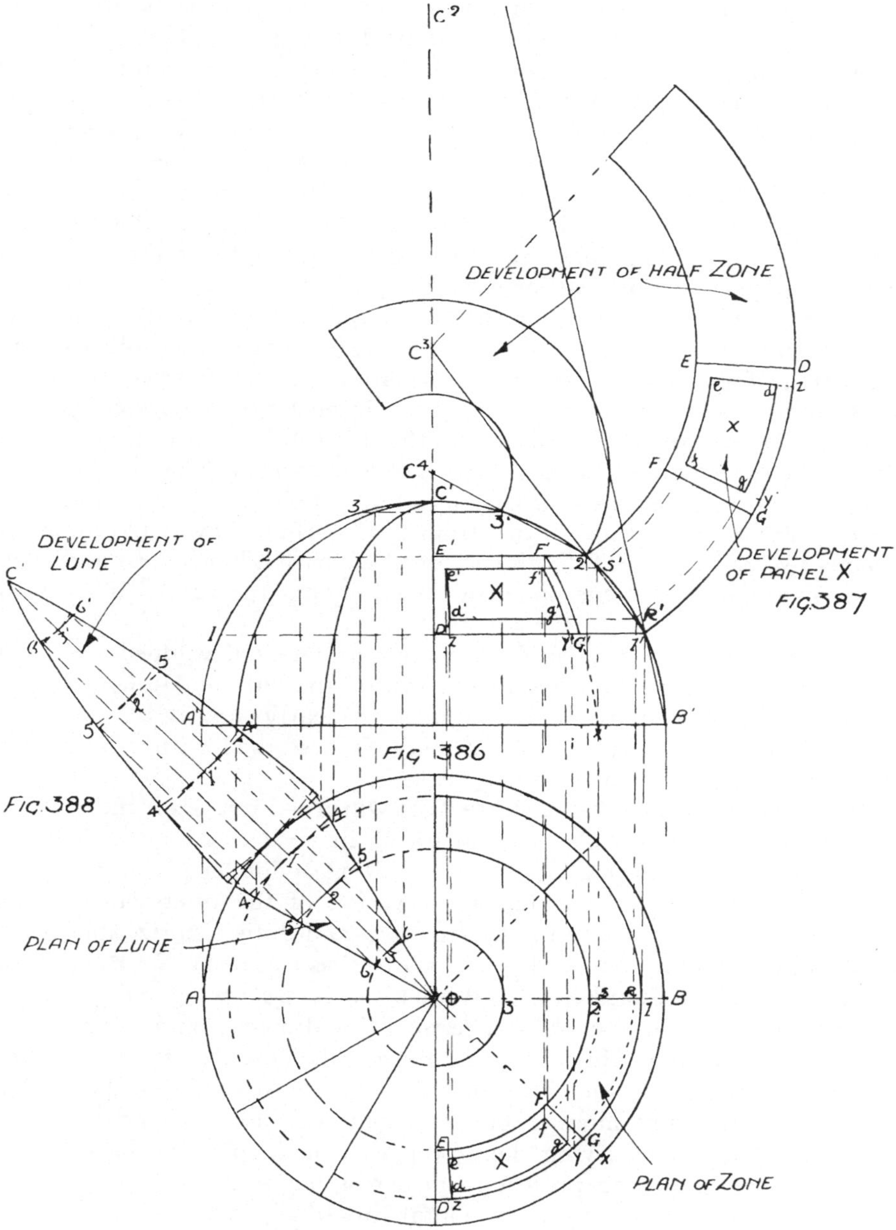

FIG 385

APPROXIMATE DEVELOPMENT OF SPHERE.

given in Figs. 385 and 386. On the right-hand side the hemisphere is divided into a series of *Conical Zones*, whilst the left-hand side is divided into a series of *Lunes* or *Gores*.

To obtain the development of the Zones, divide the outline in sectional elevation into any convenient number of parts, and draw horizontal lines representing their elevation. Draw chords between the points B′ 1′, 1′ 2′, 2′ 3′, producing them to meet the axis line C C′ produced in points C^2 C^3 C^4.

To develop the *Zone* represented by points 1′ 2′: with C^3 as centre and C^3 1′ radius, draw an arc. With C^3 as centre, C^3 2′ radius, draw another arc. These arcs are the development of the horizontal arris lines of the *Zone*. Step off in plan a series of equal divisions along the line representing the horizontal projection of the arris line, and transfer these divisions along the corresponding line in development. This will give the length of the development required. Repeat this process for each *Zone*.

Assume panelling is required on the internal surface of the sphere, as in Fig. 387, then the projection of the panelling is obtained as follows: Mark the position of the panel lines on the elevation outline between points 1′ 2′, as at S′ and R′, and project horizontal lines in elevation from these points, as shown. Also drop vertical projectors to the line C B in plan in points S and R. With C as centre, C S and C R radius, draw arcs to their position in plan. Mark in the position of panel, as at *d e f g*, and project these points up to elevation to meet the horizontal lines drawn from S′ R′. Now draw a development of the panel by drawing arcs from the points S′ R′, with C^3 as centre, and measuring the correct length of the panel by stretching out the curved lines in plan, as shown.

Sometimes the spherical surface is approximately developed by dividing the surface into a series of *Lunes*, as shown on the left-hand side of Fig. 388. The method for obtaining this development is clearly shown.

As a practical application of the methods described with relation to the sections of solids, the following is given:—

Weathering Stones for Bridge Cut-water.—First draw the plan (Fig. 389) and project up to the elevation, marking in curve line intersection against the wall face between B′ and C′ (Fig. 390). Next divide the plan into a suitable number of stones, and determine an elevation of the intersection line C D. To do this, project from B′ an axial line B′ *d* and draw a horizontal projector from D in plan to *d* on the axial line. With B′ as centre, B′ *d* radius, draw an arc cutting D′ B′ produced in *d′*. With B′ as centre, swing point B into the horizontal line D′ *d′* in point *b′*. A vertical line drawn from *b′* to meet a horizontal line drawn from C′ in point C′ determines the position of the wall line and the points *c′* and *d′* in side elevation (Fig. 391). Next divide the wall line, C B in plan, into any number of convenient divisions, as at *e f g h j* K *l*. Then, with O as centre, swing these points round in plan to the plan of the intersection line C D, and project them across horizontally to the axial line B′ *d*, and swing them up to the horizontal line D′ *d′*, as shown.

Draw projectors from these points on the wall line in plan up to elevation, as at points *e′ f′ g′*, etc., and from these points draw horizontal projectors to meet the vertical projectors, drawn from their corresponding points in the line D′ *d′*. The intersection of these lines will determine points through which to draw the curve, representing the side elevation of the intersection line C′ *d′*. To obtain the elevation of the joint lines, project direct up to elevation the points L *e f g h j* K, which are on the plan of the normal joint, to intersect horizontal projectors from *e′ f′*, etc., in E′ F′ G′ H′, etc. A fair curve drawn through these points is the elevation of the normal vertical joint K L.

To obtain the side elevation of this joint line, draw horizontal projectors from the points L *e f*, etc., in plan to the axial line B′ *d*, swing them round as before, with B′ as centre, into line D′ *d′*, and project them vertically to meet horizontal projectors, drawn from their corresponding points in elevation. Next project up the joint lines K M and K K in a similar manner, thus obtaining their elevation and side elevation. Obtain point M″ on the intersection line C′ *d′* in side elevation, and project from this point a horizontal line to the elevation in point M′ on the line C′ D′. Take any number of points between K M in plan, as at 1 2, swing them round to the wall line with centre O, and project up as before explained. To obtain the side elevation of the joint line between K K in plan, mark any number of points, as at 3 4 5; project these points across to the axial line B′ *d*, swing them round as before, then, with O as centre, swing these points in plan to the wall line C B, and project them up to the elevation of the wall line intersection curve C′ B′. From these points draw horizontal projectors to meet their corresponding projectors in side elevation at points K′ 3′ 4′ 5′ K″.

The side elevation of the joint between K M will be a vertical line drawn as shown between K″ and M″. The elevation of the joint line between K K will also be a vertical line between K″ K′.

Next obtain the true shape of the section made by the cutting plane K L; or, in other words, obtain the true shape of the joint mould to mark on the vertical joint represented in plan by the line L K and K M.

Produce the joint line L K in plan to meet the centre line C B in point *l*; then, with *l* as centre, swing round the points K *h g f e* and L into the vertical plane as shown, and project them up to meet horizontal projectors drawn from their corresponding points in elevation in points L′ E F G H K (Fig. 392). To project the portion of the joint between K M, with K^2 as centre, swing the points M 1 2 into the joint plane L *l*, then, with centre *l*, swing them into the vertical plane as before, and project up to elevation to meet horizontal projectors from points M′ 1′ 2′ in points M 1 2 K.

The true shape of the joint line between K K is given in the side elevation between K′ and K″.

A sketch of the Cut-water is given in Fig. 393.

The Oblique Plane.—Figures lying in *oblique* planes usually present difficulty to the student, but if the problems are solved in an orthodox manner, no serious difficulty should arise.

In solving the problem of the inclined plane, we assume first of all that a cutting plane is inserted through the inclined plane at right angles to its traces, giving the *dihedral angle*, or the angle that the plane makes with one of the planes of projection. The following rule is important in finding the angle that any plane makes with either plane of projection:—

Insert a cutting plane at right angles to the trace of the plane, and perpendicular to the plane of projection intersected by the plane represented by the *Trace*, as in Fig. 394.

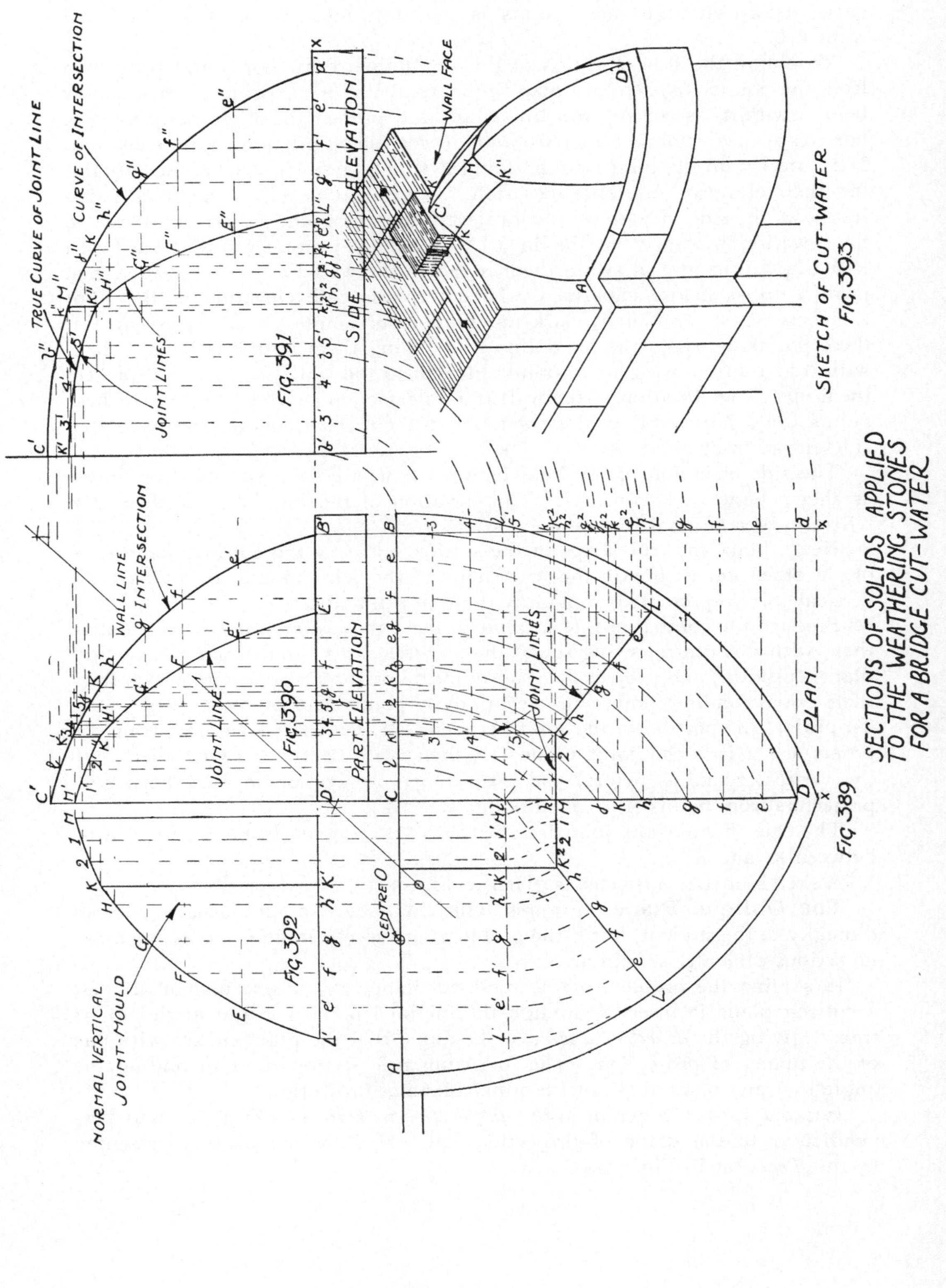

SECTIONS OF SOLIDS APPLIED TO THE WEATHERING STONES FOR A BRIDGE CUT-WATER

Then construct a right-angled triangle, which, by the process of rabatment, can be rabatted into one of the planes of projection, thus exhibiting its true inclination to the plane.

An oblique plane is given by its traces V T H; determine the inclination of the plane with H.P. (Fig. 395). From point *a* draw a line perpendicular to H.T., cutting *x y* in *b*. From *b* erect a perpendicular to cut V.T. in *b′*. With *b* as centre, *b b′* radius, draw an arc to cut a line drawn from *b* perpendicular to *a b* in B. Join B *a*, then the angle *b a* B is the inclination of the oblique plane to the H.P.

To find the inclination that the plane makes with the V.P. (Fig. 396): From any point *b′* in V.T. draw a perpendicular line to V.T., meeting *x y* in *b*. From *b* draw a line perpendicular to *x y*, cutting H.T. in *a*. With *b* as centre, *b a* radius, draw an arc cutting a line drawn from *b*, perpendicular to *b′ b* in point A. Join *b′* A, then A *b′ b* is the angle of inclination required, as shown in pictorial view (Fig. 397).

Having thus obtained the true angle made by the oblique plane with H.P., we may now proceed, by the process of rabatment, to determine the true shape of any figure lying in that plane.

Imagine the H.T. to be an axis, and the oblique plane made to revolve about H.T. until it coincides with the H.P. The path of any point or lines in the plane will be at right angles to H.T. Because *b′* (Figs. 394 and 395) is perpendicular to *a*, the plan of its path will travel through *a*, at right angles to H.T. The position of point *b′* is determined by transferring the length of *a* B along the projection of the path travelled by *b′* from point *a*. The following rule should be noted: In solving the problem of the oblique plane, *convert* the *oblique plane* into an *inclined plane*, and rabat it about its H.T. Fig. 398 shows a moulded stone standing on the H.P. cut by an oblique plane V T H. To obtain the true shape of the section made by the oblique plane, place a cutting plane T′ H′ at right angles to H.T., and rabat the cutting plane into H.P., as shown by its traces V′ T′ H′. Project all the points of the figure lying on the oblique plane into the rabatted inclined plane V′ T′ H′, then project the points perpendicular to the inclined plane V′ T′, and measure distances for the various points along these lines from T′ H′ to their respective points in plan, thus obtaining the true projected section (Fig. 399).

To rabat the section into H.P., with T′ as centre, swing the points on V′ T′ to meet T′ H′ produced; then draw projectors from these points parallel to H.T. to intersect projectors from their corresponding points in plan, drawn at right angles to H.T., as in Fig. 400.

When the traces of an oblique plane are given, as in Fig. 371; to determine the *dihedral angle* and the true form of any figure on the plane, proceed in a similar manner to that before described by *inserting a cutting plane at right angles to the H.T.*, as in Fig. 401. Extend H.T. behind the V.P., and from any point *a*, in H.T. produced, draw a line *a b* at right angles to H.T., cutting *x y* in *b*.

From *b* erect a perpendicular to *x y* to cut V.T. in *b′*. From *b* draw a line at right angles to *a b*, and along this line mark off distances *b* B, equal to

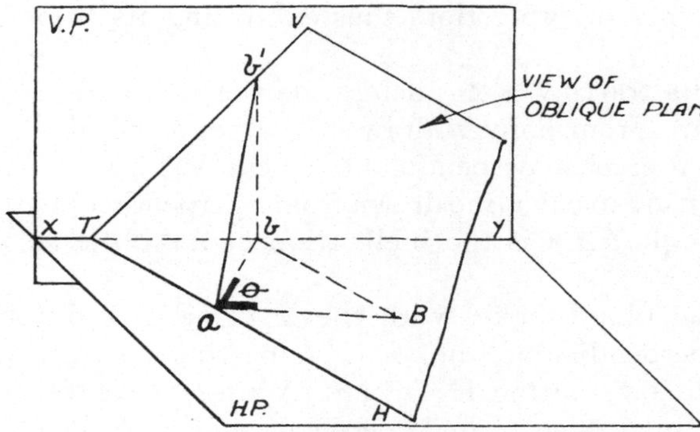

FIG. 394

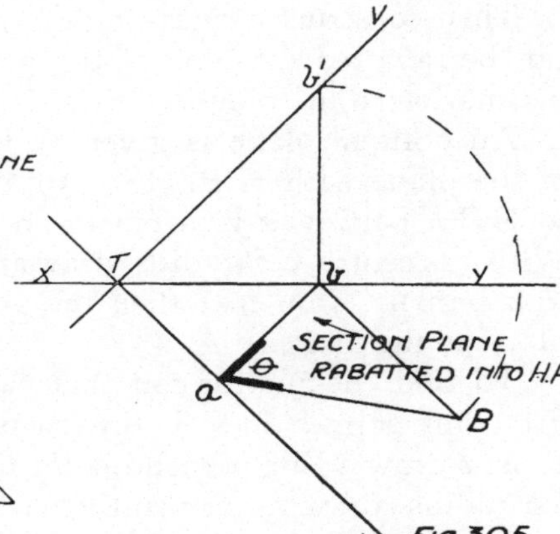

FIG. 395

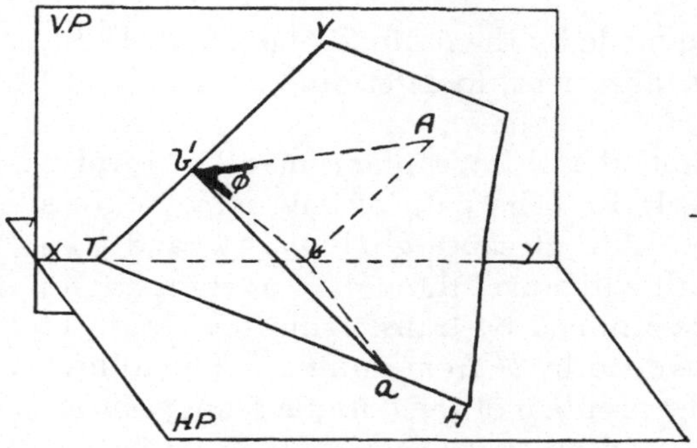

FIG. 397

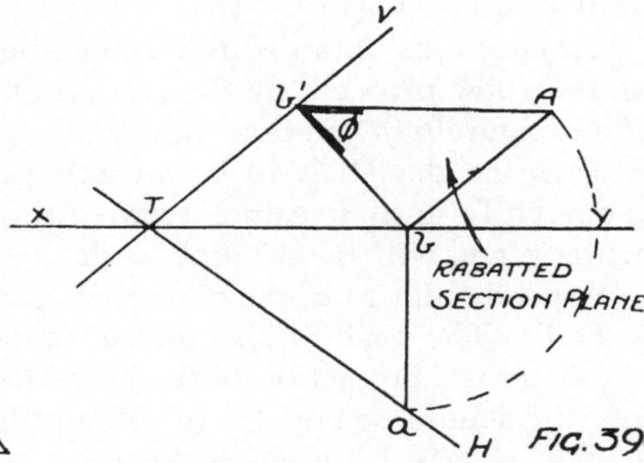

FIG. 396

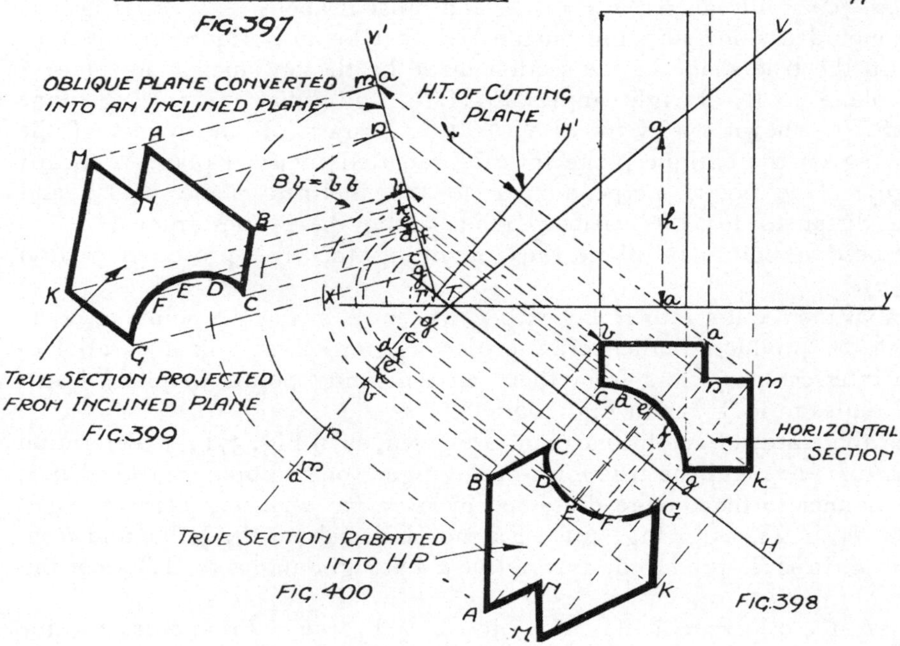

FIG. 399

FIG. 400

FIG. 398

THE OBLIQUE PLANE.

b b'. Join *a* B and produce *b a* to C, then the angle B *a* C is the *obtuse angle* the plane makes with H.P.

Applications of the above principle are given in Figs. 409 and 416.

Intersection of Planes.—When two non-parallel planes meet or cut each other, the line along which they meet is termed the line of intersection, and is common to both planes. The angle between two planes is the angle between two lines, one in each plane drawn from any point on the intersection of the planes and perpendicular to the line of intersection.

To determine the angle between two planes, pass a plane perpendicular to the line of intersection of the two planes.

Draw the traces of the two planes and determine the line of intersection (Fig. 402); also draw the assumed rabatted section of both planes made by a cutting plane perpendicular to H T and H' T', as at L M N and L' M' N'. Draw a rabatted elevation of the intersection line *a* B. Insert a cutting plane C D at right angles to *a b*, cutting *a b* in O. From O erect a perpendicular to *a* B, intersecting *a* B in E. With O as centre, O E radius, draw an arc cutting *a b* in F. Join C F and D F. Then the angle C F D is the angle between the planes.

This angle is not often required in general masonry, but its application frequently occurs in marble masonry.

Marble Window Lining.—Apply the above principles to the following:—

Marble slabs form the inside lining for the splayed jambs and head of a window opening, as in Figs. 403 and 404. Determine the face moulds and bevel for working the mitre joints between the jambs and head.

To obtain the face moulds, find the true length and inclination of the intersection line *a' b'*, as described for Fig. 372, by drawing the sections as at *a" b"* (Fig. 405). With *b"* as centre, *b" a"* radius, draw an arc cutting a vertical line drawn from *b"* in a^3. From a^3 draw a horizontal line to meet a vertical line drawn from *a'* in A^3. Join *b'* A^3, thus obtaining the face mould for the slab forming the head. With *b* in plan as centre, and *b a* radius, swing the line *a b* into a line drawn horizontally from *b*, meeting this line in a^4. Draw a vertical projector from a^4 to intersect a line drawn horizontally from *a'* in elevation, in A^4.

Join *b'* A^4, thus obtaining the face mould for the jambs.

To obtain the bevel for the mitre joint, when the shift-stock is held perpendicular to the intersection line *a' b'*, find the true length and inclination of this line, as at *a'* B. Insert a cutting plane C D at right angles to *a' b'*, cutting *a' b'* in O.

From O erect a perpendicular to *a'* B, intersecting *a'* B in E.

With O as centre, O E radius, draw an arc cutting *a' b'* in F. Join C F and D F. The angle between these two planes is now obtained by inserting a plane perpendicular to the line of intersection *a' b'*. As the joint surfaces are to form a mitre joint, it is necessary to bisect the angle already obtained, and produce the bisector to X. Then the angle C F X, or D F X, is the bevel required for working the joints.

The following example is given as a further application, showing the method of determining the bevel for the joint surfaces between two planes. Marble slabs forming the inside splayed jambs for a window opening are in position,

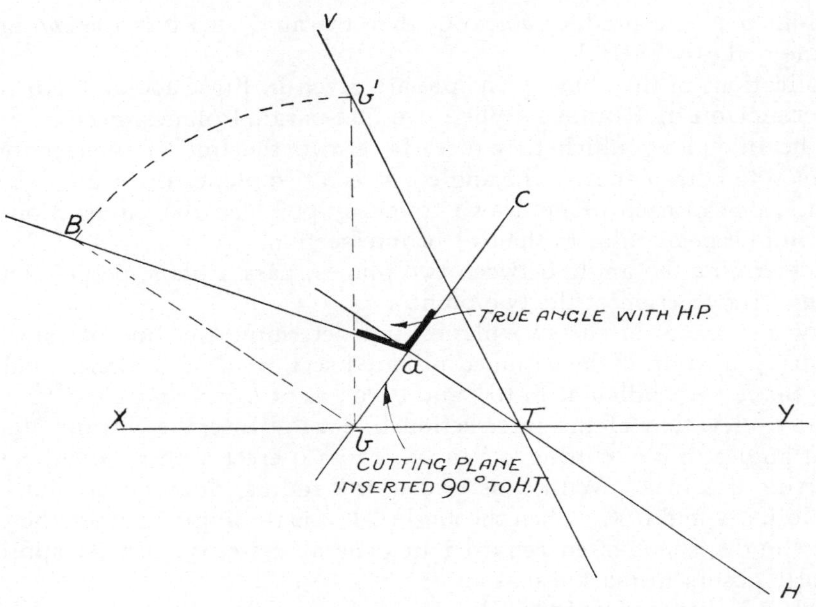

FIG. 401.

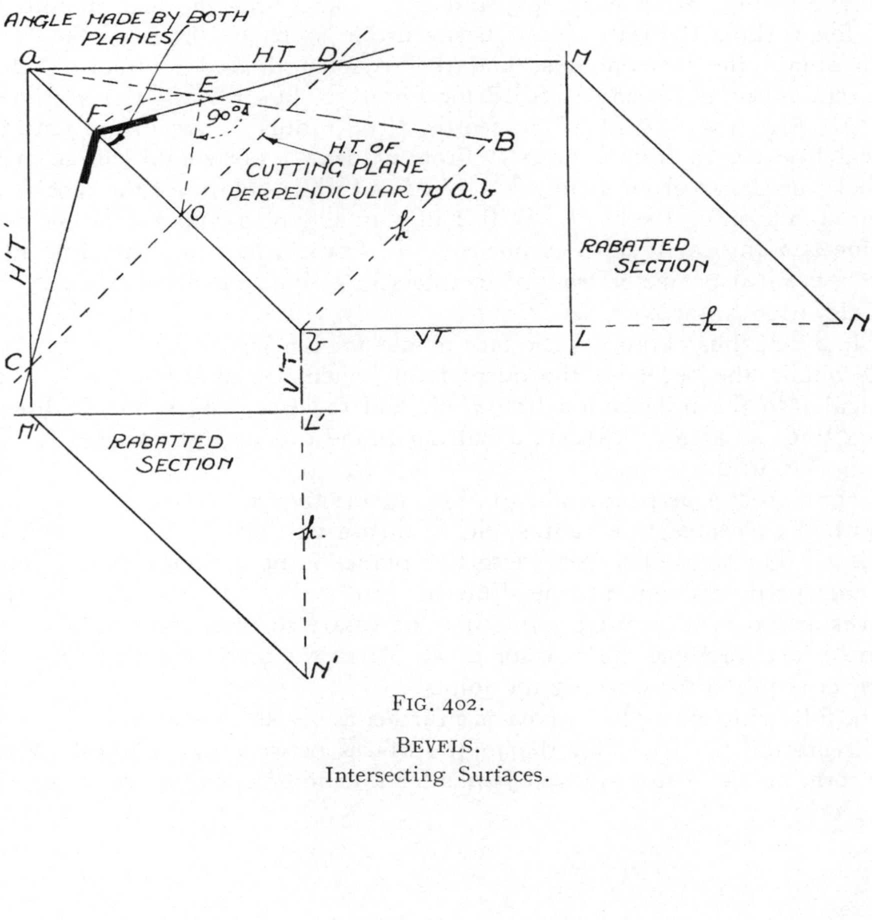

FIG. 402.

BEVELS.

Intersecting Surfaces.

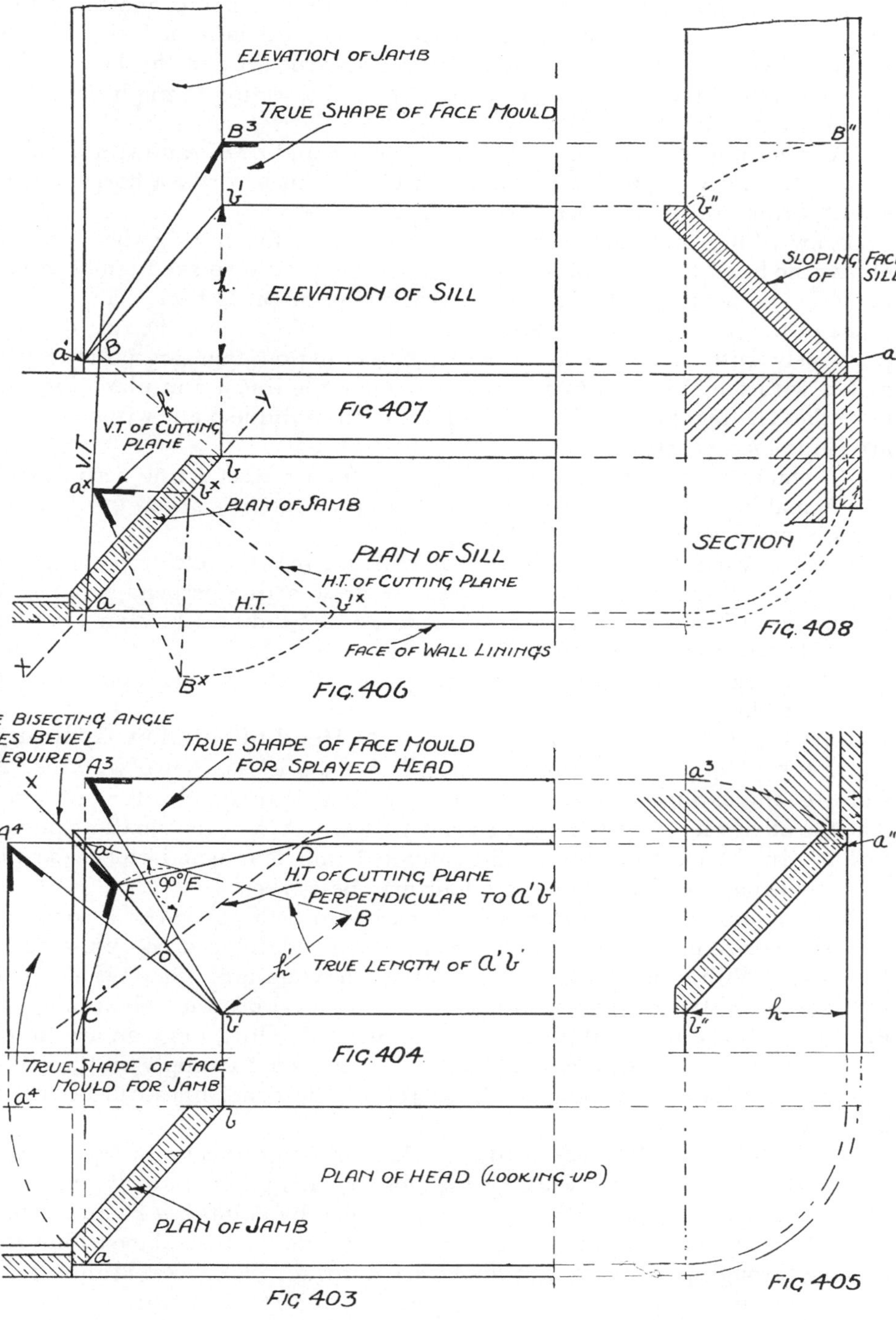

Bevels for Marble Linings.

as shown in Figs. 406 and 407. It is required to fit a slab of marble between the jamb linings to form a splayed window board. Obtain the face mould and bevel for working the joint between the window board and the jamb. Draw the elevation of the line of intersection a' b' and a section through the window board, as at a'' b'' (Fig. 408).

To determine the face mould, with a'' centre and a'' b'' radius, draw an arc cutting the vertical projector from a'' in B″. From B″ draw a horizontal line to cut a line drawn vertically from b' in B^3.

Draw a' B^3, thus determining the face mould for cutting the joint. To obtain the bevel for the joint surface against the vertical splayed jambs, assume the face of the jambs to be the vertical plane of projection V.P., then the plane of the window board is an oblique plane. Now apply the rule discussed previously with reference to finding the inclination that an oblique plane makes with the V.P. In this instance, assume the line a b in plan (Fig. 406) to be the X Y line, then the horizontal trace of the oblique plane will be the front arris line marked H.T.

To obtain the vertical trace of the plane, from b erect a line perpendicular to a b, making b B equal to the height of b' in elevation marked h. Join a B; this gives the vertical trace (V.T).

Select any point on the vertical trace, as at a^x; insert a cutting plane at right angles to V.T. to cut X Y in b^x. From b^x draw a line perpendicular to X Y to cut the H.T. in b'^x. With b^x as centre, b^x b'^x radius, draw an arc to cut a line drawn from b^x perpendicular to a^x b^x in B^x. Join a^x B^x, then the angle b^x a^x B^x is the angle required for setting the shift-stock when applied on the stone at right angles to the arris line a b.

Setting Out Marble Lining to Conical Head of Window Opening.—Draw the plan and elevation of the marble work, as shown in Figs. 409 and 410, indicating the position of the panelling if required. It is important that the stones forming the head should be worked from the smallest possible stones. In the example the head is divided into four equal stones, so that, by the setting out of one stone, the four may be worked.

Now assume the face surface of the marble to be a portion of a cone, with its axis horizontal. The surface can then be developed as explained in Fig. 383. With O as centre and O a radius, draw an arc. Step off along this arc from A, distance equal to the arc a' b' in elevation, thus obtaining point B. Join A and B to O. With O as centre and O d radius, draw an arc to meet A O and B O in D and C, then the developed face mould to apply on the conical surface of the stones is complete (Fig. 411). The panel lines can be marked if desired.

To Obtain the Plane Mould.—A plane is assumed containing the points a' b' c' d' in elevation for working the stone. This plane is really an oblique plane, its horizontal projection being represented by points a d c b in the plan. To find the true shape of the plane, convert it into an inclined plane by inserting a plane through point b to meet the horizontal trace in X. Rabat this cutting plane into H.P. by drawing a line perpendicular to b X from point b, and measuring along this line distance equal to h in elevation, thus obtaining b^2.

Join b^2 X, this line being the true length of the line b X on the plane.

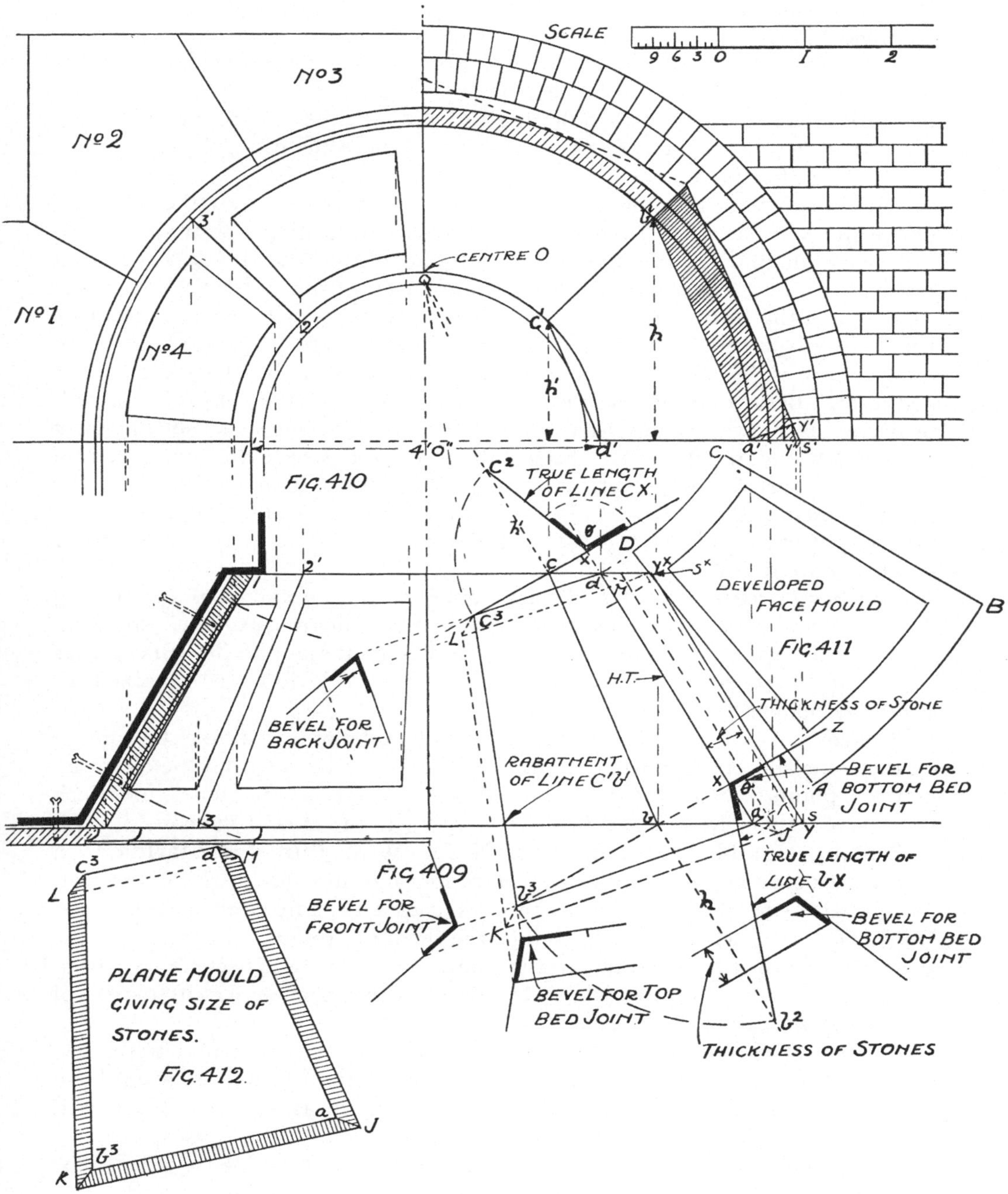

APPLICATION TO MARBLE LININGS FOR CONICAL HEAD OVER WINDOW OPENING.

With X as centre, radius b^2, draw an arc to cut b X produced in b^3. Join b^3 a. Repeat this process at point C as shown. The true shape of the plane is now obtained.

It is now necessary to find the bevels for working the splayed joints, and to determine the overall size of the stone. To obtain the bevel for the bottom bed joint, produce b X to Z, then the angle b^2 X Z is the bevel required. Because the joints are all normal, or radiate from the axis of the cone, the same bevel may be used for the top bed joint.

The thickness of the stone should now be determined. Draw the thickness of the stone in elevation, represented by the shaded portion. From a' draw a line perpendicular to line a' b' to y'. Rotate y' into the line a' s' in point y'', then project y'' down to plan in point y, and from y draw y y^x parallel to a d, then the lines a d, y y^x, represent the thickness of the stone. Produce line a d clear of the drawing, and draw two lines representing the thickness of the stone, at right angles to a d produced. Mark in the bevel for the bottom bed joint. Between these two lines on a d produced, the line J M is determined, projected as shown. Draw lines S J and S^x M at right angles to SS^x, to meet the line J M in points J and M. From J draw line J K parallel to a b^3. Produce c^3 b^3, and mark in the bevel for the top bed joint and the thickness of the stone. Draw K L parallel to c^3 b^3. Draw line y^x M at right angles to Y Y^x to meet J M in M, and draw line M L parallel to d c^3. This completes the cover mould required.

The bevels for the front and back joints may be obtained by producing line a b^3 and drawing lines representing the thickness of the stone at right angles to a b^3. Produce line J K to cut these parallel lines, thus obtaining the bevel for the front joint. The back joint is obtained in a similar manner.

The cover mould is shown in Fig. 412. A templet of the curve to be applied on the front and back joint is required. These may be cut from the elevation between points a' b' and d' c'. Mark these templets on the stone, and work the curved surface in straight drafts converging to the axis of the cone.

Setting Out Marble Lining for Niche, with Joints Radiating to a Centre Boss.—Draw the plan and elevation and add thickness for the wall linings, also plot the moulding to be worked round the front arris of the hood (Figs. 413 and 414). The wall linings are separate from the niche stones, the joint between the two occurring at A in plan. Draw the semicircle representing the outline of the centre boss in elevation, so that the semicircular arris of the boss-stone will lie in a vertical plane.

Project down to plan the position of the boss-stone, and complete the bed mould for same. Next mark, in elevation, the position of the radiating hood joints, and obtain their horizontal projection in plan, as at 1 2. These will be elliptical curves.

Next draw the vertical section through the centre line of the niche (Fig. 415), and project the side elevation of the radiating joints as shown.

The stones for the hood may be obtained from stones represented by their size in plan and elevation, similar to the method which would be adopted if worked in solid stone. This method would be expensive in material such as

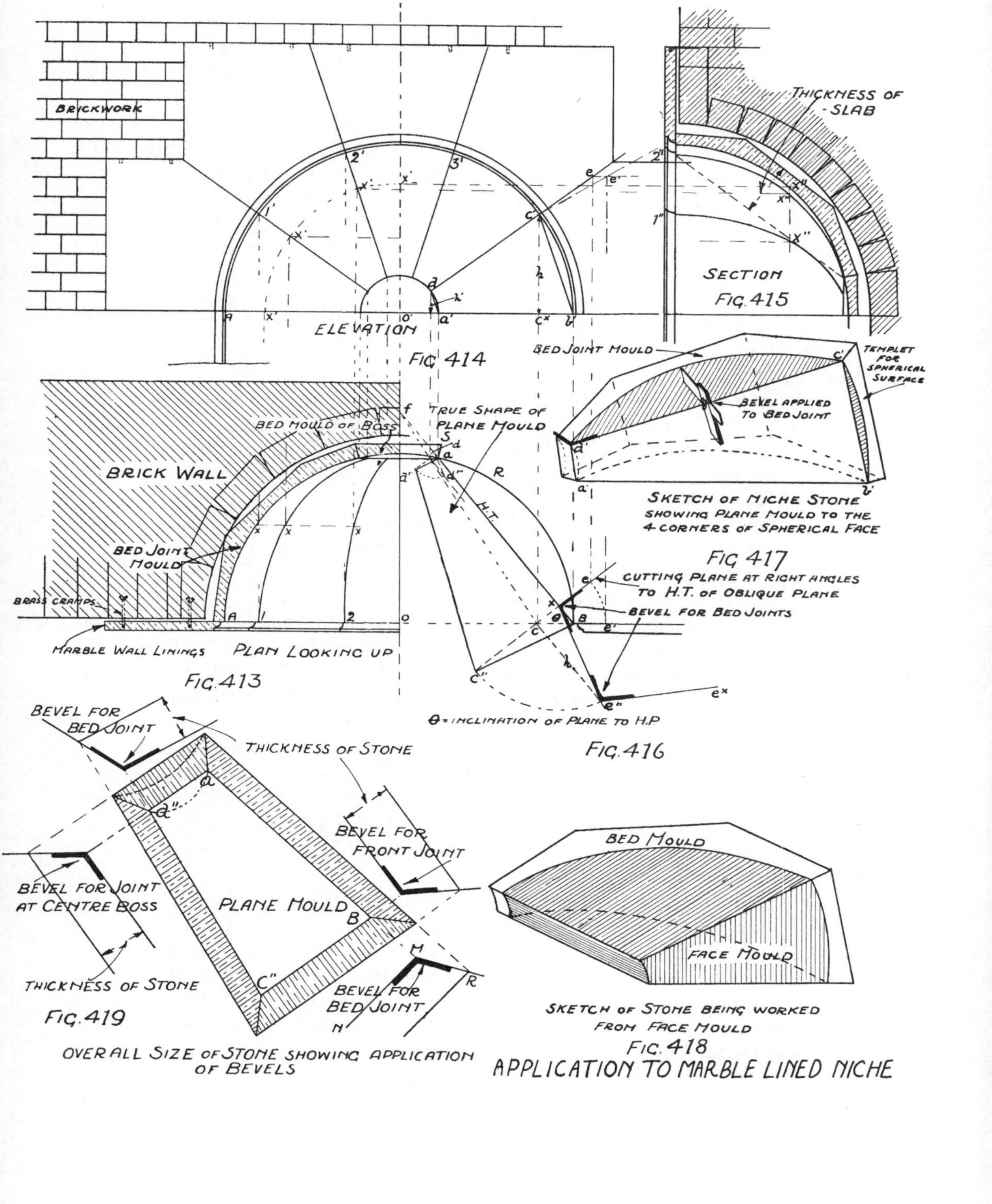

APPLICATION TO MARBLE LINED NICHE

marble, so it is important to economise in material by adopting a method, to be explained, whereby a minimum amount of material may be used.

It will be noticed that, by the application of the geometrical principles already dealt with, moulds and bevels may be obtained for working the stones. As the hood is divided into a number of equal stones, the various moulds will apply to all the stones comprising the hood.

It is first necessary to assume an oblique plane, which will contain the four points of the spherical surface, as at *b′ c′ a′ d′* in elevation (Fig. 414). This plane is leaning forward and also inclined to the vertical plane, as shown in plan by the horizontal trace of the plane, by the line *a* B. To obtain the true shape of the figure on this plane, it is best to convert the plane into an inclined plane by inserting a cutting plane at right angles to the H.T. and rabatting it into the H.P., as shown in Fig. 416. First obtain the horizontal projection of the line *d′ c′* in elevation by projecting it into plan, to intersect the front line at *c*, and the H.T. produced to the centre line at F. This will give the projection of the line required.

Now rotate the plane into the horizontal plane; this will cause the points *c* and *d* to travel along a path at right angles to H.T. Through *c* and *d* draw lines at right angles to H.T. Produce a line through *c* to *e*, cutting the H.T. in X. This line will represent the H.T. of a new plane inserted at right angles to the H.T. of the oblique plane.

It is now necessary to obtain the true length of the line which is marked in plan as C X and the dihedral angle made by the cutting plane.

To do this, erect a perpendicular to C *e* from point C, and from C mark off the height C *e″* equal to the vertical height *h* in elevation.

Join *e″* to X, then *e″* X is the true length of the line CX. With point X as centre, radius X *e″*, draw an arc to cut the line C X produced in C″. Repeat the same process at *a d*. Join points B C″, *a d″*, and C″ *d″*. This will give the true shape of a plane mould touching points *a′ b′ c′ d′*. The stone may now be marked out to the shape of this plane mould, but as the various joints are not at right angles to it, it is necessary to obtain the bevels for these in order that the overall size of the stone may be determined. The bevel for the bed joints is obtained by producing the line C X to *e*, then the angle *e* X *e″* is the bevel required for the bottom bed joint. The angle for the top bed joint is the same as that for the bottom bed joint, because all these joints are normal to the spherical surface.

Another method of obtaining this bevel is as follows:—

With C as centre, C *e* radius, draw an arc to cut the front line in *e′*. Project point *e* up into elevation to cut the normal bed line in *e*. From *e* in elevation draw a horizontal line to intersect a projector drawn from *e′* in point *e′*. Join C′ and *e′*, then the angle *b′* C′ *e′* is the angle required. The bed joints of the stone may now be worked and tested by the application of the shift-stock set to the angle obtained from the drawing.

The bevel for the front edge is taken from plan, as at X B *e′*.

When the bed joints have been worked true, apply the bed mould to the points *a′ b′* for the bottom bed and *d′ c′* for the top bed. The application of this mould gives the bevels for the front joint, and also the joint between the hood stones and the centre boss.

Next mark the templet for the spherical surface on the front joint between *c′ b′*. The spherical surface may now be worked to the lines marked on the stone, and may be tested with the templet cut to the reverse of the spherical surface. To work the conical bed joint between the hood stones and the centre boss, apply a reverse templet cut to the curve between *a′ d′* in elevation, and apply a bevel taken from the plan at S *a* R. This templet should be applied on the spherical surface normal to the curve.

A sketch of the stone, showing the plane mould and the application of the shift-stock for working the bed joints from the plane surface, is given in Fig. 417, whilst Fig. 418 is a sketch showing the size of the stone required if worked from a face mould.

To obtain the size of the stone required, decide upon the thickness of the slab. This can be measured approximately from the section (Fig. 415), which is a cutting through the centre of one of the stones, sufficient material being allowed for the amount of curve of the spherical surface, as shown by the dotted line in the section from point 2′.

Draw the plane mould as shown in Fig. 419. Produce *a* B to M, draw M N at right angles to *a* B, and draw a line parallel to M N to represent the thickness of the stone. At M mark in the bevel for the bed joints, thus obtaining point R. From R draw a line parallel to *a* B. Repeat this process at each arris line of the plane mould, marking in the bevels as shown. In practice the overall size of the stone would be reduced, without detriment, by omitting a portion of the front and back splayed joints. An alternative method for working the joint against the centre boss is shown in Fig. 425. This necessitates the introduction of the interpenetration of a Cone and Pyramid, which will be discussed later.

Intersection of Solids.—When two surfaces meet or penetrate, the line in which they meet is termed the line of intersection.

On the line of intersection of two surfaces there are generally certain important points. These points should always be determined first, and afterwards such intermediate points as will allow of the complete intersection being correctly drawn may be determined.

Problems connected with the interpenetration of solids frequently occur in masonry drawing, their correct solution requiring a knowledge of the principles involved. Groin lines in vaulting are the lines of intersection of two or more surfaces; the determination of these lines is essential for obtaining the various moulds for working the stones.

The Geometry of Intersecting Vaults.—The method of determining the *line of intersection*, or *groin line:* Draw the axes or centre lines of the large and small vaults in plan, and plot the plan of the vault, thus determining the intersection of the vaults at the wall line, as at A and B (Fig. 420). Now erect sectional elevations of both vaults, as shown in Figs. 421 and 422. Divide the soffit line of the small vault into a convenient number of divisions, as at 1′ 2′ 3′ C′, and through these points draw normal lines to the curve representing the vertical traces of the joint planes. From the points 1′ 2′ 3′ C′ draw horizontal projectors to the axis of rotation, and swing them round into the line A′ B′ produced. From the points in A′ B′ produced, draw vertical projectors to cut the curve of the soffit line of the large vault (Fig. 422) in points 1″ 2″ 3″ C″. From these points draw horizontal projectors to meet vertical projectors drawn from 1′ 2′ 3′ C′ in points 1 2 3 C. A fair curve drawn through these points will give the plan of the *groin line*, or *line of intersection*.

The points 1 2 3 on the groin line are also the plans of the points where the joint planes intersect the groin line.

To Obtain the Horizontal Projection of the Joint Plane.—These planes should be considered as *oblique section planes* passing through the large cylindrical vault. The horizontal projection of these planes produces an ellipse. Produce the vertical traces of the joint planes to meet the line C^x, which represents the highest part of the soffit of the large vault, in points 1^z 2^z 3^z. From these points draw vertical projectors to the plan of this line, cutting it in points 1^z 2^z 3^z.

We have now determined the half-major and half-minor axes for drawing the elliptical curve. These curves are the horizontal projections of the joint planes, and may be drawn by the trammel method. It will be found that these curves pass through the points 1 1^y, 2 2^y, 3 3^y, as shown in the plan (Fig. 420). O C^z is the semi-axis for each elliptical curve.

An outline of the finished plan, looking upwards, and the internal elevation are shown in Figs. 423 and 424.

Intersection of Cone and Pyramid.—The following is given as an alternative method for working the conical bed joint against the centre boss (Fig. 416), by projecting the curve *a′ d′* on to the plane mould and using a templet cut to the bevel B *a* S, which would be applied on the surface of the plane mould (Fig. 417).

The projection of this curve involves the application of the geometrical problem of determining the lines of intersection made by the interpenetration of a cone and pyramid. The geometrical diagram for determining this line of intersection is given in Fig. 425.

Let V *b* C be the horizontal projection of the slant side of a pyramid and V′ *b′* C′ the vertical projection of the slant side. Let V′ V^2 E be the horizontal projection of the cone and V′ E E′ part of the vertical projection of the cone, V′ being the elevation of the common vertex, the axes of both solids lying in the horizontal plane.

Point *a* is the point of intersection of the generator V^2 E of the cone, with the slant arris V *b* of the pyramid. (This line we will call the horizontal trace.)

Project *a* up to *a′* in X Y, then with V′ as centre and V′ *a′* radius, draw an arc cutting V′ C′ in *d′*. Project *d′* into plan to meet the slant arris V C of the pyramid in *d*. A line drawn from *a* to *d* should be parallel to X Y. Now divide the arc E F′, which represents the part base of the cone, into any convenient number of divisions, as at 1^2 2^2 3^2, and draw generators through these points from V′. Produce these lines to cut the base line of the pyramid *b′* C′ in points 1′ 2′ 3′. Project the points 1^2 2^2 3^2 down to X Y, and connect these points to V^2, thus obtaining the horizontal projection of these generators. Now

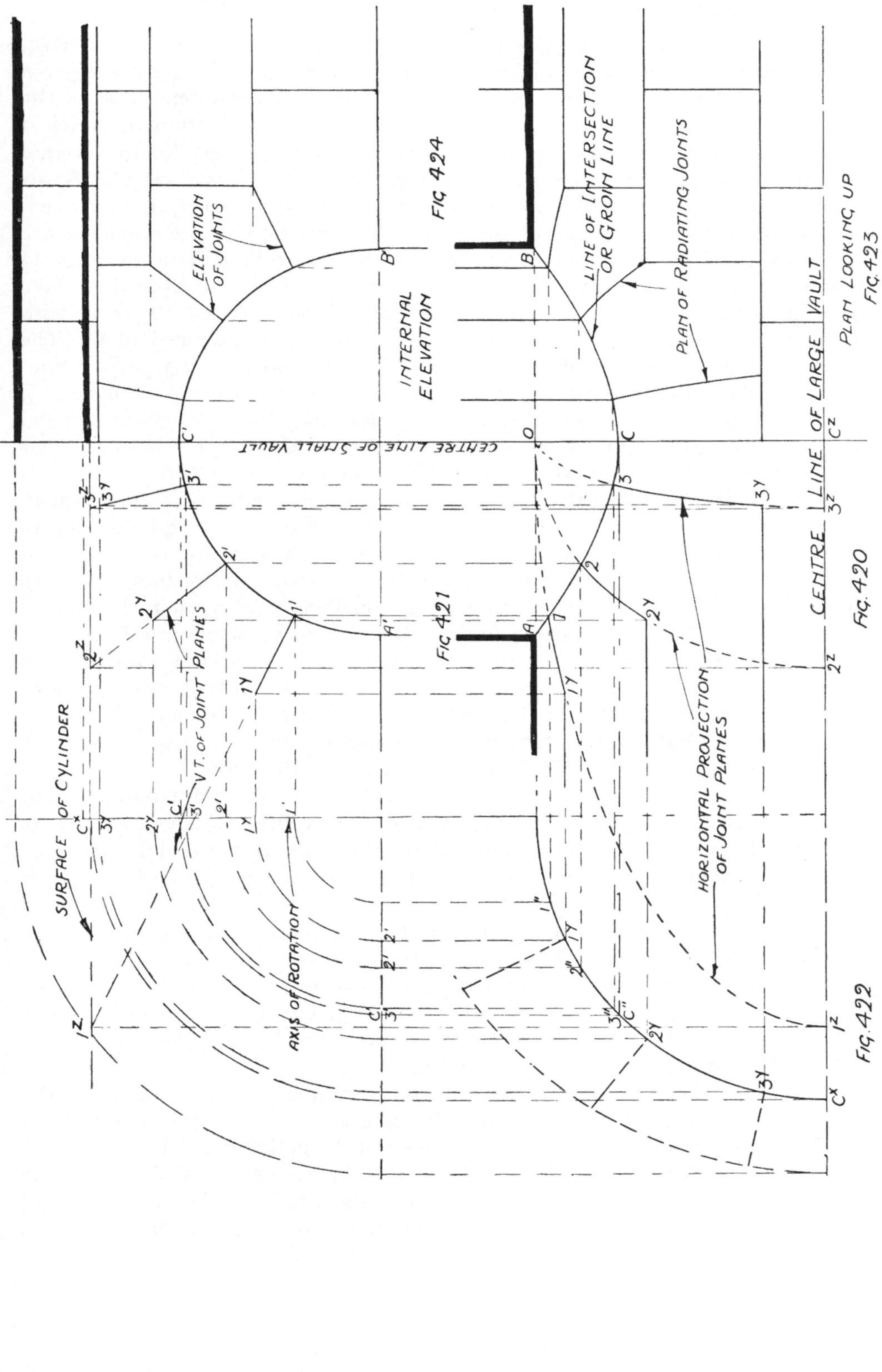

PROJECTION OF INCLINED BED JOINTS CUTTING CURVED SURFACES.

project the points 1′ 2′ 3′ to the horizontal projection of the base of the pyramid, in points 1 2 3.

Connect these points to the vertex V. Where these generators intersect the generators of the cone, as in points 1 2 3, a fair curve may be drawn from *a* to *d*, thus determining the horizontal projection of the intersection of the conical surface and the slant surface of the pyramid. To obtain the vertical projection of this intersection, project points 1 2 3 on the intersection line up into elevation to meet their corresponding generators, as at points 1^x 2^x 3^x. A fair curve drawn through these points from *a′* to *b′* will be the projection required.

To obtain the true shape of this curve with relation to the slant surface of the pyramid, rabat the slant surface into the horizontal plane by rotating it about its horizontal trace. Insert a cutting plane *d* X, and rabat this plane into the horizontal plane, as shown at *d* X d^3. Produce *d* X to D, then, with X d^3 radius and X as centre, draw an arc cutting *d* X produced in D. The line *a* D is the rabatment of the line *a d*. From the points 1 2 3 project lines perpendicular to *d* X to cut the rabatted section plane X d^3 in points 1 2 3. Also from points 1 2 3 on the plan of the intersection line, draw lines parallel to *d* X D. Where the generators of the pyramid meet the line *d* X in plan, draw lines perpendicular to *d* X to cut the rabatted inclined line d^3 X in points 1 2 3. With X as centre swing these points round to meet the line *d* X produced in points 1 2 3. Also place a cutting plane through point C at right angles to H T, and by similar procedure obtain the true shape of the slant side of pyramid. Draw the generators as shown in Fig. 426. Where these cut lines drawn parallel to *d* X, from the points 1 2 3 on the curve in plan, determines the several points through which to draw the true curve of intersection.

The Helix.—The helix is the curve generated by a point which moves along the surface of a cylinder in such a way that a constant ratio is maintained between its travel round the cylinder and parallel to its axis.

The Axial Pitch of a Helix is the distance travelled by the describing point along the cylinder, whilst moving once round the cylinder.

From the above remarks it will be evident that if the point moves round the arc of one division which, in Fig. 427, is one-twelfth of the circumference, then it will, at the same time, move along the cylinder parallel to the axis, a distance equal to one-twelfth of the pitch. The construction of a helix is shown in Fig. 427.

Divide the circumference of the circle representing the section of the cylinder into twelve equal divisions, as at 1 2 3 4, etc.

Next decide on the pitch, dividing this also into twelve equal divisions. Draw vertical projectors from the points 1 2 3 4, etc. in section, to intersect the divisions of the pitch. Through these intersections draw the curve which will be the vertical projection of the helix.

To obtain the development of the helix, stretch out the surface of the cylinder occupied by the pitch, thereby forming a rectangle. Join points 1 1′; then the straight line 1 1′ is the development of the helix (Fig. 428).

A Helical Surface is generated by a straight line, which slides along the helix, always remaining perpendicular to the axis of the cylinder, and radiating from it, at the same time revolving about that axis with a uniform motion.

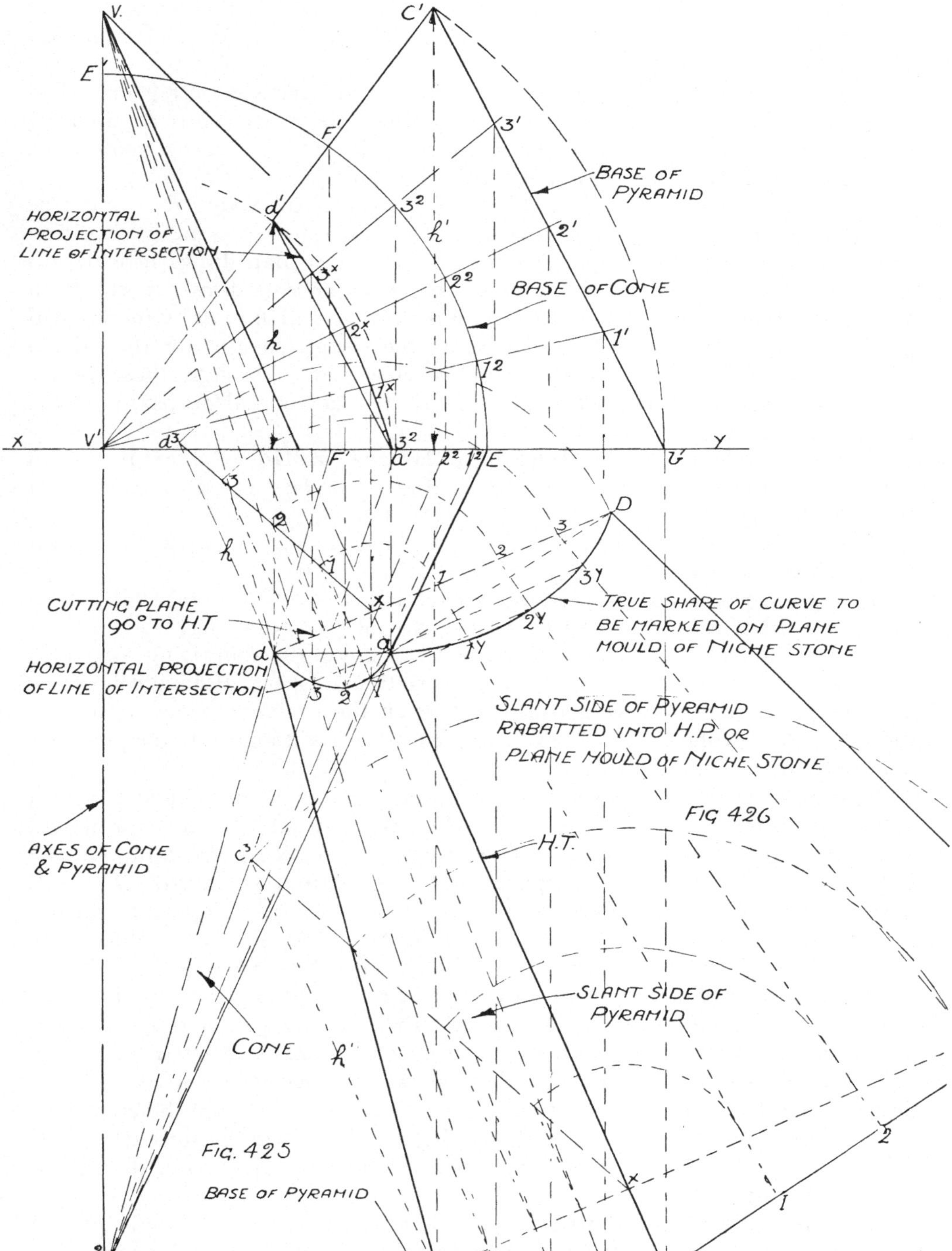

INTERPENETRATION OF CONE AND PYRAMID.

Diagram showing Method for Determining Curve Joint Line on Plane Mould of Marble Niche.

Any point in the radiating line will generate a helix. Helical surfaces occur frequently in masonry problems, and their solution requires a clear understanding of the principles mentioned.

Take, for example, a stone plinth, carried on the free ends of a geometrical stair. Let 4′ 10′, 4″ 10″ be the plan of the plinth and O the centre for striking the curves, or common axis of the cylinders (Fig. 429). Draw a centre line 4 10, and divide this curve into any number of equal divisions, as at 5 6 7, etc. Through these points draw normal lines to the centre O, cutting the curve line 4″ 10″ in points 5″ 6″ 7″, etc. Assume these lines to represent the position of the riser lines of the steps. From these points in plan draw vertical projectors to meet horizontal lines representing riser heights drawn in elevation (Fig. 430), thus obtaining a series of points through which to draw the vertical projection of the helices. Next decide on the vertical height for the plinth, adding this height above each point in elevation. Through these points draw another series of helices, thereby obtaining the vertical projection of the four arris lines of the plinth.

Axial and Normal Sections of a Helical Solid.—An axial section means a section made by a plane containing the axis. The axial section of the solid in Fig. 430 is a rectangle, as at R S T U.

A normal section is a section made by a plane at right angles to the central helix line contained in the solid.

The application of normal plane sections of helical solids does not usually occur in masonry, but it is of great importance in joinery handrailing, where plane-cut surfaces are used as joint surfaces between the units comprising a wreath.

Masonry should be viewed from a constructional aspect, although the piece of work forms a decorative feature, so that in the problem under discussion the joint surfaces should be as nearly as possible at right angles to the pressure, or, in this instance, the *pitch*.

This necessitates *twisted*, or *helicoidal* joint surfaces, having top and bottom arrises horizontal. To obtain the projection for a suitable joint for a helical stair plinth, draw a development of the centre line helix by stretching out the centre line in plan to pass through point 7 in elevation, as shown. Through point 7 draw a line perpendicular to the centre line development, cutting the top and bottom lines of this development. Through these points draw horizontal lines cutting the elevation of the centre line helix on the top and bottom helical surfaces in points M and N. Project these points down to the centre line in plan in points M and N; through the points thus obtained draw lines converging to centre O, cutting the inner and outer curves in points M^2 M^3 and N^2 N^3. Project these points to meet the horizontal lines drawn through M and N in elevation. These points should be on the elevations of the arrises of the helical solid, as shown. Draw curves through points M^3 7 N^3 and M^2 7 N^2, then the horizontal and vertical projection of the helical joint is obtained.

The method for determining the developed face moulds to apply on the cylindrical surfaces, thereby fixing the correct position of the joint arrises, is shown in Fig. 431. First draw the development of the centre line helix, and place in the vertical height of the plinth, then draw a perpendicular line to

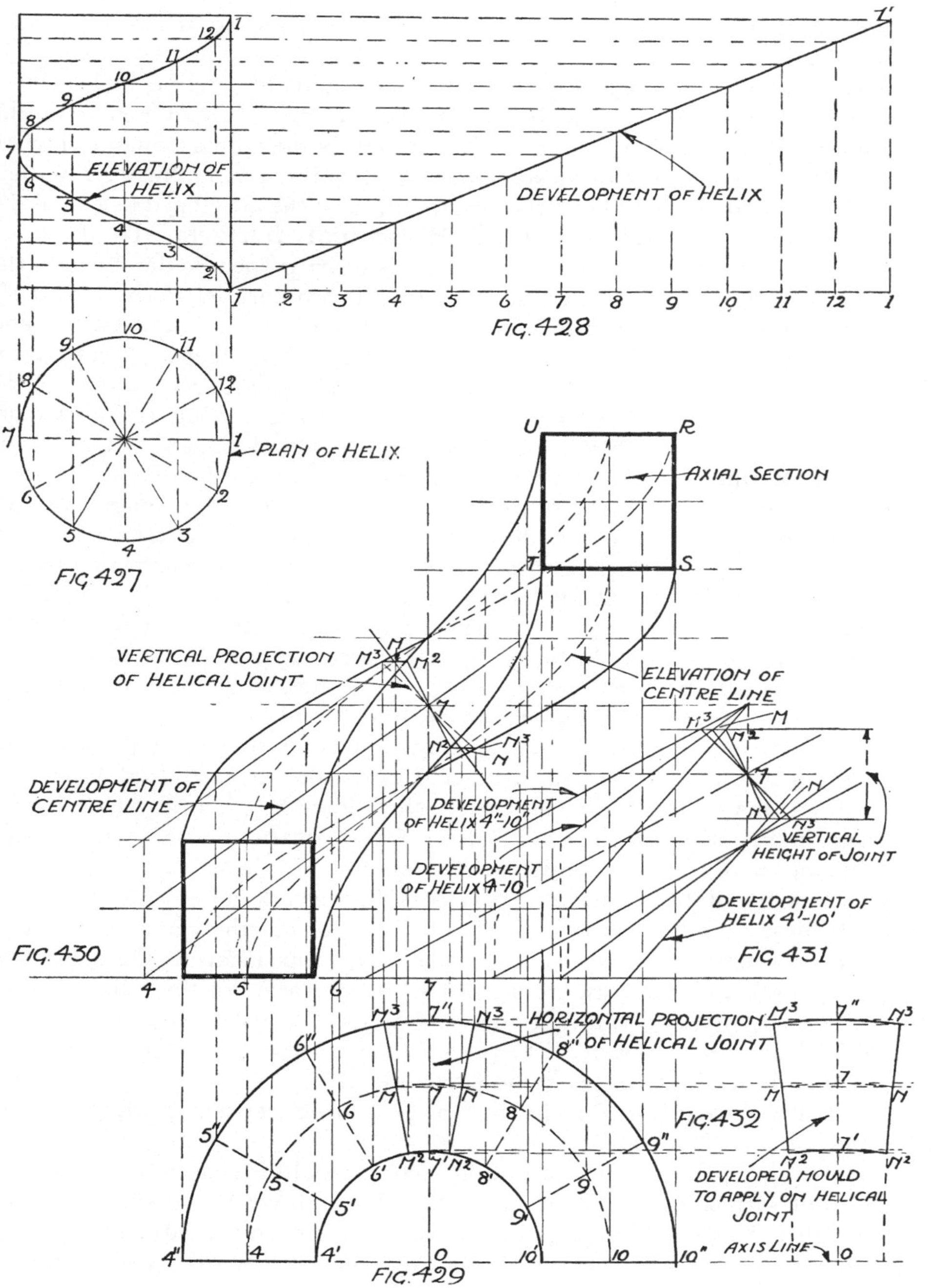

THE HELIX AND HELICAL SOLIDS.

meet the top and bottom arrises in points M and N. Horizontal lines drawn through these points determine the vertical height of the joint along the axis of the cylinder.

From point 7 draw the development of the inner and outer cylindrical surfaces as shown. Where the top and bottom arrises of these developments cut the horizontal lines in points $M^3 N^3$ and $M^2 N^2$ determines the points through which to draw the lines representing the development of the joint arrises, thus giving their correct length and bevel. These lines are required on the developed face moulds.

Approximate Development of Helicoidal Surface.—To draw an approximate development of the joint surfaces draw perpendicular lines to the axis line O 7″ in plan, to cut a new axial line drawn parallel to the axis O 7″. Make this new line the centre of the joint. Project the points M^3 7″ N^3 and M 7 N, also M^2 7′ N^2 in plan, as shown, and measure the length of each joint line in development, as at 7 M, 7 N in Fig. 431. Then, with the compasses on point 7 (Fig. 432), and radius 7 M, or 7 N, draw arcs cutting the projectors drawn from M and N in plan in points M and N in development. Repeat this process for the lines $M^3 N^3$ and $M^2 N^2$. Because the vertical height of the joint is the distance travelled along the axis of a cylinder, this length can be transferred to the axis line in development, as shown. A series of curves drawn through the points thus obtained will give the development required.

Setting Out Pediment in Straight Wall.—In setting out full size, it is necessary to draw only sufficient in elevation to obtain the face mould for the *apex stone* and the bed mould for the *springer stone* in plan.

First draw the bottom bed line in elevation and the centre line at right angles to the bed line. Determine the return wall line W, and draw in the true section of the cornice mould between N and W (Fig. 433). From N draw the line N C, representing the rake of the pediment, and draw lines from N′ and X parallel to N C to cut the centre line in E and F. Now draw the plan of the pediment, as shown in Fig. 434, by projecting down from elevation. Next draw a line at right angles N C in any position, and on this line from X′ mark off the vertical heights of the horizontal members between X and W, as at X′ H′. Through these points draw lines parallel to N C to cut the centre line. Terminate these lines on the weathering formed on the top surface of the horizontal cornice as shown. Now draw from point H′ the elevation of the joint lines for the springing stone, and continue the vertical joint to the bottom fillet line of the raking cornice. Draw H′ J′ at right angles to N C. Then the outline for the face mould of the springer stone is complete.

To obtain the outline for the bed mould, drop vertical projectors from H′ J′ into plan as shown.

To work the springer stone, a raking section mould is required for marking on the joint surface J′ H′. This mould may be obtained by an auxiliary plan of the figure on the inclined plane of the joint, or by rabatment. In this example it is obtained by rabatment (see Fig. 435). First draw the *horizontal projection* of the figure lying on the inclined joint plane by dropping vertical projectors into plan, as at J and H. Between points N′ X in elevation mark any number of points. Draw lines from these points parallel to N C to the vertical trace of the joint plane. Also from N′ X drop vertical projectors to cut the

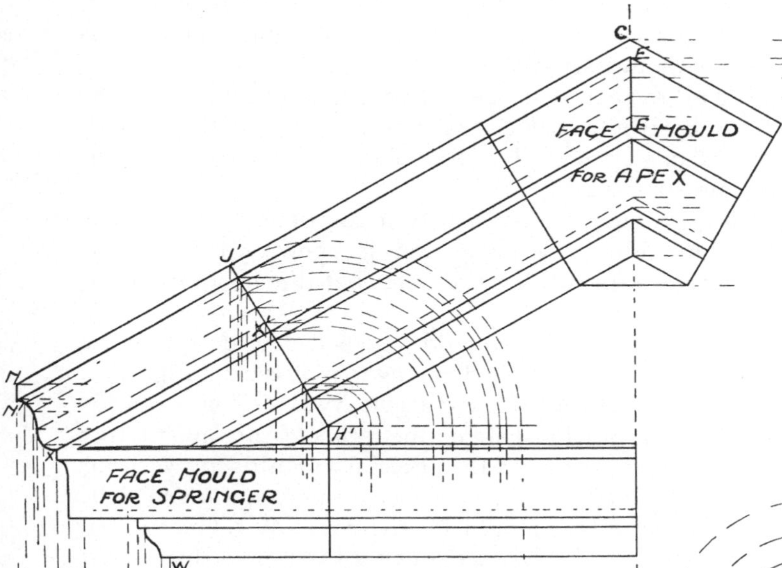

ELEVATION of PEDIMENT
FIG. 433

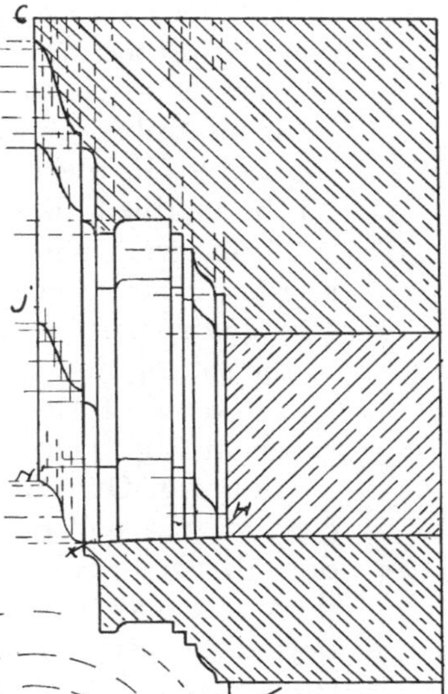

SECTION ON CENTRE LINE
FIG. 437

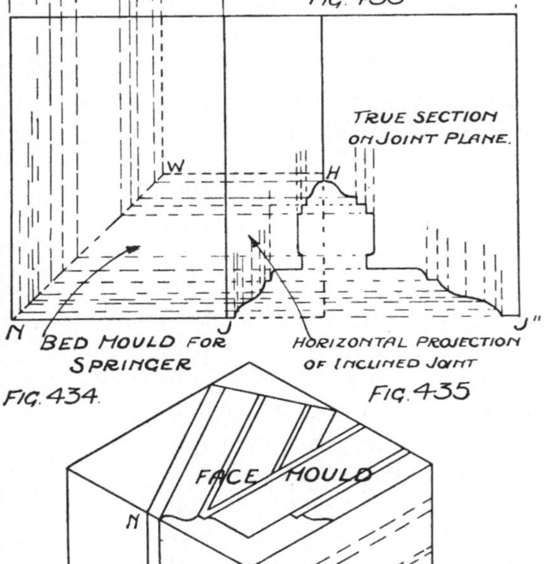

BED MOULD FOR SPRINGER
FIG. 434

HORIZONTAL PROJECTION OF INCLINED JOINT
FIG. 435

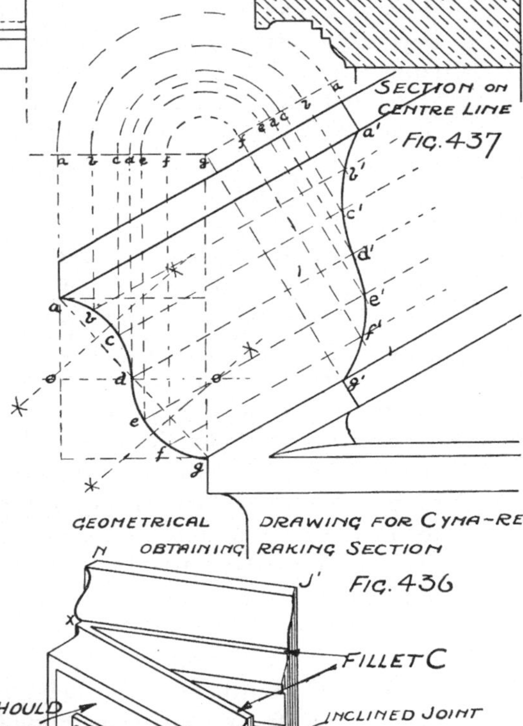

GEOMETRICAL DRAWING FOR CYMA-RECTA OBTAINING RAKING SECTION
FIG. 436

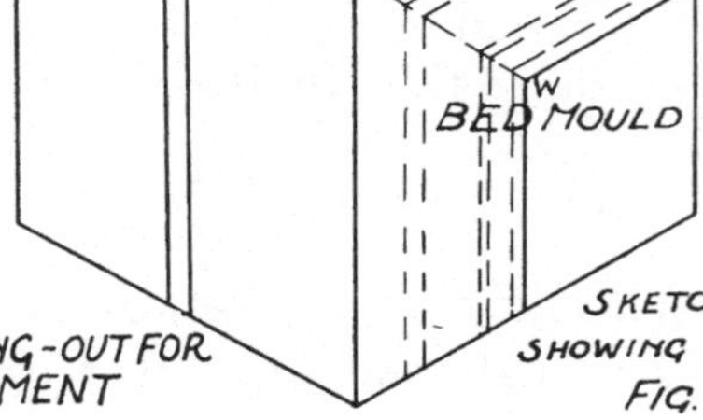

SETTING-OUT FOR PEDIMENT

SKETCH OF SPRINGER SHOWING MOULDS APPLIED
FIG. 438

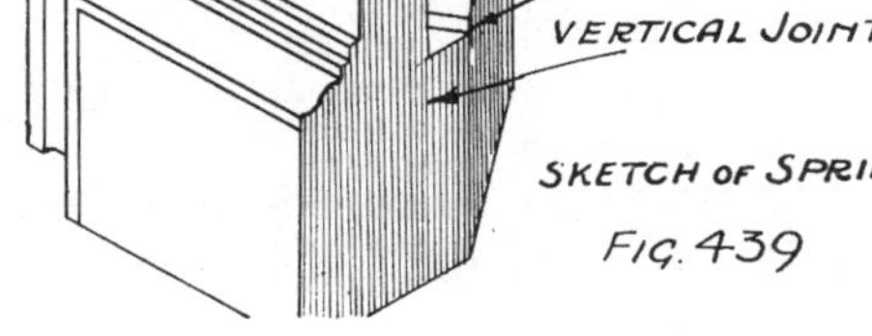

SKETCH OF SPRINGER
FIG. 439

mitre line N W in plan. From these points in plan draw horizontal projectors to meet vertical projectors drawn from the V.T. of the joint plane, thus completing the outline of the moulding as shown.

With H′ as centre, swing all the points in the V.T. of the joint plane, into the horizontal plane to intersect horizontal projectors drawn from their corresponding points in the horizontal projection. The true shape of the section lying on the inclined plane is now obtained.

A diagram showing the geometrical construction for obtaining the true section of the cyma-recta moulding by projection is shown in Fig. 436. Fig. 437 is a section on the centre line of the pediment, but this is not essential to the setting out.

A sketch of the stone showing the face mould and bed mould applied is given in Fig. 438, and a sketch of the finished springer is given in Fig. 439.

Setting Out Segmental Pediment in Straight Wall.—The example chosen is taken from Fig. 191. As previously explained, this feature must be set out full size in order that the correct sizes and the necessary moulds for working the stones may be obtained. It is only necessary to draw in elevation sufficient beyond the centre line to obtain the face mould for the *apex stone*, and in plan sufficient to obtain the bed mould for the *springer stone*. First draw the bottom bed line and the centre line in elevation (Fig. 440). Determine from the scale drawing the position of the centre line of shaft at X, and place in the lines representing the top bed line of the frieze immediately over the capital. This determines points *w* and *y* in the elevation set out. Mark in the true section of the cornice mould from point W, thus determining point N′. Next fix the height of point C above the bottom bed line, and obtain the centre for striking the segmental raking nose line N′ C, and draw the lines representing the member lines of the raking cornice as explained in the previous example.

Draw the plan as shown in Fig. 441, by projecting from elevation, the position of the wall lines being determined from the scale drawing or setting out of the doorway (Fig. 191).

From point Y in elevation, place in the section of the lower members of the cornice, giving the outline for the return of the nosing, as at point E. This point determines the position for the vertical joint, which should now be drawn to intersect the bottom bed line of the raking cornice at F. From F draw the normal joint line F H. Drop points E and H into plan, thus determining the outline of the face mould and bed mould for the springer stone.

The method of obtaining the raking joint mould is shown by projecting lines from the various points in the V.T. of the joint plane at right angles to the V.T., and measuring their projection from the wall line, these distances being taken from the true section of the cornice in elevation between N′ and W. This raking section could be obtained by drawing a section on the centre line, as shown in Fig. 442.

The geometrical construction for obtaining the raking section of the ogee is shown in Fig. 443.

Setting Out Semi-elliptical Skew Arch.—Draw the elevation of the arch on the outside wall line, placing in the normal joint lines for the voussoirs (Fig. 444).

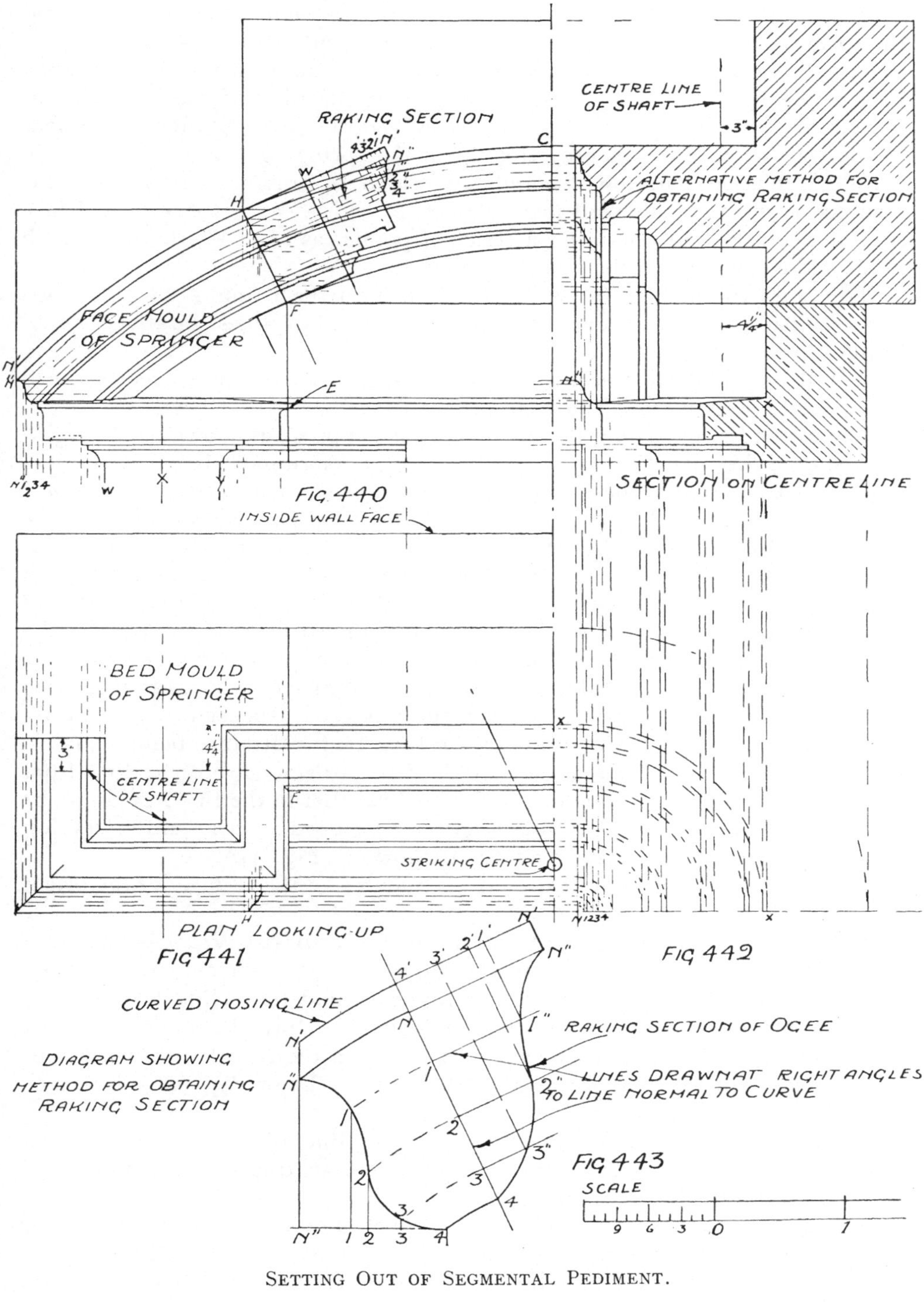

Setting Out of Segmental Pediment.

Next draw the inside and outside wall lines in plan, and project vertical lines from the points A E F, etc., in elevation to cut the outside wall line in points A E F, etc. (Fig. 445). At point A in plan, place in the *angle of skew* determined by the line A A′, which represents the plan of the arris line at the bottom bed of the springing stone. Where this line cuts the inside face of the wall at A′, draw a projector up to the springing line in elevation in point A′, thus determining a point from which to draw the elevation of the arch curve on the inside wall face. Points E′ and F′, etc., are projected in a similar manner, thus obtaining a series of points, as at E′ F′, through which to draw the elevation of the inside arch curve.

To obtain the moulds for stone No. 2, project the points E F G H D down to the outside wall face, and from these points draw lines parallel to A A′ in plan to the inside wall face.

It will be found most economical to work the stone from a prism instead of a block viewed at right angles to the wall face.

The method of obtaining the bed mould of the prism is shown on the right-hand side of the plan (Fig. 446). When this is determined, it is necessary to project a true section of the prism. This may be done by drawing an *auxiliary elevation* at right angles to the long edges of the prism, thus: Produce the plan lines of the arrises of the prism to cut a new X′ Y′ line drawn at right angles to the plan lines. Measure the heights of the various arris lines in elevation and mark their heights in front of the new X′Y′ line, as shown in Fig. 447. It is now necessary to obtain the development of the surfaces of the prism. To do this, unfold the surfaces commencing from the arris line G G′, then all the arris points will travel along paths at right angles to G G′. From H draw a line at right angles to G G′. Project the line G G′ to meet this line in G^3. From G^3 mark point F^3 on the line drawn from H, the distance being equal to G″ F″ measured from the section mould of the prism. Project a line from point F in plan to meet a line drawn from F^3 parallel to the line A A′ in point F^2. Join G F^2 and G′ F′, thus obtaining the developed joint mould. Develop the other surfaces in a similar manner, as shown in Fig. 448.

A sketch showing the application of the moulds on the surfaces of the prism is given in Fig. 449.

Setting Out Semicircular Arch in Cylindrical Wall.—The arch curve in this instance is assumed as semicircular in development, so that when the development of the arch curve is wrapped round the cylindrical surface, the elevation of the curve is not a true semicircle. Draw the plan of the wall and set out the reveals of the opening, drawing in the reveal lines normal to the plan curve of the wall, as at A A′ (Fig. 450). Draw the centre line C C, projecting it up to the V.P.

Stretch out the arc A C in plan on the springing line, then, with C A as radius and O′ as centre, draw a quadrant of a circle representing the development of the arch curve in elevation (Fig. 451).

Now divide the curve A C into an odd number of divisions, as at 1 2 3 4 5 6, and select points 2 4 6 as the points of intersection of the joints with the arch curve. Place in the normal joints and arrange the bonding to suit the surrounding ashlar courses.

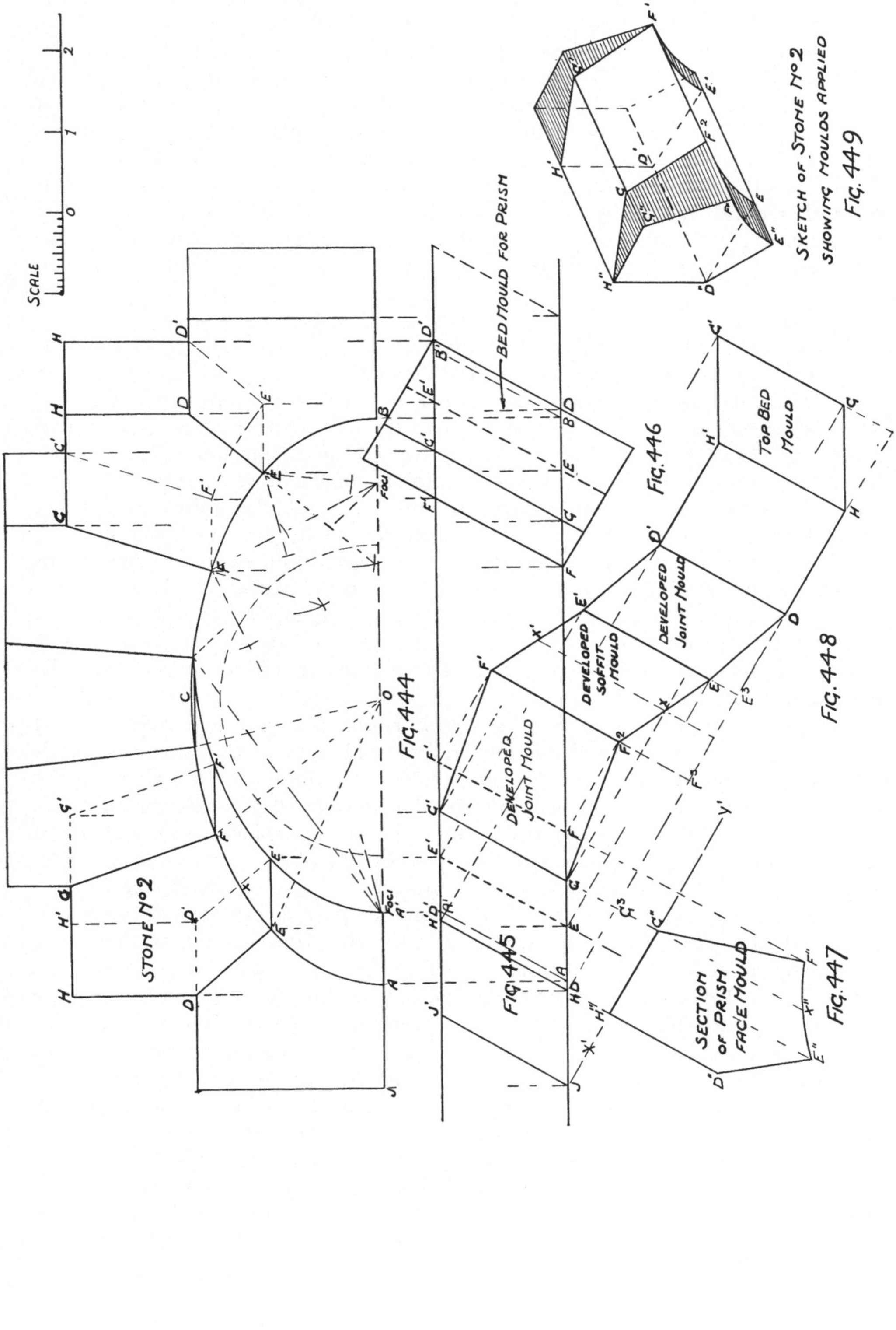

SEMI-ELLIPTICAL SKEW ARCH.

Project points 1 2 3 4, etc., down to the development line, and with the dividers, step these divisions round the arc A C in plan. Through these points draw normals to cut the inside wall face in points A′ 1′ 2′ 3′ 4′, etc.

Now set up the inside developed elevation (Fig. 452) by stretching out the points on the inside wall face to a development line, on the right-hand side of the setting out. From these points in the development line, draw vertical projectors to meet horizontal lines drawn from points 1 2 3 4, etc., in the outside developed elevation. A curve drawn through these intersections determines the developed arch line on the concave or inside wall face. Complete this development by drawing the joint lines through points 2″ 4″ 6″; these converge to centre O′. The position and heights of the horizontal beds and vertical joints are transferred from the outside development, as shown at D′ E′ F′.

The outside and inside developed face moulds are now complete.

To obtain the bed moulds, select any stone, as, for example, stone No. 2. Project the points D 2 4 E E down to the development line, and transfer them to their correct position on the outside convex surface in plan, as at D 2 E 4. Through these points draw lines normal to the curve, cutting the inside wall face in points D′ 2′ E′ 4′, thus obtaining the bed mould for stone No. 2. The horizontal projection of the helicoidal joint is shown in plan, as at E E′ 4 4′, points X and X′ being the centre horizontal line of the joint. To assist in working the stones, a vertical *boning line* should be placed on the moulds; this is shown in the elevation of stone No. 2 between points 2 2. Project point 2 down to the development line and transfer it to the curve wall line in plan. Through point 2 in plan draw a normal line, cutting the inside wall face in 2′. Transfer this point to the inside development line, and project up to the inside developed elevation, as shown in 2″. These lines should be marked on the face moulds and the bed mould.

In this example a moulding is shown running round the arch, so that developed joint moulds would be required to apply on the joint surfaces. It is necessary to obtain the horizontal projection of the joint surfaces showing the outline of the moulding on these, in order that developed moulds may be cut to apply on the joint surfaces of the stones. To do this, place in the outline of the section mould in the plan of the piers outlined from point A, as at A K L X. Draw the normal lines through these points to the outside curved face line.

Transfer the position of these points on the wall face to the development line, and from the points projected to the springing line, draw arcs, centre O′, in the developed outside elevation, cutting the joint lines in points K L X. Draw vertical projectors from these points to the development line, and transfer them to their correct position in plan, and draw normal lines through these points in plan to the inside wall face. Now draw concentric lines in plan from points A K and L in the plan of the true section, cutting the normal lines just drawn in points A K and L. This determines the horizontal projection of the joint surface between the stones. The bed mould for stone No. 2 is shown projected out from the plan in Fig. 453, the horizontal projection of the joint surface between stones Nos. 2 and 3 thus being clearly defined. To develop these surfaces, refer to the geometrical diagram (Fig. 432), which shows the approximate development of helicoidal surfaces, and apply the principle there explained to the present example. This is also shown in Fig. 454. Obtain the centre of the outside developed joint line 4 E, as at X in elevation, and transfer this point to its correct position in plan. Through X in plan draw a line to centre O. Now draw a line O^2 X^2 parallel to the line O X, and project the points 4 X E and 4′ X′ E′ at right angles to the line O X, to cut the line O^2 X^2 in points X^2 and X^2. Produce the lines beyond O^2 X^2. Next obtain the centre of the joint line 4″ E′ on the inside developed elevation; then, with X′ 4″ radius, and with X^2 in the joint development as centre, draw arcs to cut the perpendicular lines drawn from E′ 4′ in points E^2 4^2. With X 4 radius and X^2 in development as centre, draw arcs cutting perpendiculars drawn from E 4 in points E^2 4^2. Curves drawn through these points determine the outline of the developed joint mould. The outline of the moulding in development is obtained by measuring the distances of points L K X in the inside and outside developed joint lines (*i.e.*, in elevation), and transferring them round the arcs 4^2 X^2 E^2 in the developed joint section, and drawing lines K K′, L L′. Measure the lengths of these lines to the moulding in the horizontal projection of the joint, and transfer the lengths thus obtained to the developed section, as shown.

To Develop the Soffit.—As the surface of the soffit forms the surface of a cuneoid, it is not actually capable of development; but a close approximation may be obtained by the following method. Let A A′ and B B′ be the plan of the soffit (Fig. 455), the semicircle 1 9 being the section in development, taken through the cuneoid at right angles to the axis 5 F at point 5. This section will be a close approximation for the development of the arch face.

Divide the semicircle into a number of equal parts, as at 1 2 3, etc., and project these points into the plan line as shown.

Draw the line F 2′ at right angles to F 1, and equal to the vertical height of points 2 2 in section. Now measure the length of the arc 1 2 in section, and from point 1 in plan draw an arc. Measure the length of the line F^2 in plan; then, with the point 2′ in development as centre, draw an arc to cut the arc drawn from point 1 in point 2^x. Join points 1 2^x and 2^x 2′. Repeat this process until the development is complete, as in Fig. 456. To obtain the development of the arch lines, mark off on the lines 2′ 2^x and 3′ 3^x, etc., the lengths of the lines F A and F A′, as shown.

The projections of stones Nos. 2 and 3 are given in Figs. 457 and 458.

Fig. 459 is a sketch of stone No. 2, showing the face and bed moulds applied to the stone.

Setting Out a Flat Arch with Projecting Key and Wing Stones in a Cylindrical Wall.—First draw the outline of the wall in plan (Fig. 460), marking in the reveal lines; also draw the soffit and top bed lines in elevation. Next draw the section through the centre line (Fig. 461), thereby fixing the projection and profile of the key and wing stones. Determine the width of the key and wing stones at the top bed front wall line in plan, and draw the normal joint lines to cut the inside wall line. Produce these lines beyond the front wall line to the projection of the key at A, which is found by projecting from the centre line section and drawing concentric arcs A O A and B P B, representing

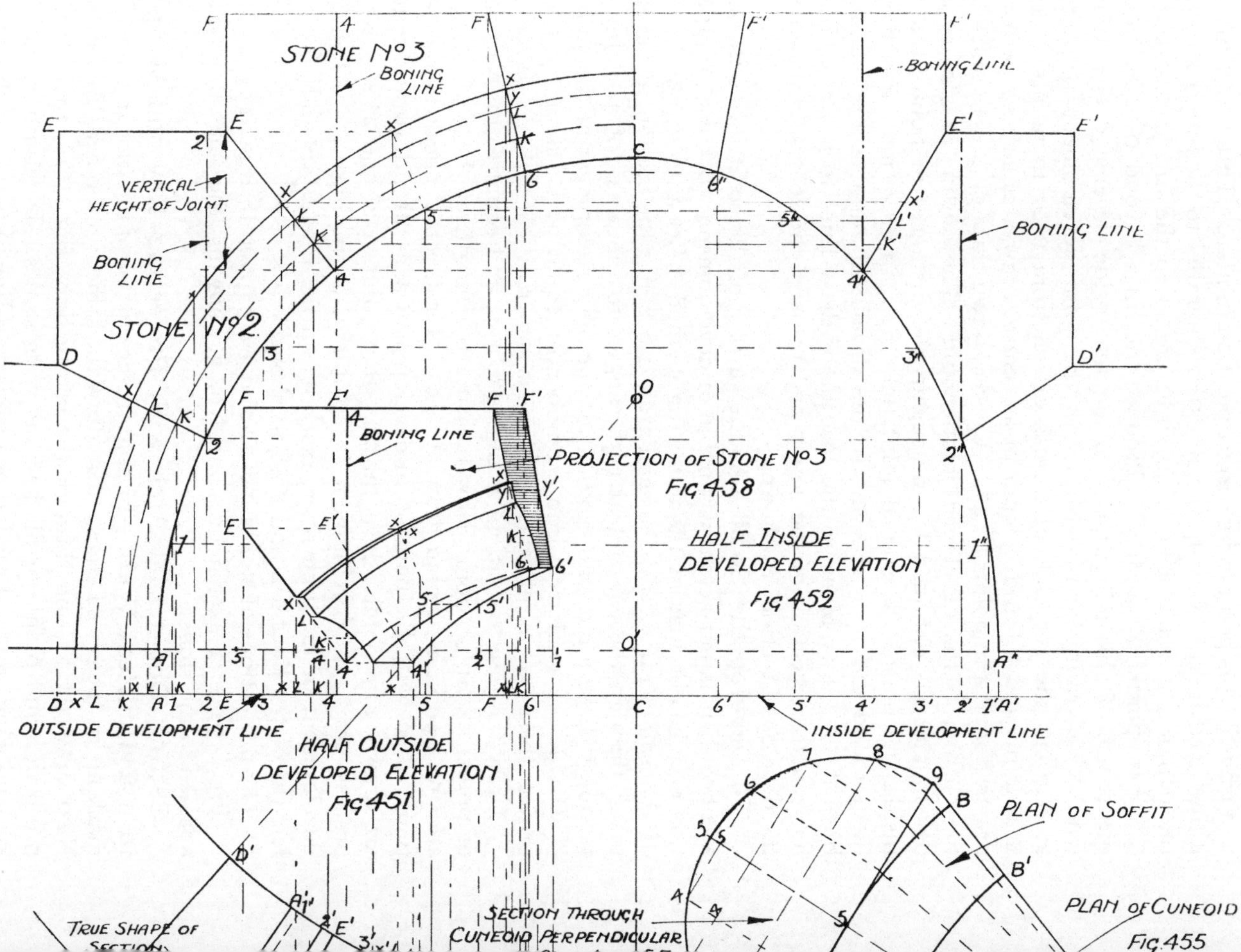
STONE No 3
BONING LINE
VERTICAL HEIGHT OF JOINT
BONING LINE
STONE No 2
BONING LINE
BONING LINE
BONING LINE
PROJECTION OF STONE No 3
FIG 458
HALF INSIDE
DEVELOPED ELEVATION
FIG 452
OUTSIDE DEVELOPMENT LINE
INSIDE DEVELOPMENT LINE
HALF OUTSIDE
DEVELOPED ELEVATION
FIG 451
PLAN OF SOFFIT
PLAN OF CUNEOID
FIG 455
SECTION THROUGH
CUNEOID PERPENDICULAR
TRUE SHAPE OF

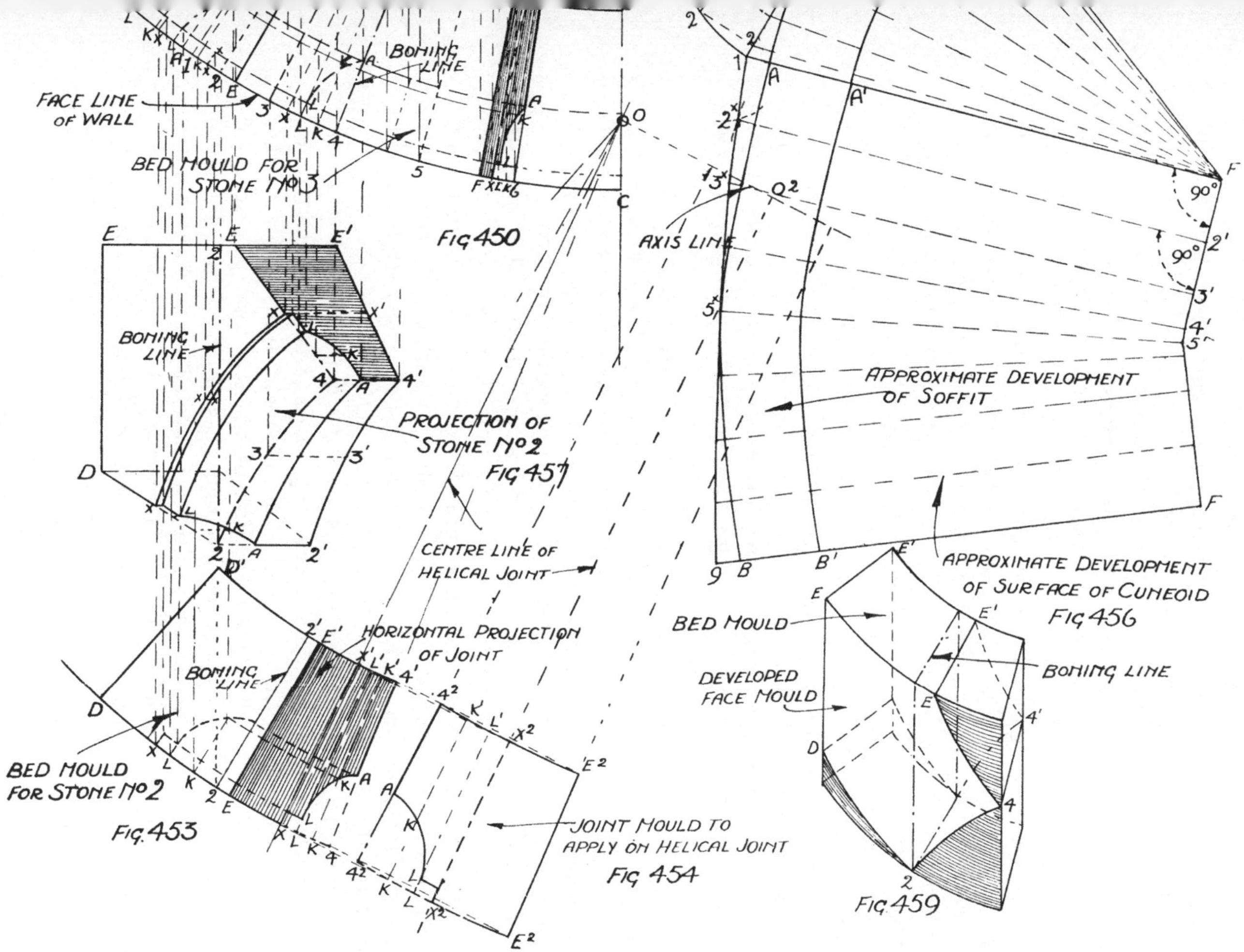

Semi-arch in Wall, Circular on Plan.

the projections of the key and wing stones. Stretch out the inside width of the key *o′ a* on to a development line, and project up into the right-hand developed elevation of the inside wall face, at *a* (Fig. 462).

Draw a radiating line from *a* to *a′*, and project this point on to the development line, and transfer it in plan at *a′*.

Draw a normal line to cut the front line of the key at A′; this is the plan of a point at the angle of the front bottom arris of the key.

Join A′ A, and produce the normal line *a′* A′ to A″. Thus far we have obtained the bed mould of the keystone. Next obtain the bed mould for the wing stone by determining the width of the wing stones on the inside wall surface, as at *b*, and transfer this distance from the centre line O′ to the development line, and project up to the inside development at *b*; then from *b* draw the radiating joint line to *b′*. Project *b′* to the development line and transfer it to the inside wall line in plan, as at *b′*. Through this point draw a normal line to the front, terminating on the bottom arris of the wing stone at B′. Produce the normal line *b′* B′ to X. Join B to B′ and B^2 to B^3. The bed mould for the wing stones is now obtained, the outline being shown by the points B B^2 B^3 a^v *b*, as in Fig. 463. The next bed mould required is for the side stones, resting on the piers, and is obtained by transferring the shape of the wall in plan to the mould, there being no projecting front surface to these stones.

We have now obtained moulds for the beds and the inside developed face. It must be remembered that the front faces of the key and wing stones are conical, so that it is necessary to develop the outside face moulds on the vertical cylindrical surface. This surface must be worked and the face moulds applied to it, the stone being then worked to the required shape, the face being afterwards dressed to the correct conical shape as a last operation. The lines for working the face are obtained by applying the joint moulds to the joint surfaces. In the drawing, the developed radiating joint lines are shown straight, but when these moulds are wrapped round the cylindrical surfaces they will not produce straight lines, but helical curves. This curve, being very slight, would be scarcely noticeable in practice.

Next develop the outside face moulds. This must be done on a cylindrical surface containing the front top arris line in each stone. Stretch out the distance O A on to the development line, and also the distance O A″, which is the bottom arris line of the keystone produced to the cylindrical surface at A″. Transfer these distances on to the development line, and project them to their correct position in development—that is, to the lines representing the top and bottom surfaces of the key from A A″, thus obtaining the outline for the developed face mould of the key (Fig. 464). To obtain the developed face mould of the wing stones, stretch out the cylindrical surface represented by the top arris line, as at B B^2, which should be produced to meet the bottom joint arris B^3 a^v. Produce the bottom arris of the joint B′ *b′* to intersect the cylindrical surface in X, and transfer these points to the development line, thus obtaining the face mould required, as in Fig. 465.

A developed face mould is also required for the side stones, thus giving the position of the radiating joint lines. This mould is developed on the cylindrical

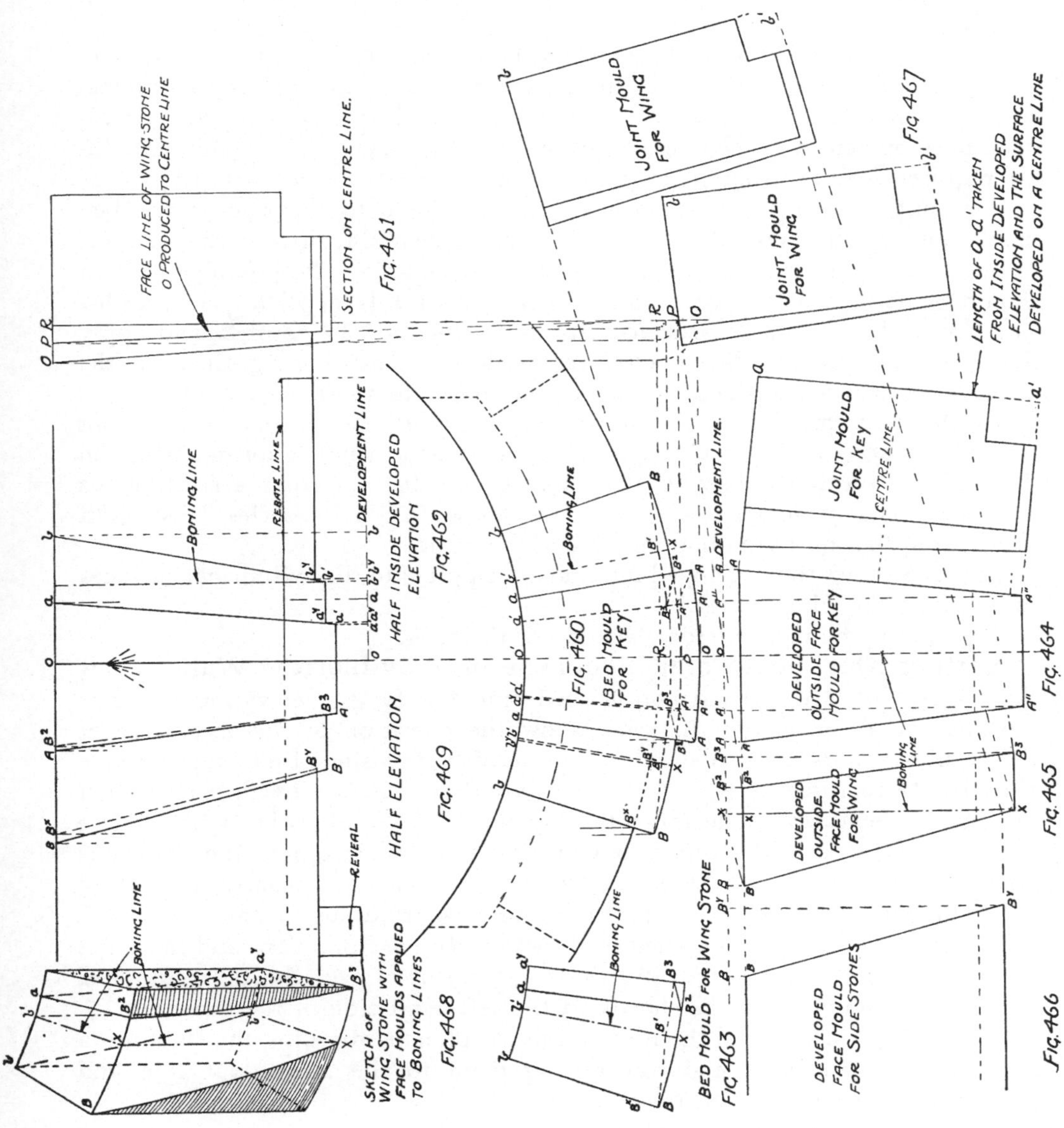

FLAT ARCH WITH PROJECTING KEY AND WING STONES IN WALL, CIRCULAR ON PLAN.

wall face, and will extend horizontally from the top joint arris line $B^x b$ to the bottom arris line $B^y b^y$. Stretch this distance out on the development line and project to the top and bottom bed lines in elevation, as shown in Fig. 466. Only sufficient length is required for this mould to be marked correctly on the surface.

To obtain the joint moulds, rotate each joint about its front arris line and project in each case, as shown in the drawing. These are only approximately correct. Their true shape should be determined as in the case of the approximate development of a helicoidal surface (see Fig. 432). Two joint moulds are required for the wing stones, and one for the keystone, as in Fig. 467.

To work the wing stones, select a piece of stone to the size of the bed mould and to the vertical height taken from elevation, then work the stone to the cylindrical surfaces covered by the bed mould. The boning line b' X should be marked on the stone when the bed mould is applied. Square this line down the cylindrical surfaces from the top bed to point X, and apply the inside and outside developed face moulds to these boning lines on the concave and convex surfaces. Shape the stones to these face moulds, and apply the joint moulds to the joints, marking on the outline of the front conical surfaces, and the check for frame. Cut a templet for marking on the soffit of the stone, giving the curve of the bottom front arris line, as at B' B^3. The conical surface may now be worked in a series of straight drafts from the top to the bottom edge of the stone.

Fig. 468 shows the developed face moulds applied to the cylindrical surfaces for working the wing stones.

Fig. 469 is the half outside elevation of the arch.

Setting Out Segmental Pediment in a Cylindrical Wall.—Draw the centre line Y Y and the bottom bed line X X for the elevation. Draw the plan at a convenient distance below the elevation to the measurement and radius required, also draw the plan of the nosing line, and assume this line to represent the surface of a cylinder. It is on this surface that the developments are made for obtaining the face moulds (Fig. 470). There are two methods, either of which may be adopted, unless the choice is governed by data already supplied by the architect; but usually the setting out of the unit is left to the discretion of the masonry draughtsman.

The elevation may be obtained by direct projection from plan, and the curve of the pediment on elevation, drawn as a segment of a circle, then stretched out into development, or as shown on the left-hand side of the drawing. A development is made from the nosing line in plan by dividing the line into a number of equal parts and transferring them to the stretch-out line and projecting up into elevation (Fig. 471).

The divisions are numbered 1 2 3 4, etc. With your dividers step these along the stretch-out line, numbering them as shown. Project vertical lines from these points up into elevation. From point 7 mark off the projection of the cornice along the stretch-out line, and draw the section of the cornice in elevation to the details supplied. Between the top arris of the nosing N thus obtained, and the central height of the pediment point C—this height should be taken from the drawings—draw an arc as before described.

With radius more than half N C and with N and C as centres, draw intersecting arcs, through which draw a line cutting the centre line Y Y in point O. With O as centre, draw the arc between N C, which forms the outline of the pediment in development. Also swing round the lower nosing line and the fillet line at the lower part of the ogee.

Project vertical and horizontal ordinates from all selected points on the cornice section to the lines N N and N 7, lettering them as shown. Transfer points *a b c d e f* N to the centre line, which is a normal to the curve, and through these points draw parallel lines representing the lower members of the cornice for the pediment on the nosing face. Also from *a b c d e f* N draw horizontal lines which will represent the horizontal cornice members. Next place in the jointing of the pediment as required; simple jointing is shown for clearness. We have now the developed face moulds required for working various stones.

Now consider the treatment for the plan. Assume that the pediment is independent, as, for instance, a feature over a dormer window. In this case the horizontal members of the cornice on the return ends would run through to the inside face of the wall. Circumstances govern the treatment of these return mouldings, for geometrically they should converge to the centre of the plan curve of the pediment, which means that the lines of the members would not be parallel. This treatment is shown on the left-hand side, whilst another treatment is demonstrated on the right-hand side of the drawing. In the latter the return members have all been drawn parallel to the return wall line. This method complicates the intersection of the moulding of the return with that of the front moulding, thereby producing a crippled mitre line, unless the projections of the cornice on the return ends are treated as being projected from tangents to the plan curve, thus obtaining a true mitre. This construction will be explained later.

On the left-hand side we will set out the plan on geometrical principles, giving first consideration to the plan of the mitre line. As the mitre line is formed by the intersection of curved and straight lines, a true intersection cannot be a straight line, but must be slightly curved on plan. Transfer the points of projection already obtained, as at N *g h f k l m n o p* 7, and mark them off on the nosing line in plan from point 7, and draw normal lines from these points to represent the plan of the mouldings. The intersections of these lines with their respective lines in plan produce the slightly curved line as shown, but this curving is scarcely visible in actual practice. Draw the joint lines of the springer in plan as shown at D D′, E E′, thus determining the horizontal projection of the inclined bed joint. Next obtain the face moulds for application to the inside face of the wall (Fig. 472). This is done by drawing normal lines in plan from points 1 2 3 4 5 6 7 to the inside wall face in points 1′ 2′ 3′ 4′, etc., and also the points *p*′ O′ *n*′, etc., representing the members of the moulding on the return end. Step all these points on to the stretch-out line as shown, and erect perpendiculars. Transfer the heights on these lines from the corresponding heights on the lines in the developed outside elevation. This is necessary in order to obtain the curve at the inside wall face. The heights are the same, because all normal lines are horizontal. Complete the development

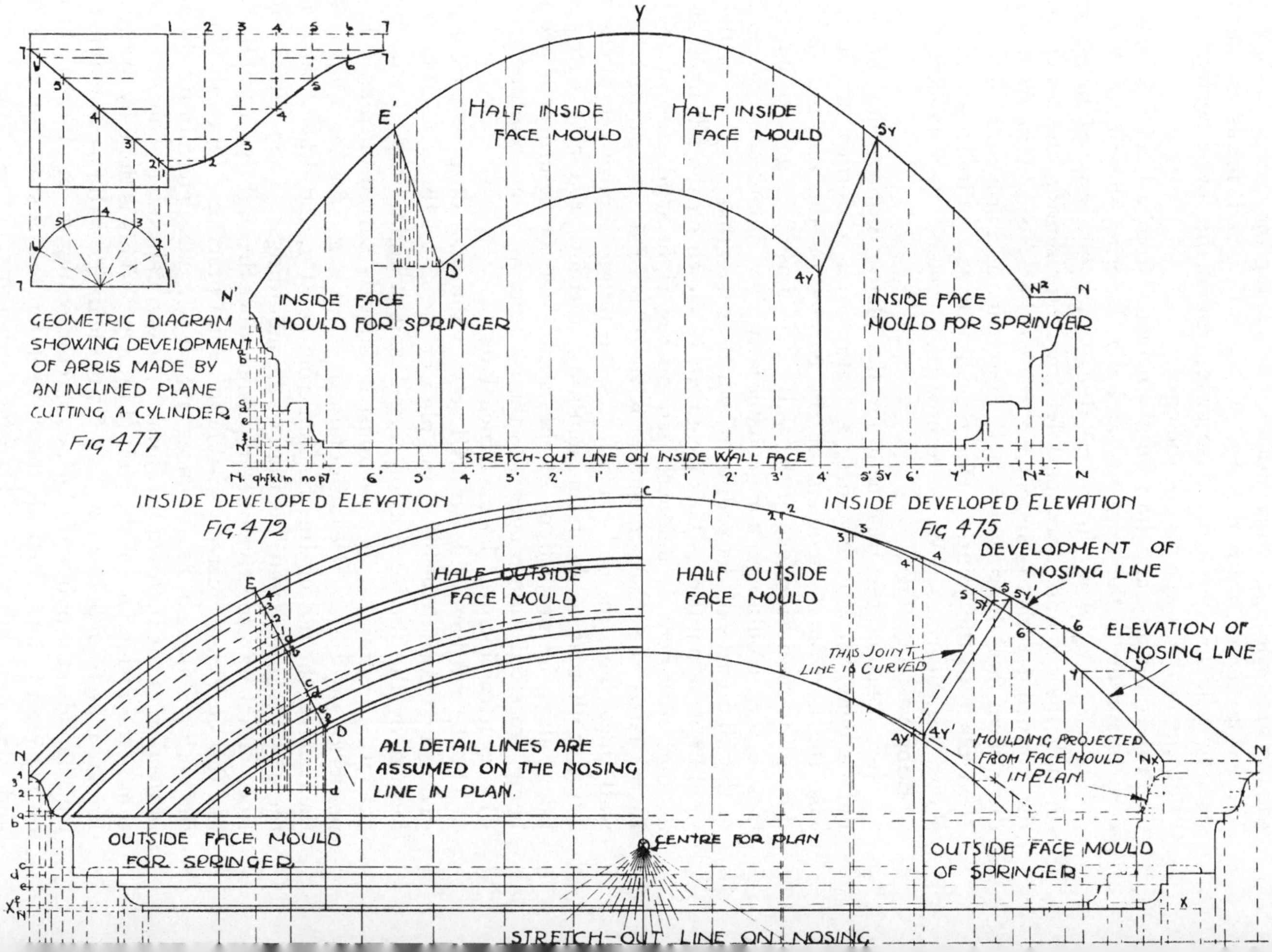

GEOMETRIC DIAGRAM SHOWING DEVELOPMENT OF ARRIS MADE BY AN INCLINED PLANE CUTTING A CYLINDER

Fig 477

INSIDE DEVELOPED ELEVATION

Fig 472

INSIDE DEVELOPED ELEVATION

Fig 475

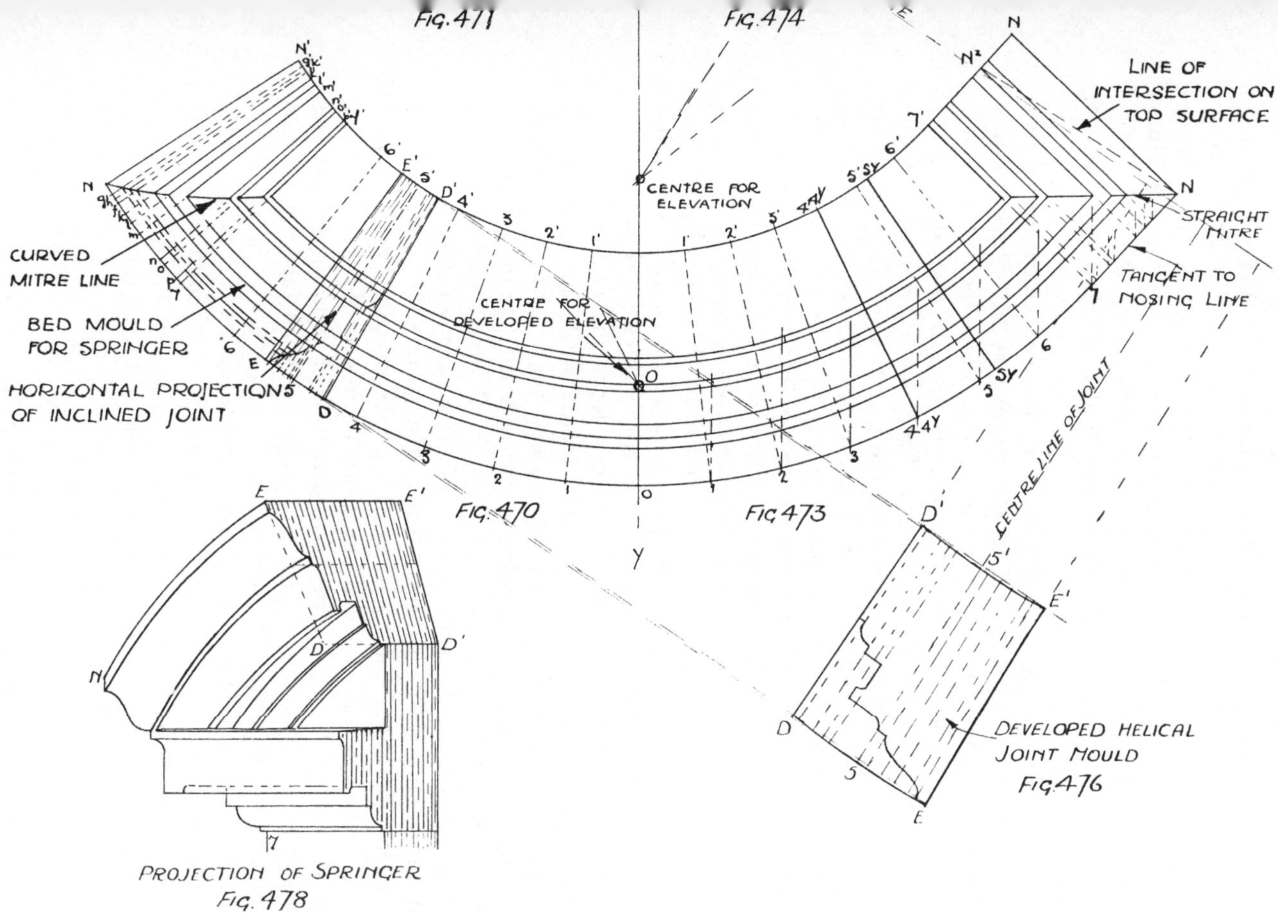

Setting Out for Segmental Pediment in Wall, Circular on Plan.

on the inside wall face as shown on the drawing. A second method of setting out the pediment is shown on the right-hand side of the drawing.

Draw the plan to the return wall face as before, produce the line representing the return wall face, to the curve line of the nosing at point 7 (Fig. 473). Now erect a perpendicular from this line, as a tangent to the plan curve at point 7.

Mark on this tangent the projection of the cornice, and draw the lines representing the plan of the moulding members parallel to the wall line. The intersections of these lines with those of the front members will produce a true mitre line in plan.

Divide the plan as before in points 1 2 3, etc., and project these points direct into elevation, including the projection of the cornice at N. Mark on this line in elevation the height of the cornice, as at N^x. Then locate the centre for an arc to be drawn through N^x and C, and draw the arc representing the curve of the pediment in elevation (Fig. 474). From the intersection of this curve with the vertical projectors, drawn from the points 1 2 7 in plan, draw horizontal lines. Now mark off on a stretch-out line the development of points 1 2 3, etc., and erect from these points perpendiculars to cut the horizontal lines just drawn. These intersections are points through which to draw the development curve of the pediment. Determine the position for the joint in plan at 4^y 4^y. Project the joint 4^y into elevation to meet the elevation of the lowest member of the pediment cornice, and draw the normal joint line as at 4^y 5^y. The line between these points is really a helical curve. Now draw horizontal lines from these points to meet the curve lines in development, thus obtaining the position of the joint lines in development, as at $4^{y'}$ $5^{y'}$.

Next set up the developed elevation of the inside wall face (Fig. 475). To produce a good curve line for the top arris along the wall face, it is best to draw a radiating line in plan from point N to the striking point. This line will represent a horizontal intersection of the curved upper surface with a horizontal surface running in from the nosing of the return cornice.

Stretch out the inside wall surface in plan, also points N^z and N. The height to be marked on these lines will be the corresponding heights taken from the outside elevation. *Note*.—All the lines for the projected elevation from plan are viewed as being on the cylindrical face represented by the nosing line in plan, and all points required for the developed face moulds are drawn horizontally from this projected elevation.

Next draw the horizontal projection of the inclined joint E D (Fig. 470). All the points are dropped on a horizontal line *e d* (Fig. 471), and transferred to the nosing line in plan, between points E D.

Draw radiating lines from these points to the centre of the plan curve; then, where these intersect, their corresponding lines drawn in plan concentric with the nosing line will give the outline of the moulded section projected into the horizontal plane.

To obtain the joint mould (Fig. 476), refer to Fig. 432, which will explain the method for developing a helicoidal surface. To determine the outline of the moulded portion, transfer all the points between E and D in the developed joint line (Fig. 471) round the arc E D in the development (Fig. 476). Also transfer the points on the developed joint line E′ D′ (Fig. 472), marking them round the arc E′ D′ in development, and draw the lines as shown. Measure along these lines distances taken from their corresponding lines in the horizontal projection of the joint, thus obtaining a series of points through which to draw the moulded outline.

In the setting out, a helicoidal joint surface has been assumed, but in practice a straight-cut joint is sometimes arranged; but although this is cheaper to execute, it does not produce the best results. If a straight-cut joint is required, the joint lines in the developments will be curved instead of being straight lines, as shown. A geometrical diagram showing the reason for this is given in Fig. 477. A projection of the springer stone is given in Fig. 478.

The Setting Out Corbelling at Splayed Angle of Building.—Draw the axial line O O′ and the elevation of bed line A A. Plot the plan to form a splayed angle of 45° to the two face lines of the building produced, which are 90° to each other, the length of the splay being 8 ft. 6 in., as in Fig. 479.

To set out the plan of the circular portion above the corbelling, draw a normal line from one of the points of intersection of the splayed face with the wall face, as at B, and produce this normal to intersect the centre line at C. Project point B into elevation (Fig. 480). With C as centre, C O radius, rabat the line C O into the V.P., as at O″, and project into section, to the height of the top course (Fig. 481). Set out the thickness of the wall in plan (18 in.). Select any point D′ in line A A, making this the point of intersection of the conical face with the splayed vertical face. This point determines the slope of the surface of the inverted cone, or first decide upon the slope of the cone surface to determine the point D′.

Project D′ into plan, as at D, and rabat it into the V.P. at D″. Determine point 5′ in section by measuring down, say, 15 in., and making this point the position for the commencement of the conical surface. Join 5′ and D″, this line being the true slope of the surface of the inverted cone, on the centre line, or the vertical section through the axis of the cone, thus giving its true angle with the horizontal beds for the various courses. From D on the wall line in plan draw a normal line through C, producing it to cut the curve B O in point D^3.

Next draw the heights for the courses in elevation, and plot the circular window opening. Produce these bed lines to cut the line 5′ D″ in section. Now imagine this inverted cone being cut by a vertical plane represented by the splayed vertical face line of the wall, as shown in the diagram (Fig. 482). The section thus made would be a hyperbolic curve. A portion of this curve is to be seen in elevation, and various points to be obtained on this curve are required for the bed moulds. This curve of intersection is obtained by producing the horizontal bed lines to cut the section of the cone in points 2′ 3′ 4′ 5′. From these points drop projectors into plan to intersect the vertical splay face B D^2. Project these points up to elevation to meet their corresponding bed lines already drawn. Through these points a fair curve may be drawn, which is the elevation of the line of intersection of the conical face with the vertical splayed face. The next curve to obtain is the one forming the arris of the hooded arch over the circular opening. This curve is obtained by a section plane cutting through the inverted cone. This plane passes through point O in plan to the splayed wall face at D^2. The section of this cutting plane is obtained by rabatting these points O and D^2 into the V.P. and projecting up, as shown in Fig. 481. As this section cuts the cone obliquely, its projection is a portion of an ellipse, which may be obtained by projection as before described. Draw this line in section, as at D^2 2^2 3^2 4^2 5′, and swing them round to the axial line C O; then project them horizontally to intersect the plan lines projected from the section of the conical face in points 4 3 2, etc. A curve drawn through these points will give the plan of the arris of the arch curve from D′ to 5′. Project these points in plan into the elevation, to meet their corresponding bed lines. This will give the elevation of the section made by the cutting plane between points D′ and 5′. Now arrange for the coving of the arch, thereby obtaining the intersection line D′ 6′ R Q P S. To do this, select any point S as the intersection of the coving with the vertical wall face on the centre line.

Project this point across to the section and place in the shape desired for the coving on a vertical section through the centre line.

In this example the coving is assumed to be a similar curve on any vertical cutting plane.

Insert vertical planes through the points 2 3 F in the elevation of the arris line D′ 5′ (Fig. 480), and project these points to cut the section line of the arris, as in points 2^2 3^2 F′. Draw vertical projectors from these points to cut the section of the coving on the centre line.

Measure the heights of these ordinates from a horizontal line drawn from S, and transfer these distances below their corresponding points in elevation. This will give a series of points through which to draw the intersection curve line between points D′ and S. Any number of points may be obtained in a similar manner. Next decide upon the position for the radiating bed joints: these could be placed so that the lower portion of the joints are normal to the curve of the centre circular window, whilst the top portion of the joints, from the groin line, may be arranged normal to that curve, as from points 3^x 4^x.

In this problem the joint surfaces are arranged in one plane, for if the joint planes are normal to both curves, the upper portion of the joint against the keystone will not form a good key joint, owing to the angle made by the inclined joint plane and the horizontal plane being too small. In addition to this, the effect produced would not improve the appearance of the work.

Next determine the projections of the curves made by the radiating joints passing through the coving as between points 3^x Q, so that their horizontal projections may be obtained in plan to complete the outline of the bed moulds. Select any point, such as T, between points 3^x Q (Fig. 480). Through point T insert a vertical cutting plane, cutting the groin line D′ 5′ in point F, and the intersection line D′ S in point G. Project these points horizontally to their correct position in section, as at points G′ F′ (Fig. 481). Through these points draw the curve of the coving, taken from the section of the coving on the centre line.

Now project point T across to the section to cut the curve between G′ F′ in point T′. This determines the point through which to draw the side elevation of the radiating joint line between 3^x, Q, these points being projected to their correct position in section, as shown.

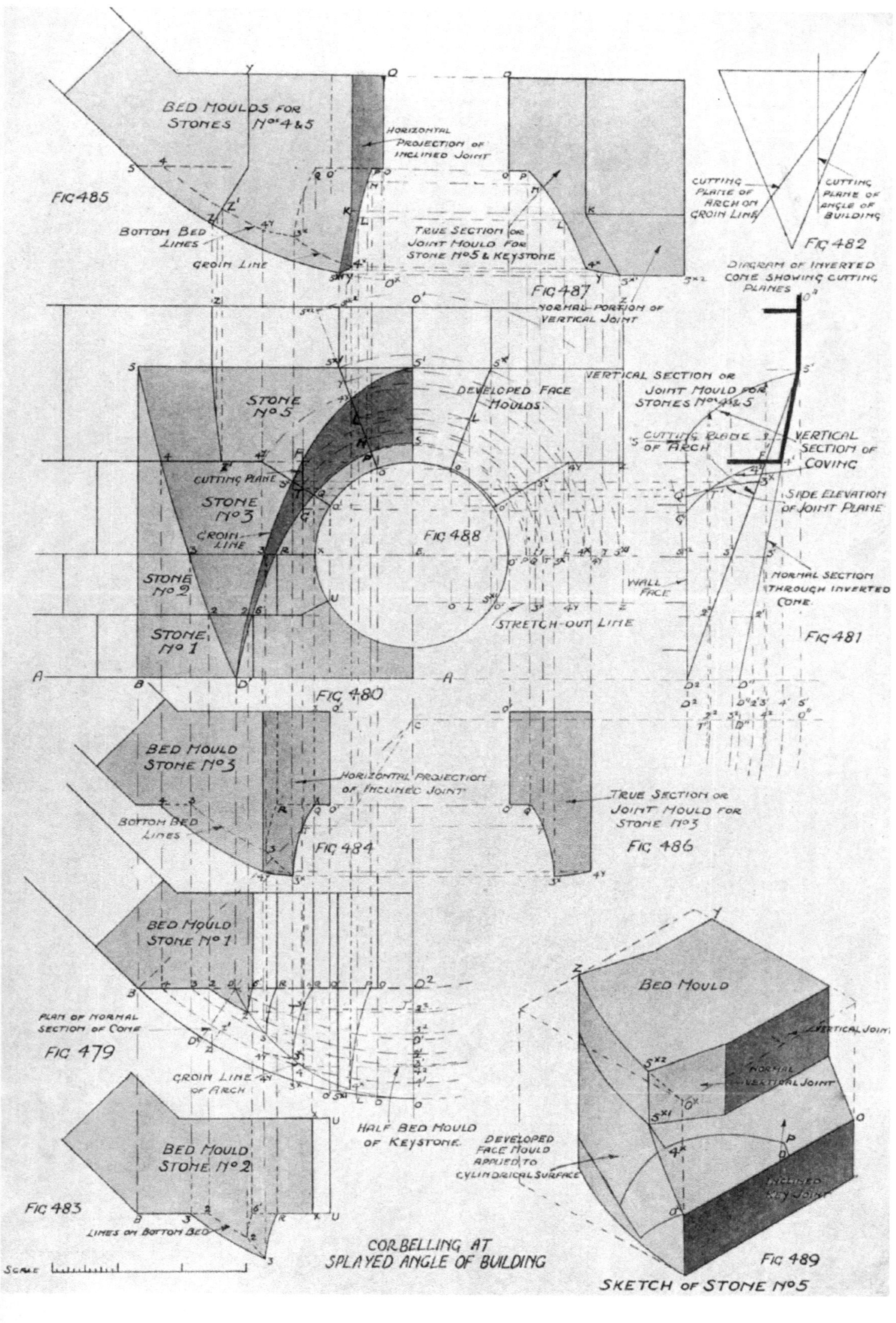

CORBELLING AT SPLAYED ANGLE OF BUILDING

Now obtain the plan of point T, as already explained, by dropping a projector from point T′ into the axial line C O″, then swinging it round into the axial line C O in plan, and projecting horizontally to meet a vertical projector drawn from point T in elevation, in point T^3. A point is thus determined through which the plan of the curve, between points 3^x Q, may be drawn. Points for drawing the plan of the other joint lines may be obtained in a similar manner.

The curve between points 3 and R, which is a horizontal section through the coving, may be obtained by inserting a series of vertical cutting planes, then projecting the true shape of these cutting planes in section in a manner similar to that already explained for points on the radiating joints.

To obtain the bed moulds for stone No. 1, which is outlined on the plan, project the points 2 2 6′ into their position in plan, point 2 being on the plan of the groin line and point 6′ on the wall line.

The bed mould for stone No. 2 is obtained in a similar manner.

Project down to plan points 3 3 R X, point 3 being on the plan of the groin line. This mould is shown projected out from plan in Fig. 483, the points 2 2 6′ on the bottom bed being marked on this mould.

The bed mould for No. 3 stone is shown projected out from plan in Fig. 484. In order that the correct shape of this mould may be obtained, it is necessary to draw the horizontal projection of the bed joint between points O′ and 4^y. Project into plan points 4^y 3^x Q O′, point 3^x being on the plan of the groin line and point 4^y being on the plan line of the horizontal bed joint No. 4. The curved portion between 4^y and 3^x in plan is a portion of an ellipse. The projection for the bed mould for stones 4 and 5 is shown in Fig. 485.

It is now necessary to obtain the section moulds to apply on the inclined joint surfaces. Take any points on the V.T. of these joint planes, such as O′ Q T 3^x 4^y, and rabat them into the horizontal plane about the centre E. Project the points to plan to meet horizontal projectors drawn from their corresponding points on the bed mould, as in Fig. 486.

Similarly with the inclined joint O $5^{x\prime}$, rabat all the points into the horizontal plane, and project them to plan, which in this drawing is shown above the elevation for clearness. The projections of all the points may easily be traced through their projections until they finally produce the true shape of the section mould, as shown in Fig. 487.

The vertical portion of this joint from $5^{x\prime}$ to 5^{x2} should be first rabatted into the plane of the inclined joint about point $5^{x\prime}$ and then rabatted into the horizontal plane about point E.

Although a complete true section of this joint is given, it could not be applied to the stone, but the inclined portion is necessary.

Developed face moulds are required for stone No. 5 and the keystone. To obtain these, produce all the points in plan parallel to the centre line to intersect the cylindrical face line, as at points Z 4^y 3^x O′ $5^{x\prime}$ L O (Fig. 479), and transfer them to the stretch-out line on the right side of the elevation. Project vertically from these points to intersect horizontal lines drawn from their corresponding points in elevation, as in Fig. 488. These moulds will wrap round the cylindrical surface as shown in the sketch of stone No. 5 (Fig. 489).

Internal face moulds should be cut to the projected elevation for marking on the back surface of the stones to assist in working the stones to the required shape. The joints may now be worked to these face moulds, after which the joint moulds may be applied.

Setting Out Hemispherical Hooded Niche.—Draw the plan (Fig. 490), including the outline of the niche on the line A A. Project these points from plan, and draw in the elevation (Fig. 491). Place in the horizontal and normal bed joints where required, and project the points C and E down to the wall line in plan. With centre O, swing these points round, thus obtaining the horizontal projection of the arris lines.

Now divide each course in plan into a convenient number of stones, and draw the plans of the vertical joints as lines converging to the centre O, as at R L for the 1st course and P N for the 2nd course. The back bonding for each course may be placed as shown. The bed moulds for the three stones in the 1st course are now obtained, and the face moulds are shown in elevation. To draw the elevation of the normal vertical joint lines, project up to elevation points K L in plan to K L in elevation, K being on the horizontal bed line A A and L on the bed line C C.

The curve may be trammelled as for an ellipse, with O′ G as semi-major axis and O′ K semi-minor axis. The curve for these joints may also be drawn through various points obtained by projection. Select any number of points on the curve outline of the niche in elevation, as at 1. Project down to the wall line in plan in 1, and swing this point round with centre O to cut the horizontal projection of the vertical joint K L in point 1. Now project this point into elevation to intersect a line drawn horizontally from point 1 on the curve line in elevation. One point on the curve is thus obtained. Repeat this process, thus obtaining the number of points required.

To draw the section (Fig. 492)—this is not actually required in the setting out—project all the points in plan to a vertical axis, then swing the points projected round into a horizontal axis and project them to the vertical plane V.P., to meet lines drawn horizontally from the elevation. The side elevation of the vertical joints is obtained by projection in a manner similar to that already explained for the front elevation of the joints. Project points K and L parallel to the wall line in plan, to meet the vertical axis line, swing them round into the horizontal axis line, and project them up to K and L in section. Take any number of points on the curve line in elevation and plan, as at 1, and project in a similar manner to intersect horizontal projectors drawn from the points in elevation.

The bed moulds for the 2nd course are obtained in a manner similar to that explained for the 1st course. The plan of the 2nd course is shown separately in Fig. 493.

The centre stone is provided with a horizontal seating to prevent it from slipping out of position. This seating forms a key, its construction being clearly shown in the drawing and the sketch.

The method adopted does not increase the dimensions of the stone, but involves extra labour.

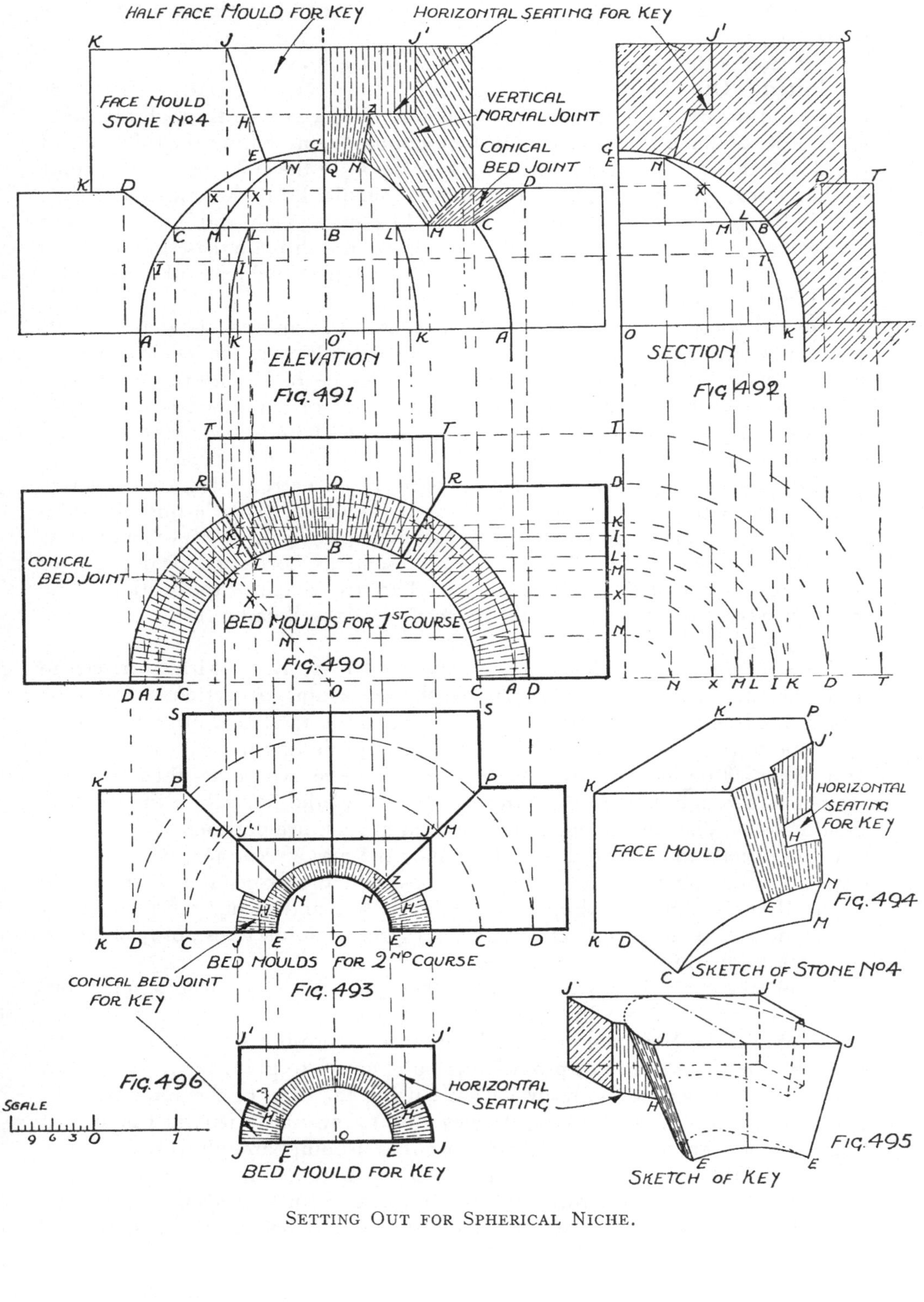

Setting Out for Spherical Niche.

Sketches illustrating the notching for the horizontal seating for the centre stone are given in Figs. 494 and 495.

Fig. 496 is the bed mould for the key or centre stone.

Setting Out Dome.—Draw the axial lines A O and C O. With O as centre and O A radius, draw the curve representing a section through the inner surface of the dome (Fig. 497). Mark off point A′ representing the thickness of the dome at the base, also from C mark C′ equal to the thickness of the dome at the crown, which, for good construction, should be of less thickness than at the base. With A′ C′ as centres, radius more than half the distance between A′ C′, draw arcs intersecting each other, and through these intersections draw a line to meet the axis line C O produced in O′, which is the centre for drawing the extradosial curve line of the dome.

Next divide the section into as many courses as required, as at D E F G H, and from centre O draw through these points normal lines to the inside spherical surface, also the horizontal bed joints of the dome in elevation. As all the joints converge to the axis of the dome, the surface contained is conical, so that the bed joints of all the stones, excluding the bottom bed joint A A′, are portions of an inverted cone, with O as the vertex. Now project point A A′ into plan to meet the axial lines, then, with O^2 as centre, swing round points A A′, which will represent the plan of the bottom bed. Divide this plan into the number of stones required (eight in this example), as in Fig. 498, and draw normals to the centre O^2. These lines represent the horizontal projections of the vertical joints of the various stones. The true shape of the section plane made by these joints is shown in the section of the dome already drawn, as at A A′, D D′.

Project point D into plan to meet axial line O^2 A′ in D; with O^2 as centre draw the curve as shown, which will be the horizontal projection of the horizontal bed line D. At D^2, where this line meets the joint line in plan, project up a line to the elevation of bed line D, meeting this line in D^2; also from A^2 project up a line meeting the base line in A^2. The two points thus obtained are points through which the vertical projection of the joint lines in elevation should pass. The curves representing the joint lines are portions of an ellipse, O A^2 being half the minor axis and O C half the major axis, for, if a circle be revolved about its axis into an oblique position, the projection of the circle is an ellipse. Similarly drop down points E F G and swing them round in plan, and where these curves cut their respective joint lines, project up to elevation to obtain points as before explained.

To obtain the minor axis for the other joint lines, project points from the plan of the bed line A in each case, to the base line in elevation, as at X X′.

To obtain the elevation of these joint lines by projection, take any points, such as Y between D and E, project this point down to plan, then, with O^2 as centre and O^2 Y as radius, draw an arc cutting the plan of the joint lines in Y and Y. Project these points up to elevation to meet a horizontal line drawn from Y in section, as shown. This determines a point through which to draw the curve of the joint line. Other points may be obtained in a similar manner.

Setting Out Dome Stone by Rectangular Block Method.—Draw the vertical line representing the axis of the dome, and lay down the section of one of the stones to a larger scale (Fig. 499). If in practice, draw to full size, and plot the stone in plan by projecting points 1 2 3 4 down to the centre line in plan, and swing them round to meet the joint lines, as at 1^x 2^x 3^x 4^x (Fig. 500). Project these points again into elevation to intersect lines drawn horizontally from points 1 2 3 4. These points represent the distance the stone passes round the curve of the dome, and thus give the thickness of the stone required. From point 1 draw a line parallel to 3^x 2^x, which would represent the thickness of a slab of stone required for this stone.

This line should be drawn to intersect a line drawn horizontally from point 4 in 4^y. Then the prism thus obtained and shown by points 1 2 2^x 3^x 4^x 4^y is a cover mould enclosing the stone. It should be noticed that, because the bed joint represented by 3 4 is conical and the line 3 3^x is horizontal, the line 3^x 4^x radiates from the same centre.

If, when drawing the parallel line 1 4^y, it is impossible to draw from point 1 without cutting through the curve 1 4, this line must be drawn as a tangent to the curve, meeting the line 1 2 produced from point 1. This method of projection for the dome stones is economical, because the stone may be machined to the prism section and to the length to be obtained later.

It is now necessary to develop two plane moulds to lie on the two inclined planes 1 4^y and 2^x 3^x. The horizontal line 1 1^x represents the curve 1 1^x in plan, which is an arc of a circle, so that the mould to be marked on the inclined plane 1 4^x must have the projection of this curve marked on it in order that the bottom conical bed may be cut to the curve thus projected. The curve required on this mould is really the arc of the circle lying in the horizontal plane, projected into the inclined plane 1 4, or the oblique section through a cone; this curve projected on to the plane may be either an elliptical or parabolic curve. Similarly, a curve is required on the other plane mould at 3 3^x.

To obtain the true shapes of these moulds, it is necessary to rabat them, or swing them round into the horizontal plane. This is done by using point 1 (or the point obtained by the intersection of the plane with the line drawn from point 1).

Draw a horizontal line from point 1, and, with 1 as centre and 4^y as radius, swing this point into the horizontal line. It is now necessary to project the points on the conical bed surface up to the plane surface, as shown at 1^y 5^y and 3^y 6^y. A central point has been taken in the drawing, but as many points as desired may be used. With points 1 and 3^x as centres, swing these points round to the horizontal line, and project them into plan as shown.

The length of the plane mould at 2^x is the length shown in plan between points 2^x 2^x. Project these points horizontally in plan to meet the line 2^x projected in rabatment, meeting this line in 2^z 2^z. Because the point 3^x, round which the plane rotates, does not move, its projection in plan will remain stationary, so that a line drawn from 2^z to 3^x will represent the position for the joints along this plane mould.

To obtain the projection of the elliptical or parabolic curve, it is necessary to project down from points 6^y 5^y and 1^y in the inclined planes to intersect their lines in plan, continued normal to the curve in plan, as at 1^y 5^y and 6^y, and then projected horizontally to meet their corresponding projectors in rabatment

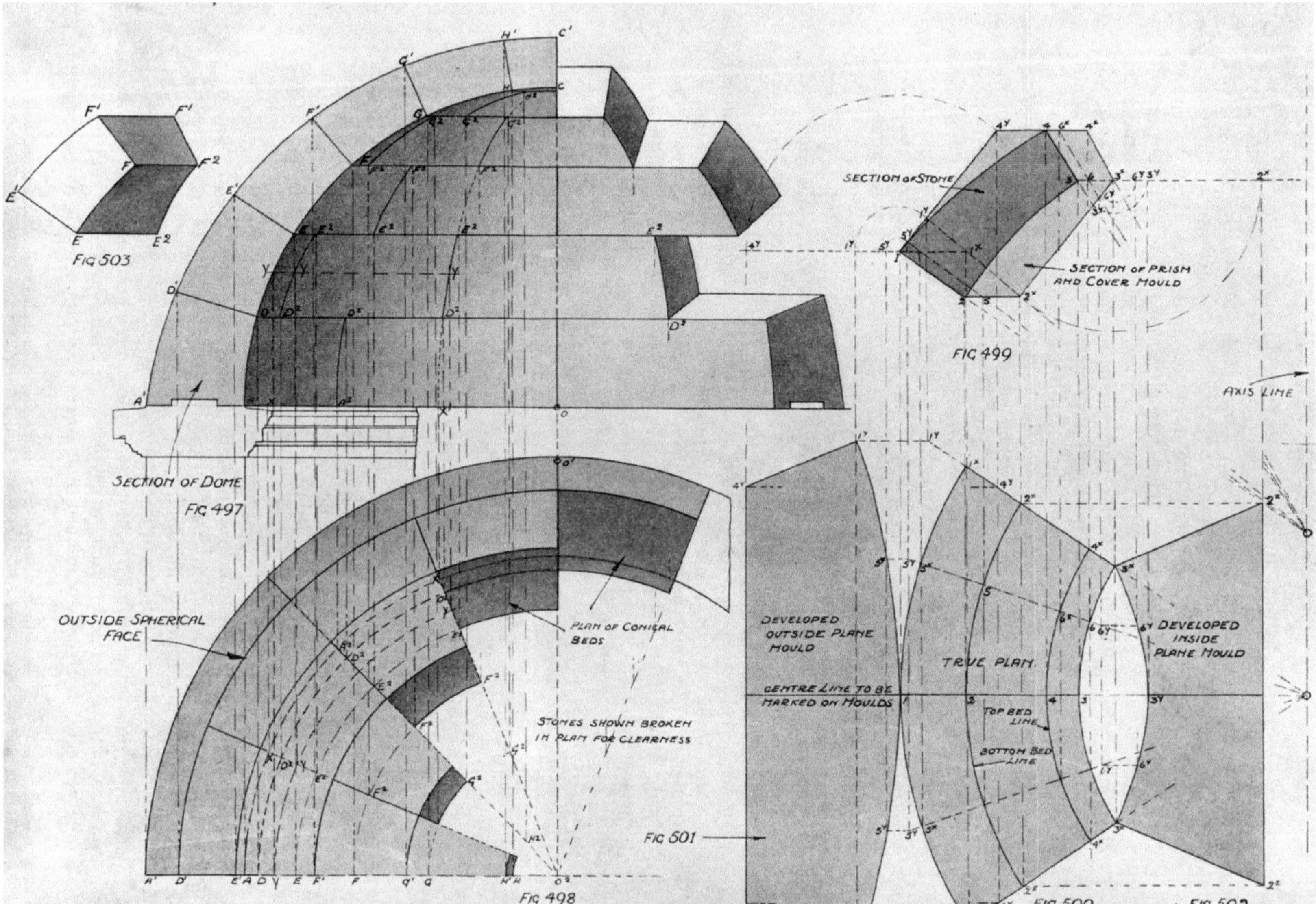

SETTING-OUT OF DOME-STONES

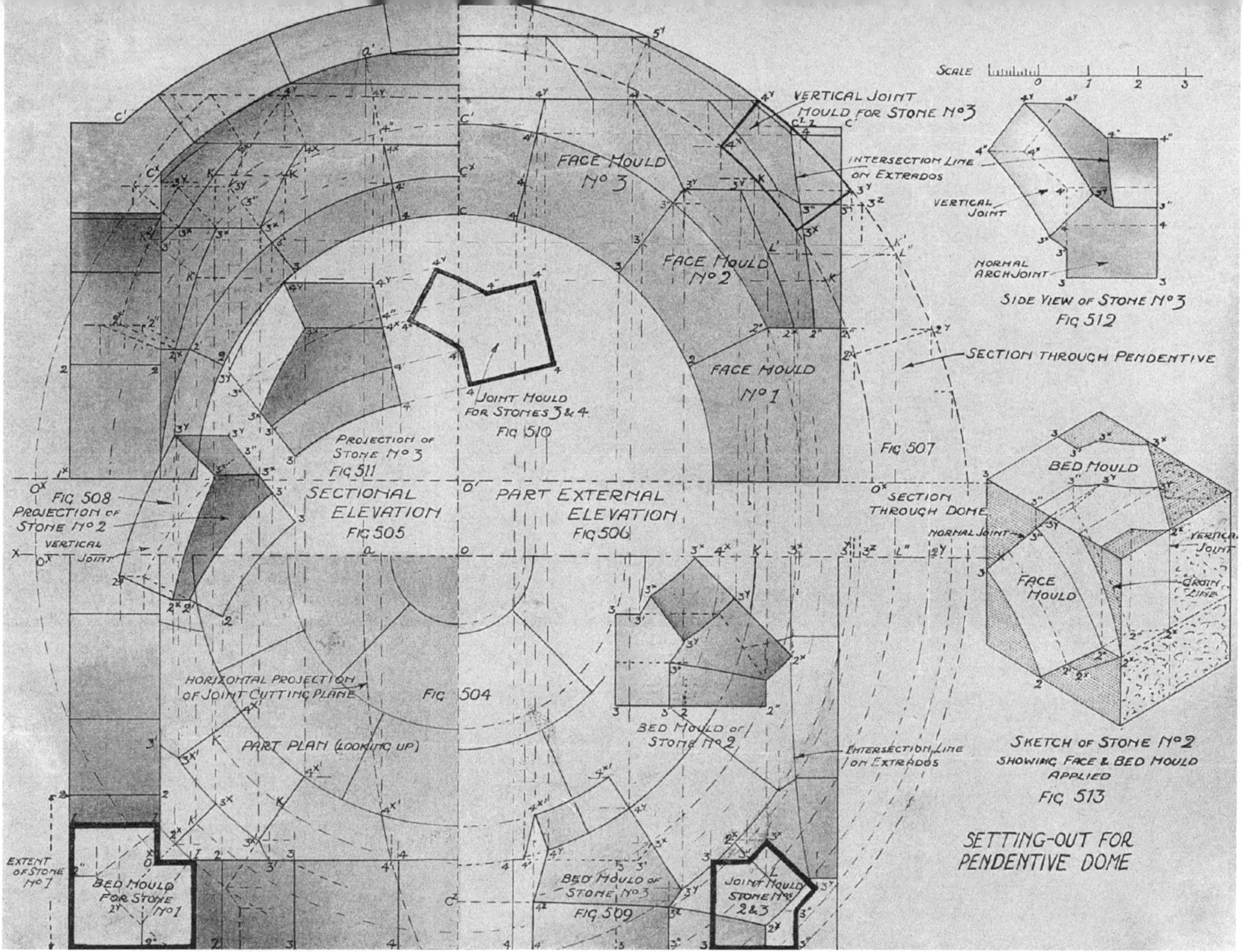

SETTING-OUT FOR PENDENTIVE DOME

in $1^y\ 5^y\ 6^y$. Notice that this increases the length of the stone, as shown between points 1^x and 1^y in plan.

A curve drawn through these points gives the curve required on the plane mould.

To obtain the outside plane mould, project point 4^y to the joint line in plan and horizontally to meet the vertical rabatted line in 4^v. Join $1^y\ 4^v$, and this is the joint line on the plane mould. The central point 1 remains stationary, so that point 1 is also a point on the curve.

The other moulds required are taken from the true plan, being the curve 4^x 4 4^x for marking on the top surface, and the bottom bed line at 2^x 2 2^x for marking on the bottom surface. A true section of the stone is required, as shown by points 1 2 3 4. Two reverse templets are required, one cut to the external spherical surface and one to the internal spherical surface.

The application of the moulds is given in Figs. 277 to 281.

Fig. 501 is the developed outside plane mould and Fig. 502 is the developed inside plane mould. The projection of one of the stones is given in Fig. 503.

Setting Out Pendentive Dome.—First draw the axial lines and plot the walls in plan to the size required, as in Fig. 504. Now project up for the elevation and determine the radius for the intersecting arches which are semicircles, the centre of these being at point O′. On the left-hand side of the elevation draw a quarter internal view of the dome (Fig. 505), and on the right, a quarter external view (Fig. 506.) Make the internal face of the arch such that the extradosial curve of the arch springs from the corners of the square compartment. On the right, draw the curve representing the external face of the arch from the same centre, O′. Draw horizontal lines from the crown of these arch curves, to intersect projectors from the walls in plan, as at C′ and C^x. These points give the outline of the vertical section through the centre of the arch.

To draw the outline of the dome with centre O′, radius O O^x, which is the half-diameter of the square compartment, swing round the curve into the horizontal line X O, and project this point up to the springing line of elevation. With O as centre, O′ O^x radius, draw the curve as shown in dotted lines. This curve represents the inside curved surface of the dome, springing from the corners of the compartment.

Next determine the thickness of the dome at the crown and at the springing. The point at which a line bisecting these two points cuts the centre axial line, determines the centre for striking the curve representing the extrados of the dome.

Next divide the cross arches into a convenient number of stones, as at 2 3 4, and draw the normal joints as shown. From the point where the arch joint 2 2′ meets the spherical face in point 2′, draw a horizontal projector to the inside dotted spherical section, as at 2^x (Fig. 507). From 2^x draw the first normal bed joint of the dome, as at $2^x\ 2^y$. From 2^y draw a horizontal projector to meet the normal bed joint on the outside face of the arch in point 2″. This determines, for convenience, the width of the outside face of the arch. Draw in this curve. Now divide the dotted section into a convenient number of stones, arranging them to suit the position of the normal arch joints. It is best to arrange these joints so that the arch joints are required to be extended to the level of these horizontal bed joints, as at 3^x on the intrados and 3^y on the extrados.

Draw horizontal lines from the intersection of the bed joints of the inside spherical surface. These lines represent the horizontal bed lines on the internal elevation, and are stopped where they meet the radiating joint lines of the cross arches produced in points 2′ and 3^x 4^x. Now continue the radiating bed lines on the extrados to meet the horizontal lines representing the bed lines on the external surface of the dome, as at 3^y 4^y. Next project the inner and outer horizontal bed lines from the dotted section, as at 2^x 2^y and 3^x 3^y, etc., into plan, and draw them concentric with each other from centre O, thus representing the horizontal projection of these bed lines in plan. The intersections of these bed lines on the internal spherical surface with the radiating joint lines are obtained by dropping down projectors from points 3^x and 4^x in elevation to points $3^{x\prime}$ $4^{x\prime}$ in plan.

The horizontal projections of the radiating joint lines cutting across the spherical surface in plan are portions of elliptical curves. The curves in plan are obtained by producing the V.T. of the joint plane to the spherical surface, as at a', and projecting it into the H.P. at a. Then O a is the semi-minor axis, and distance O O^x is the semi-major axis for drawing curve 4′ $4^{x\prime}$, similarly for the curve 3′ 3^x.

To draw the projection of the vertical normal joints of the dome in the elevation, divide the plan into suitable-sized stones as shown, and project the extremities of the plan of each of the joint lines to intersect their respective bed lines in elevation, as at 2^x 3^x and 3^x 4^x.

To obtain intermediate points through which to draw the curves, select any point K on the plan of the line; swing this point round to the horizontal line X O and project it up to the inside spherical surface in section, and from this point on section draw a horizontal line to cut a perpendicular line drawn from point K in plan; then the intersection gives a point on the curve. As many points as required may be drawn in a similar manner. The same procedure applies for the elevation of the joint lines on the extrados. Take any point L in plan, swing this point round, and project up to the dotted section in point L″. From L″ draw a horizontal line to meet a vertical projector drawn from L in plan, thus determining point L′ through which to draw a curve. These curves are portions of an ellipse, as may be proved by the oblique projection of a circle, which is the section through the centre of a sphere and rotated about its vertical axis, as already discussed. The chief object to the setter-out of masonry is to transfer on to the plane of the setting-out board just sufficient of the particulars to determine the necessary moulds for the working of the various stones. All unnecessary lines have therefore been omitted from this drawing, so that the principles of projection stand out clearly to the student. Now project the individual stones so that the necessary moulds in each case are obtainable.

The bed mould and face mould for No. 1 stone are clearly shown in plan (Fig. 504) and outside elevation (Fig. 506). If an archivolt moulding is required around the arches, then joint moulds would be necessary to apply on the inclined joint surfaces. To obtain the projection of stone No. 2, draw horizontal projectors from points $2''$ 3^y in the outside elevation to meet the inclined joint lines on the internal elevation produced in points $2''$ 3^y. Through the extremities of the joint lines 2^x 3^x in elevation draw normal lines from centre O′, meeting the horizontal lines drawn from $2''$ 3^y in 2^x 3^y. To obtain an intermediate point through which to draw the curve representing the joint arris on the outside spherical surface, through point K on the dotted section draw a normal line from O′ cutting the outside spherical surface in point K′. From K′ draw a horizontal projector to meet a normal line drawn through K on the internal joint arris line in point K^2. This stone is shown projected out of the drawing for clearness in Fig. 508.

To obtain the bed mould for No. 3 stone, project all the points down from the outside face mould of the stone to meet their corresponding lines drawn in plan, which are obtained by projecting them horizontally to the dotted outside section line. Project them down to the line X O and, with centre O, swing them round in plan, as at 3^y 4^y. The positions of the various points on the inside may be obtained by projecting across from the internal plan and elevation, as shown in Fig. 509.

Before the lines for this mould are complete, it is necessary to determine the plan of the intersection between the outside spherical surface and the extrados surfaces of the arches. Project points C′ $4''$ $3''$ $2''$ in elevation to the outside dotted section line, as at C^z 4^z 3^z 2^y, and project them into plan as before explained to meet vertical projectors drawn from C′ $4''$ $3''$ $2''$, as shown in points C^z 4^z 3^z 2^y. A curve drawn through these points determines the plan of the intersection line.

Draw the elliptical curves between 4^y 4^z and 3^y 3^z as before explained, then the outline of the bed mould is complete.

Joint moulds are required to apply on the joints 3 3^y and 4 4^y. These are obtained by projection, as shown on the drawing in Fig. 510. The projection of stone No. 3 is clearly shown in Fig. 511, and a vertical normal joint mould for stone No. 3 is indicated on the dotted section. A side view projection of stone No. 3 is shown in Fig. 512, and a sketch illustrating the application of the moulds for stone No. 2 is given in Fig. 513.

Setting Out Ribbed Pendentive Ceiling.—Draw the plan of the walls (Fig. 514) and draw the plan of the centre line of the ribs in their required positions, then erect a sectional elevation showing the depth of the faces of the cross arches, as in Fig. 515.

The working surface of the dome is assumed on the nose line of the ribs, the necessary projections being made on this surface. It is best to obtain first the section through the domical surface on the nose line by swinging round the half-diagonal O C in plan, then projecting it into the section at C′. With O′ as centre and O′ C′ radius, draw the outline of the curve, which will represent the nosing line. Draw a true section of the ribs as shown, and from this obtain the complete diagonal section, together with the surface C″ of the dome, as at C″ (Fig. 516). Continue the diagonal section of surface round to intersect the inside vertical face of the arch in the sectional elevation, as at point C^x. From this point draw a horizontal line to C^y on the centre line, thus determining the position of the intersection of the vertical face of

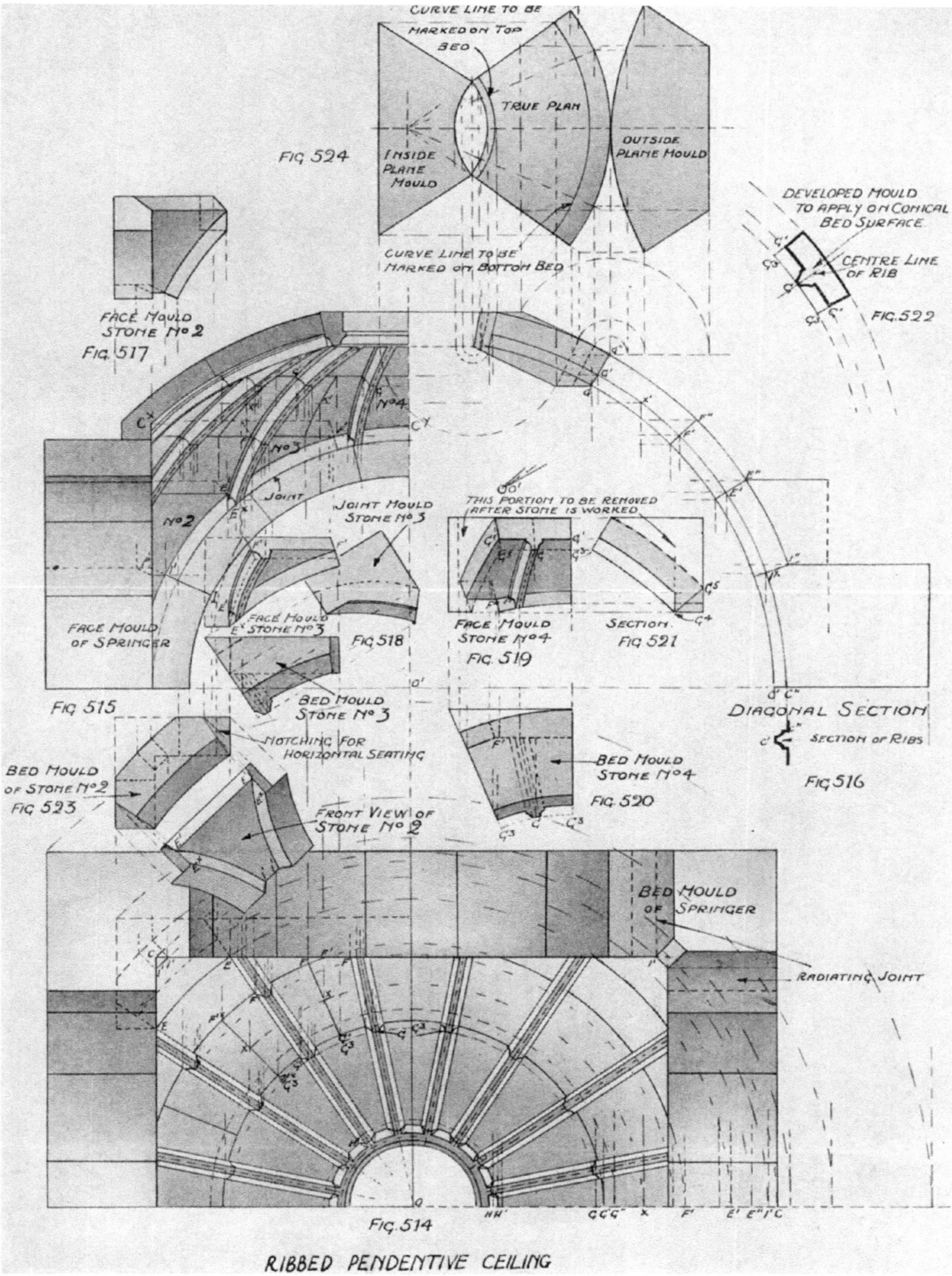

RIBBED PENDENTIVE CEILING

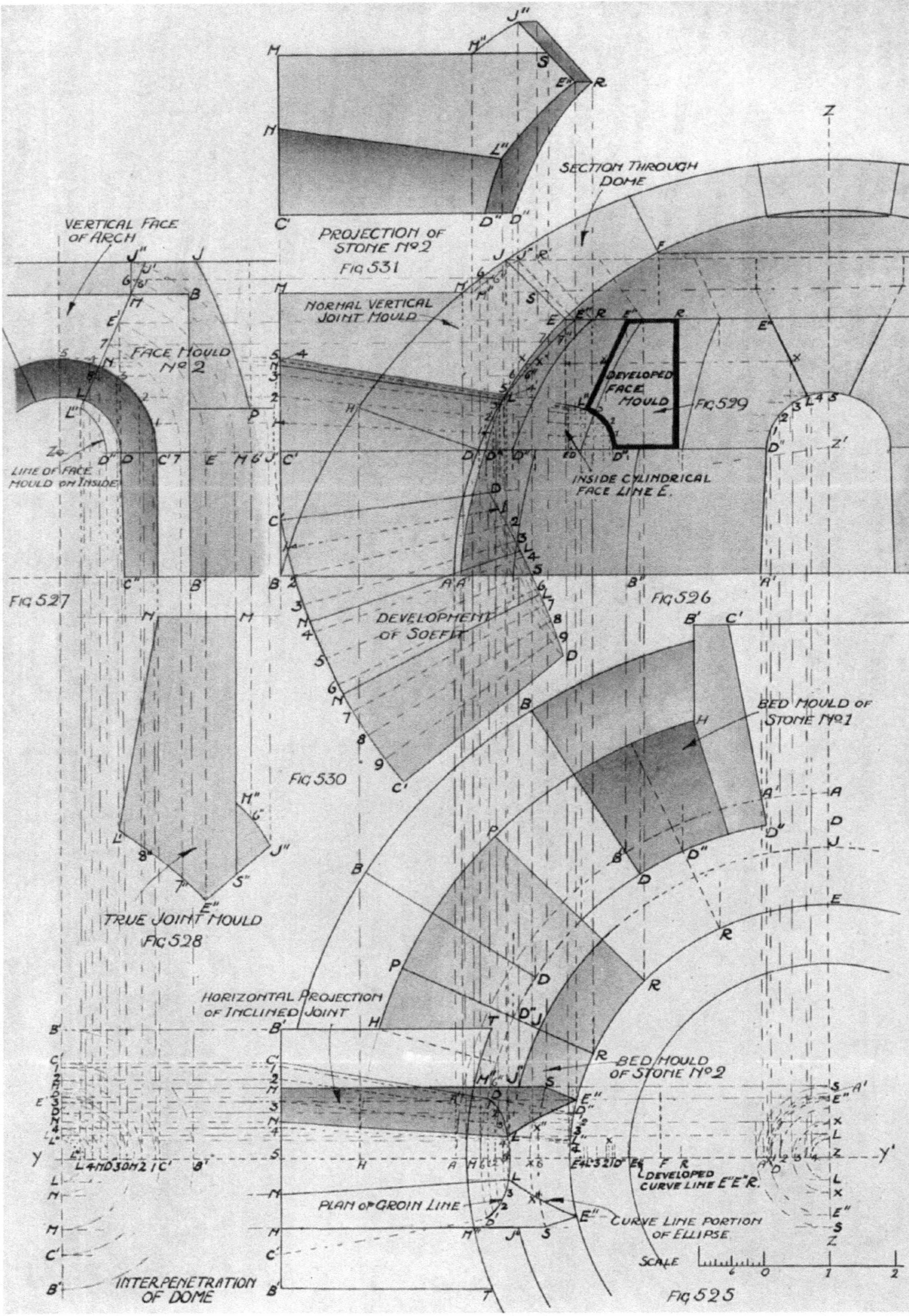
PROJECTION OF STONE No 2
FIG 531
SECTION THROUGH DOME
VERTICAL FACE OF ARCH
FACE MOULD No 2
NORMAL VERTICAL JOINT MOULD
DEVELOPED FACE MOULD
FIG 529
INSIDE CYLINDRICAL FACE LINE E.
LINE OF FACE MOULD ON INSIDE
FIG 527
FIG 526
DEVELOPMENT OF SOFFIT
BED MOULD OF STONE No 1
FIG 530
TRUE JOINT MOULD
FIG 528
HORIZONTAL PROJECTION OF INCLINED JOINT
BED MOULD OF STONE No 2
DEVELOPED CURVE LINE E"E"R.
PLAN OF GROIN LINE
CURVE LINE PORTION OF ELLIPSE
SCALE
INTERPENETRATION OF DOME
FIG 525

the arches with the spherical surface C″ of the dome. Now fix the first bed joint of the cross arches, as at point 1, and draw the normal joint line from centre O′ to meet the intersection with the outside ring of the arch face in point 1′. This point represents the position for the first horizontal bed joint of the dome, and may be drawn in elevation. Project point 1′ across to the diagonal section (Fig. 516) to cut the line C″ in point 1^x, and draw in the normal bed joint through point 1^x from centre O′. Now place in the depth of the stones in the ceiling, cutting the normal bed joint in 1″, and project this point across to elevation. This determines the height of the face mould for the springers.

Drop point 1′ into the wall line in plan; this will give the radius for drawing the horizontal bed line. Now complete the bed mould for the springer as shown on the right-hand side of the plan, also complete the plan of the ribs and project up to elevation their intersections with the wall face, as at E and F. Where point E cuts the wall line intersection of the dome in elevation in point E′ determines the position for the horizontal bed line.

Project line E′ across to the diagonal section and place in the normal joint. Now decide upon the internal diameter of the eye at the crown in plan and project up to elevation as shown. Complete the section of the horizontal rib surrounding the eye and also its projection in plan. Divide the remaining section of the dome surface into a convenient number of courses and project across to elevation, thus determining the horizontal bed lines. To project the elevation of the ribs, drop down projectors from the bed line intersection on the diagonal section into plan. Swing them round, as at E′ F′ G′, thus determining the horizontal bed lines in plan. Project up to elevation their intersection with the plan of the ribs, as at E F′ G′, to cut the horizontal bed lines of the dome in points E′ F′ G′. We thus obtain the elevation of the line of intersection of the ribs with the domical surface. Next project upwards in a similar manner the centre line of ribs, which in this case should intersect horizontal lines drawn from the diagonal section of the nosing line. These points, such as G, on the diagonal section should be dropped down to plan and swung round, thus obtaining the plan of the joint line at the nose of the ribs, as at point G.

It must be clearly understood that the nosing line of the ribs and the domical surface are concentric spherical surfaces which are cut by the four vertical planes of the face of the arches.

Next place in plan the arrangement of the joints, as at F′ G′ and F'^x G'^x, and project their elevations, which are elliptical curves and may be drawn by selecting points in the plan of the joint, as at X, and swinging them round and projecting to the diagonal section as at X′. These points projected horizontally to elevation to meet vertical projectors from the points in plan determine the necessary points through which to draw the curve, as at X′. Although the back surfaces of the stones forming a domical ceiling are to be roughly dressed concentric with the inside spherical surface, as shown in the section (Fig. 516), it will be best to work most of the stones to their required shape by applying the projected face moulds on a vertical back surface.

The surface forming the joint on the extrados of the cross arches should be worked square from the back vertical surface, so that the intersection line with the domical ceiling I′ E′ F′ C^{v} may be found by squaring and gauging from this joint surface.

The face moulds for stones Nos. 2, 3, and 4 are shown projected in Figs. 517, 518, 519.

To work No. 4 stone, apply the bed mould (Fig. 520) to the top surface of the stone, also the face mould to the back vertical surface. Work the vertical normal joints to the outline of the bed mould and mark on these surfaces the joint mould (Fig. 521). Next work the top conical bed joint and apply to this surface the developed bed mould. The application of this mould determines the position of the rib and the domical arris line.

To develop the moulds to apply on the conical bed joint for determining the position and outline of the rib section: Imagine the normal line G″ G′ and G in section to be a generator of a cone, the horizontal line from G″ to be the base, and O′ the vertex, then, with radius O′ G′, draw an arc. Step off along this arc distances equal to the points in plan along the arc G^3 G^3. The development curve for this conic surface corresponds with the section curve because the curve is drawn from centre O′.

Therefore, the points taken from the plan line G^3 G^3 may be marked along the diagonal section curve from point G, although in the drawing this development has been projected out for clearness (Fig. 522).

Now work the spherical surface on the nosing line, as at G^3, then work the bottom arch joint square from the back surface and trammel a line parallel to the back along this surface representing the arris line for the vertical face, as at G^4. Next work the notching for the conical bed joint, as at F″, either from the back vertical surface or by means of a bevel applied to the spherical face. The inclined joint at F″ can be squared from the back surface, whilst the conical bed may be worked to a bevel cut from the section mould (Fig. 521), as at G^3 G^4 G^5. This bevel should be applied from the spherical surface already worked. The intersection of the vertical and spherical surfaces should be determined by working the vertical portion square from the bottom circular arch joint, and should be gauged to correspond to the width of the vertical face from the joint arris. The position for the rib nosing on this arris may be obtained by stepping its distance along the arris from the joint. The rib can now be worked, also the spherical surface at G′ from the lines on the stone, the back being roughly dressed to the required shape as a last operation. A projected front view of stone No. 2, also the bed mould, is given in Fig. 523, whilst Fig. 517 is the projected face mould.

The stones not in contact with the cross arches may be worked as ordinary dome stones, the setting out for these being shown in Fig. 524. The inside plane mould is developed on the nosing line, whilst developed bed joint moulds determine the position for the ribs, the spherical ground surface being worked in conjunction with the working of the ribs.

Setting Out Interpenetration of Dome.—First draw the axial lines, also the plan and sectional elevation of the dome (Figs. 525 and 526). To draw the elevation of the penetrating window (Fig. 527), determine the width of the opening on the vertical face line, and draw a semicircle representing the soffit line of the arch. Divide this curve into any number of equal divisions, as at points 1 2 3 4 5. Project these points across to the vertical face of the sectional elevation, and converge them to the vertex Z′.

Next draw the normal bed lines representing the inverted conical bed joints of the dome, as at D H, E J, etc., and draw horizontal lines from the points D E representing the horizontal bed lines on the inside elevation. Place in the joint lines on the elevation of the penetrating arch (Fig. 527). Project points 1 2 3 4 and 5 into the line Y Y′ in plan from the elevation. Swing them round, and project horizontally to the face line B′ B′ in plan. Now draw converging lines to centre Z from these points.

Again project lines from the intersection of lines 1 2 3 4, with the inside spherical face of the dome at points D 1′ 2′ 3′ 4′ 5′ down to line Y Y′; then, with centre Z, swing them round to meet the plan lines in points D 1 2 3 4 5, point 5 being on the centre line. Through these points draw the curve representing the plan of the groin line made by the intersection of the arch with the inside spherical surface of the dome. Project these points up to elevation to meet their corresponding lines, which radiate to centre Z′. Then through these points a curve may be drawn representing the side elevation of the groin line. Measure the distance of the points D 1 2 3 4 in plan from the line Y Y′, and transfer these distances to the inside elevation to intersect their corresponding lines drawn horizontally from their intersection on the groin line to the inside elevation. This will produce the inside elevation of the arch. Place in all the vertical normal joint lines for the dome in plan, and indicate the bed moulds for the bottom course, as at B B′ C′ D″ D. The reveals of the opening radiate to the axis of the dome, and the soffit of the arch is conical.

Now project the points H D E J down to the line Y Y′ in plan for the 2nd course. In order to obtain the bed moulds, it is necessary to consider the radiating joint surfaces of the arch. These have been drawn as plane surfaces cutting the inside spherical surface obliquely. Therefore, a horizontal line drawn through each point on the inclined joint line up to the point M will be parallel to the axis line Y Y′ of the sphere. The upper portion of the joint should be a normal vertical surface, but if this is arranged thus, a difficulty arises in placing the keystone into position. As an alternative method, the radiating joint may be continued up to the bed line J in the same plane as the inclined joint, but this creates a sharp arris for the keystone at the intersection of the bed line, and also increases the size of the keystone considerably. In this example the joint surface between M J″ has been drawn vertical, and parallel, to the axis of the dome.

Although in correct dome construction it is necessary for all horizontal and vertical joint surfaces, or cutting planes, to be normal to the axis of the dome, it is wisest in cases such as the one under consideration to place these planes with the view of obtaining the best advantage structurally, while practical requirements are being also kept in mind. All dome penetrations are sources of weakness in the structure, except where complete rings are omitted at the crown.

As the radiating arch joints in this case are not in strict accordance with the rules governing dome construction, therefore the small joint surface between

M and J″ should not make any appreciable difference in the stability of the structure.

To obtain the horizontal projection of this joint for plotting the bed mould, project the points L N M, representing the joint line in elevation (Fig. 527), into the line Y Y′, and swing them round as before and project them into plan. As L and N are normal points on the surface of the conical soffit, a line drawn through N in plan to centre Z will give the horizontal projection of the arris line L N, the position of L being determined where the line from N cuts the plan of the groin line at L. The top arris line from M being horizontal, and parallel to the axis Y Y′ of the sphere, this line may be drawn parallel to the line Y Y′ in plan from the point M. Then, as the portion of the joint between M J″ is vertical and also parallel to the line Y Y′, the arris line from M should be continued past J″ in plan, to point S, which is on the conical bed surface of the dome.

The position of this point S is determined by producing M M′ in section to cut the conical bed surface J E in S. Drop a projector from S into the line Y Y′ in plan, then, with Z as centre, swing this point round to cut the plan of the arris line M J″ in S.

Next determine the inside elevation of the normal joint line. To do this, project the horizontal arris line E in section to the elevation of the joint in Fig. 527 at point E′. Project this point down to line Y Y′ and swing it round as before, and draw a projector parallel to the line Y Y′ to cut the plan of the arris line E in point E″.

Transfer the distance E″ from Y Y′ to the inside elevation along the horizontal arris line E, measured from the centre line, thus determining E″ in elevation. Point L may be determined in a similar manner. This projection is shown rotated about the axis line Z Z.

Join E″ to Z′, this being the elevation of the joint line across the inside spherical surface.

To complete the horizontal projection of the joint, take any point X on the joint line of the inside elevation, project this point down to the line Y Y′, and swing it round as before. From X in elevation draw a horizontal line to cut the section of dome in X. Drop a vertical projector from this point in section down to the line Y Y′; then, with Z as centre, swing the point round to meet the parallel line already drawn from the axis line Z Z in the point X″. A curve drawn through E″ X″ L is the horizontal projection of the joint arris running across the inside spherical surface. The horizontal projection of the joint surface is now complete.

This also completes the outline for the bed mould for the second stone, the mould being indicated on the drawing by the points B′ H P R E″ L N. To obtain the true shape of the mould to apply on the inclined joint surface, rotate all the points in the elevation of the joint (Fig. 527) into the horizontal plane as shown, and draw projectors from these points perpendicular to the line D C′. Draw the front face line M N perpendicular to these projectors, and measure the distance of each point from the line M N from the section of the inclined joint.

In measuring the depth of the section for this joint mould, it must be remembered that the joint is not vertical but inclined; therefore, the top portion, as at E, travels a certain distance around the curvature of the dome. Hence, all the points required travel along their respective curve. Project E″ up from plan to the horizontal arris line E in point E″, also J″ in plan to the horizontal arris line J in J″, and proceed similarly with point M″ X″ and 6″. Draw lines through these points, thereby obtaining the vertical projection of the joint surface.

Measure all these points in the vertical projection of the joint surface from the vertical face line M B, and transfer them to their corresponding rabatted projectors beyond the line M N in rabatment.

The vertical joint between M J″ should be first rabatted into the plane of the inclined joint, and then rabatted into the horizontal plane as shown. The true shape of the mould is given in Fig. 528.

For the working of the second stone an inside developed face mould is required. This mould is developed on a vertical cylindrical surface formed at the arris line E. From E in section draw a vertical line representing this cylindrical surface, and produce all the generating lines on the conical soffit. To determine the curve of the conical soffit on the developed cylindrical surface, first obtain the side elevation of the intersection of the conical soffit and the cylindrical surface by projecting up the points D″ 123L″ 4 to meet the generators of the cone in section. Also produce their corresponding generators in plan to meet the plan of the arris line E. Now stretch out the points on the arc E″ E in plan along the horizontal arris line D in elevation from the vertical line drawn from E. Vertical projectors drawn from these points to meet horizontal projectors drawn from the points in the vertical cylindrical surface determine the outline of the developed face mould, as shown in Fig. 529. The line between E″ L″ is a curve in development.

It is now necessary to develop the conical soffit of the intersecting arch, as a developed mould is required for application on the stone. The method of development is clearly shown in Fig. 530. With Z′ 5 radius and centre Z′, draw an arc 5′ C′. Step off along this arc the points on the elevation of the arch curve (Fig. 527), and through these points draw lines converging to centre Z′. To obtain the development along the groin line, mark off on these converging lines from the arc already drawn distances equal to their corresponding lines in section as at points 1 1′. The face mould should cover the extreme points of the stone, so it is necessary to determine points D″ L″ in elevation. These are obtained by drawing projectors parallel to Y Y from points D″ and L″ in the plan of the arris line E, swinging them round into the axis line and projecting them up into elevation to meet the joint line produced in L″ and the horizontal line D C′ produced in D″. Any intermediate point for drawing the curve may be obtained by projecting in a similar manner.

Projection of stone No. 2 is given in Fig. 531.

The Setting Out Intersecting Vaults of Unequal Span.—Draw the axial lines Z Z and O′ F′, also lines Y Y and X X at convenient distances. Set up the cross section of the small vault, as at A F B (Fig. 532), which is semicircular, and decide on the thickness of the vault. In doing this, arrange that the vault is thinner at the crown than at the springing. Divide the intrados curve of this vault into a convenient number of stones (seven in this case). Draw normals through these points representing the bed joints. Allow for a horizontal top bed on the springing stone at H. As the vaults are to be the same height at the crown, but of different span, it will readily be seen that, in order to obtain a straight groin line in plan, the larger vault will be projected as an ellipse. Set out the plan to the required size, and place in the groin lines (only one-quarter need be drawn on the left-hand side).

Next obtain the section and the intrados curve of the large vault (Fig. 533). Select any number of points on the inside curve of the small vault, and project them down to the groin line in plan. Draw projectors from these points to the section of the large vault, parallel to the axis line; also project the points C D E F in section horizontally to meet the axis line Y Y. Swing these points into the horizontal plane, and project them parallel to line Y Y to meet their respective lines already projected from the groin line in plan in points C″ D″ E″ F″. These intersections will give the points through which to draw the elliptical curve forming the intrados of the large vault; also these points C″ D″ E″ F″ are the positions where the bed joint lines of this vault intersect the intrados curve.

This curve may be drawn as a semi-ellipse after obtaining the major axis and semi-minor axis by means of a trammel, after which the points C″ D″ E″ should be projected around as before described.

To draw the normal bed joints for the large vault, a very important geometrical principle in normals and tangents to the ellipse may be applied and as stated in the theorem, page 140; also a diagram demonstrating this principle is given in Fig. 338. Draw a tangent to point C to intersect the springing line in C^x; project this point into plan to meet a line drawn at 45° from A in C^y. Project this point to the springing line of the large vault, thereby obtaining point C^z. Connect this point with C″; then C″ is the point of contact. A normal to the ellipse at C″ is a line drawn 90° to the tangent, and by drawing this line we obtain the normal bed joint.

Repeat the process for the remaining bed joints. It is more convenient to draw the tangents at points D and E to intersect the minor axis F O, as at points D^x and E^x. Project these points to axis line Y Y, swing them round into the H.P., and project them to intersect the minor axis in point D^y E^y. Connect these to their points of contact on the ellipse, thus obtaining the tangents. Next draw the bed joints at right angles to these tangents. The curve of the extrados for the large vault should now be determined, the shape of which may be projected from the extrados of the small vault, but this does not project a good cross section for the large vault. It is best to make this curve a semi-ellipse, making the thickness of the cross section at the base the same as that of the small vault continued down to springing level, as at A′, and at the crown the same as that of the small vault, the semi-minor axis being equal to O^2 F′. Next obtain the plan of the external groin line. This is obtained by projecting points H J, etc., into plan to meet projectors drawn from points H^y and J on the section of large vault to intersect in plan at h^y and j. Other points for the determination of this groin line may be obtained in a similar manner.

It will be noticed that the intersections of the bed joints with the extrados

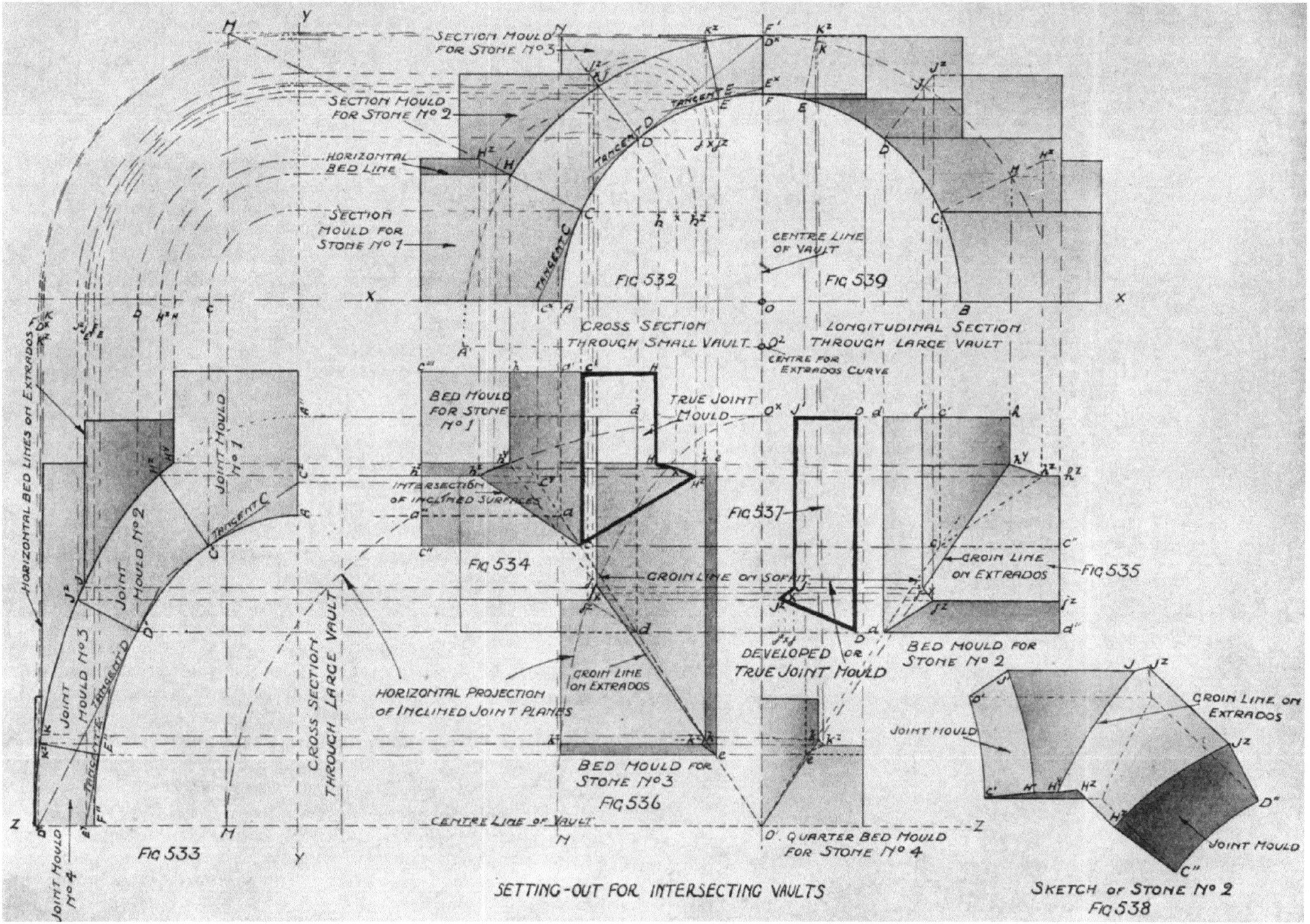

SETTING-OUT FOR INTERSECTING VAULTS

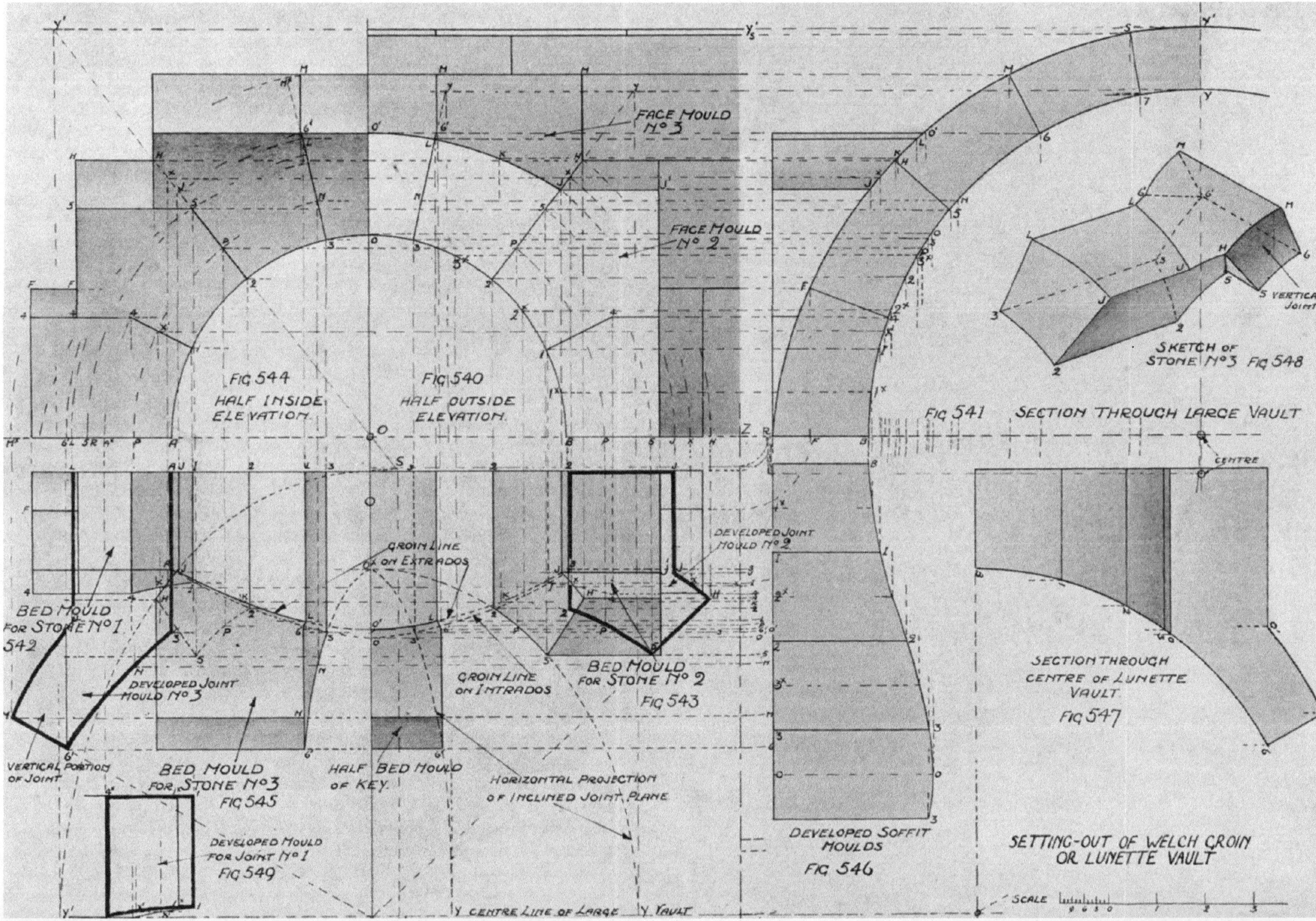

SETTING-OUT OF WELCH GROIN OR LUNETTE VAULT

of the large vault are not at the same level as those of the small one; hence it is necessary to produce the bed joint line C H to intersect the level of the large vault joint at H^z. The horizontal projection of the portion between H and H^z will be a portion of an ellipse, for actually it is cutting obliquely across the curve of the extrados, the axes of the ellipse being obtained by continuing the joint plane to the line representing the crown of the extrados and projecting it into plan, as at M. Then M O′ is the semi-major axis and O′ O^x the semi-minor axis. Repeat this method of projection for the remaining stones, the amount of difference in the level of the bed being less at each course.

Now obtain the bed moulds, commencing with stone No. 1 (Fig. 534). This is shown at the left-hand side of plan, at points C C′ C‴ C″. The arris lines for the bottom bed are indicated by points *a a′ a″*. The top bed lines are marked $h\ h^y\ h^z\ h^z$, which are also the bottom bed lines for stone No. 2 (Fig. 535). The bed mould for this stone is shown on the right side of plan, at $d\ d\ d''\ h^z\ h^z\ h\ h^y$, the lines on the top bed being $j\ j'\ j^z\ j^z$, and the line $j^z\ d$ being the horizontal projection of the intersection of the two inclined surfaces or bed joints.

The bed mould for stone No. 3 is shown projected in Fig. 536, and one-quarter of the bed mould for the centre stone is shown. The vertical joint moulds are taken from the two sections.

It is now necessary to obtain the true shape of the joint moulds for application to the inclined joint surfaces. These are shown rabatted into the horizontal plane about the points C and D, and projected into plan to meet horizontal projectors drawn from their corresponding points in plan, as shown in Fig. 537. A sketch showing stone No. 2 is given in Fig. 538, the position of the external groin line in relation to the horizontal arris lines of the bed joint being clearly indicated. A section along the axis of the large vault is given in Fig. 539.

Setting Out Welch Groin or Lunette Vault.—Draw the axes of the small and large vaults and the half-elevation of the small vault (Fig. 540), also the sectional elevation of the large vault (Fig. 541).

Divide the arch line of the small vault into any convenient number of stones (seven in this example), and draw joint lines converging to centre O, as shown. Determine the height of the horizontal bed joint at point 4, and draw a horizontal line from 4 to the inside surface of the large vault at point 4. This determines the first horizontal bed line for the large vault. Now divide the remainder of the large vault section into a suitable number of stones, as at 5 6 7, and draw normal bed lines to meet the extrados curve in points F H M S. From the point H obtained, draw a horizontal projector to the elevation of the small vault to meet the radiating joint line 2 in point H. Now fix the height of the extrados surface of the small arch from point J to O′. The centre for striking this curve is taken below centre O. This curve cuts the radiating joint 3 in point L. Project horizontally from these points, thus obtaining the sectional elevation of the small arch, showing the intersection of the arch with the extrados surface of the large vault.

Now obtain the plan of the groin line on the intrados surface. Draw an axis line Z Z′ at right angles to the springing line. From a series of points on the curve of the small vault, as at 1^x 1, 2^x 2, etc., draw horizontal lines to cut the inside curve of the large vault. Project these points down to the springing line; then, with Z as centre, swing them round into the axis line, and from these points draw horizontal lines to meet vertical projectors drawn from points B 1^x 1 2^x 2, etc. A curve drawn through these intersections gives the plan of the groin line and the position of the wall face, also the points A and B in plan.

Now rotate the centre line of the large vault into plan.

To obtain the bed moulds: The bed mould for stone No. 1 is shown on the left-hand side in plan (Fig. 542). Project points A and 1 into plan to meet the plan of the groin line on the intrados. Also project the horizontal arris line 4 in the section of large vault into plan by rotating it about Z and projecting horizontally to meet the vertical projector from point 4 in elevation. The horizontal line 4 4 is the plan of the arris line. The elliptical curve between points 4 and 1 in plan, which represents the horizontal projection of the inclined arch joint travelling obliquely across the cylindrical surface of the vault, may be obtained by projection as follows: Take any point X′ between 1 and 4 in elevation; project this point across to the section of the large vault, and rotate it into the axis line Z Z′, as before explained. From this point in Z Z′ draw a horizontal line to meet a vertical projector from X′ in elevation in point X, through which to draw the curve. This completes the bed mould for No. 1 stone.

The bed mould for No. 2 stone is shown on the right side of the plan (Fig. 543). It is first necessary to obtain the plan of the intersection line on the extrados. Project the points J K L O′ in elevation to intersect the extrados of the large vault in section, rotate them into the axis line Z Z′, and project them horizontally to meet vertical projectors, drawn from their corresponding points in elevation as shown, thus determining the points J K L O′ in plan. Draw in the line J J′ in plan. This line represents the plan of the intersection of the horizontal top surface and the extrados of the large vault. In a similar manner project into plan point H, representing the horizontal arris line on the extrados of the large vault, and point 5, representing the horizontal arris line on the intrados of the large vault, as shown in Fig. 544, which is a half internal elevation. Draw the elliptical curves between J and H and 2 and 5, thus completing the bed mould for stone No. 2. These curves are the horizontal projections of the arris lines of the inclined joint surface cutting obliquely across the inside and outside cylindrical surfaces. They may be drawn by continuing the V.T. of the joint planes in elevation to the crown of the large vault at Y and Y′. Draw vertical projectors from these points to meet the plan of the centre line in Y and Y′. The distance O Y on the centre line determines the semi-minor axis, and a line drawn from A B in plan cutting the vertical axis line O O′ in O^x determines the distance O O^x, which is the semi-major axis. An ellipse drawn through these points by the trammel method passes through the points 2 and 5. A similar procedure will determine the curve through J and H, though in this instance, the centre for the extrados being below the springing line, it is necessary for the full length of the radius of the large vault to be used as the half-major axis, as shown, by continuing the curve of the extrados in section, down to a horizontal line drawn from the striking centre of the extrados curve. Project this point up to the springing line, as at R, and rotate it into Z Z′. A line drawn from this point to cut the vertical trace of the inclined joint produced through centre O in point S determines the semi-major axis. Then, with S S′ as the semi-major axis and S′ Y′ as the semi-minor axis, the elliptical curve through points H and J may be drawn.

To obtain the bed mould for No. 3 stone, first determine the position of the joints for the keystone.

The width of the keystone is determined by continuing the radiating joint line on the intrados to meet the horizontal bed line at 6′. From this point the joint is continued vertically to the horizontal bed line M on the extrados. Measure the distance of M from the centre axis line, and transfer it to the outside elevation at M, thus obtaining point 6′ on the extrados. The outside radiating joint line should be produced across the cylindrical surface to 6′. Project points L and 6′ down to plan, point L to meet the intersecting line on the extrados, and point 6′ to meet the line 6′ rotated into plan about the axis line Z Z′.

Join 6′ to 6′ parallel to the centre line O O′, and draw the elliptical curve through L 6′ in plan, then the bed moulds for stone No. 3 and the keystone are complete, as in Fig. 545.

The next operation is to obtain the developed joint moulds to apply on the inclined joint surfaces. These are shown in the drawing, rabatted into the horizontal plane, about the axis O. The construction necessary for the rabatment of these inclined planes is clearly shown in the drawing, whilst the principles involved are explained in Figs. 420-424. Developed soffit moulds are required for application on the stones. The method for obtaining this development is shown in Fig. 546.

Stretch out on a straight line the points 1^x 1, 2^x 2, etc., on the curve line of the small vault, and drop vertical projectors from these points on the soffit line of the large vault. Where these projectors meet perpendicular lines drawn from the stretch-out line determines the points through which to draw the curve representing the development of the intrados groin. A section through the centre of the small vault is given in Fig. 547. A sketch is given of stone No. 3 in Fig. 548.

The section mould to apply on the radiating joint surface between stones 1 and 2 is shown in Fig. 549.

Ramp and Twist Work.—Ramp and twist work in masonry requires treatment in setting out entirely different from that for handrailing in joinery.

In the latter the wreaths are set out and worked by a special craftsman known as a *handrailer*, but in masonry the *setter out* is required to draft the setting out so that all lines necessary for the complete working of the stones may be transferred to zinc moulds. This is essential, because the banker mason works entirely to particulars on the moulds, and is not responsible for the setting out of the work.

The particulars should be such that all the processes necessary for the completion of the stone can be executed in a mechanical manner, by transferring the lines from the moulds to the stone and definitely working to these lines. In the tangent and normal systems which are used in joinery, certain surfaces are worked to the eye, and the face moulds in the tangent system are fastened to the surfaces of the timber, in their correct position relative to the tangent lines, the material being worked between these moulds; but in masonry it is impossible to fasten these moulds to the surfaces of the stone, so that the system becomes impracticable.

The method shown in the examples dealing with ramp and twist work in this book is both simple and practical and, in the opinion of the writer, is the most economical to adopt. Whilst geometrical methods are explained in the following examples, the writer is aware that approximate methods are sometimes used in practice, but with very poor results.

Setting Out Wing Wall for Entrance Stair.—Draw the plan of the wall and steps (Fig. 550). Divide the wall into suitable stones of length, and place in normal lines from the points at which the risers intersect the plinth curve. Select a normal line as near as possible to the centre of one of the stones, and project this line as a *boning line*. The elevation of the wall may now be projected from the plan if required by dropping projectors from the concave and the convex surface on plan at the point of intersection with the normal lines just drawn, as in Fig. 551. Where these projectors intersect their corresponding arris lines of treads and risers gives a series of points through which to draw the curve determining the intersection of the steps and the curved wall. As the plinth is 12 in. high, measure points 6 in. above the point already drawn; these will give a fair curve, which in the elevation is the helicoidal top surface of the plinth.

To obtain the vertical projection of the individual stones, thereby determining the *cover face mould*, or a mould to cover the amount of "*twist*" in each stone, produce the *boning line* on plan in either direction and select any point on this line. Mark off on the *boning line* produced, a series of points representing the number of risers contained in each stone. For instance, the first stone passes through two risers, or to riser No. 3 on plan, so that two riser heights, plus two risers for the thickness of the stone, should be marked on the *boning line*, and lines drawn from these representing the riser heights at right angles to the *boning line*, as in Fig. 552. Project up the points 1 Y X 2 3 and 1′ Y′ X′ 2′ and 3′ to intersect their respective lines in projection, remembering that all normal lines on plan are horizontal lines in elevation. The *boning line* for this stone has been taken central between risers 1 and 2, so that the position of this point, in elevation, will be midway between the riser lines already drawn. Through these intersections curves may be drawn, representing the amount of twist in the stone when viewed in the direction of the normal line X X′.

It is now necessary to determine the joint lines. Naturally the correct jointing surface should be twisted according to the twist of the stones on plan and the pitch of the helical solid. The principle here involved is shown in Fig. 430. The lines 3 3′ and 6 6′ on plan should be taken as representing the centre horizontal line of the normal joint. Develop the centre line B in plan, as in Fig. 553, making the development 12 in. in vertical height. Mark a centre line on this development and project up the points representing the joint lines taken from plan, as at points 3″ and 6″. Where these intersect the lines already drawn determines the centre lines of the joints. Erect a line perpendicular to the centre line from the point of intersection; then this line represents the joint line placed normal to the pitch on the centre line on plan. Drop projectors from the points 3^y and 3^x and transfer their distances from the centre point 3″ on the development line B to the plan, measuring their distances on each side of the centre line in plan. From these points draw normal lines in plan, and at the points where they cut the concave and convex surfaces, project them to the vertical projection of the stone. Where these projectors intersect the curved lines, the position of the joint lines when viewed, as in orthographic projection, determines the position of the joints in the vertical projection. Now draw the parallel inclined lines L M and K N to clear all points in the elevation of the stone.

To obtain the plane bed mould (Fig. 554): Imagine the stones to lie in an inclined plane, the inclination to the H.P. of the first stone being the angle made by the top and bottom surfaces of the rectangle represented as the *cover mould*. First draw the horizontal projection of the rectangle P Q R S in plan to cover the amount of curve on the first stone, and then imagine the top surface of the rectangular block rotated about an axis K N into the plane of the paper. The width of this stone will be the same as the rectangle P Q R S in plan. The curve will be the projection of the circular arc in plan, projected into the inclined plane. Mark on the plan any number of normal lines, and project them up to the plane line K N. Where these points intersect K N, erect perpendicular lines to K N, and on these lines mark off distances corresponding to those on the line S R in plan. This will produce a series of points through which to draw the elliptical curves.

This section is the same as the section produced when a cylinder is cut obliquely, as shown in Figs. 555, 556, 557, 558.

Draw the plan and mark in the X Y line, and erect an elevation of the solid, placing in the V.T. of the cutting plane (Fig. 556). Next obtain the true shape of the section made by this inclined plane passing through the cylinder, as shown in Fig. 558. These elliptical curves may be drawn by the trammel method, E′ O′ being the semi-major axis and R′ O′ the semi-minor axis for the elliptical curve of the convex surface.

To obtain the joint lines on these moulds, project points B″ B′ up to the V.T. and draw lines from these points perpendicular to V.T. to cut the elliptical curves in points B″ and B′. The curves between these points are drawn as straight lines, but, if correctly drawn, they should be slightly curved, because they are not horizontal lines, as may be seen by reference to Fig. 302. Since all the generators of the helicoidal joint surface are at right angles to the axis, this being vertical, then the generators, or straight lines, will be horizontal; therefore, because these helicoidal joint surfaces do not intersect the upper and lower plane surfaces of the stone in horizontal arris lines, they will be slightly curved.

We have now obtained the *cover face mould* and the developed plane

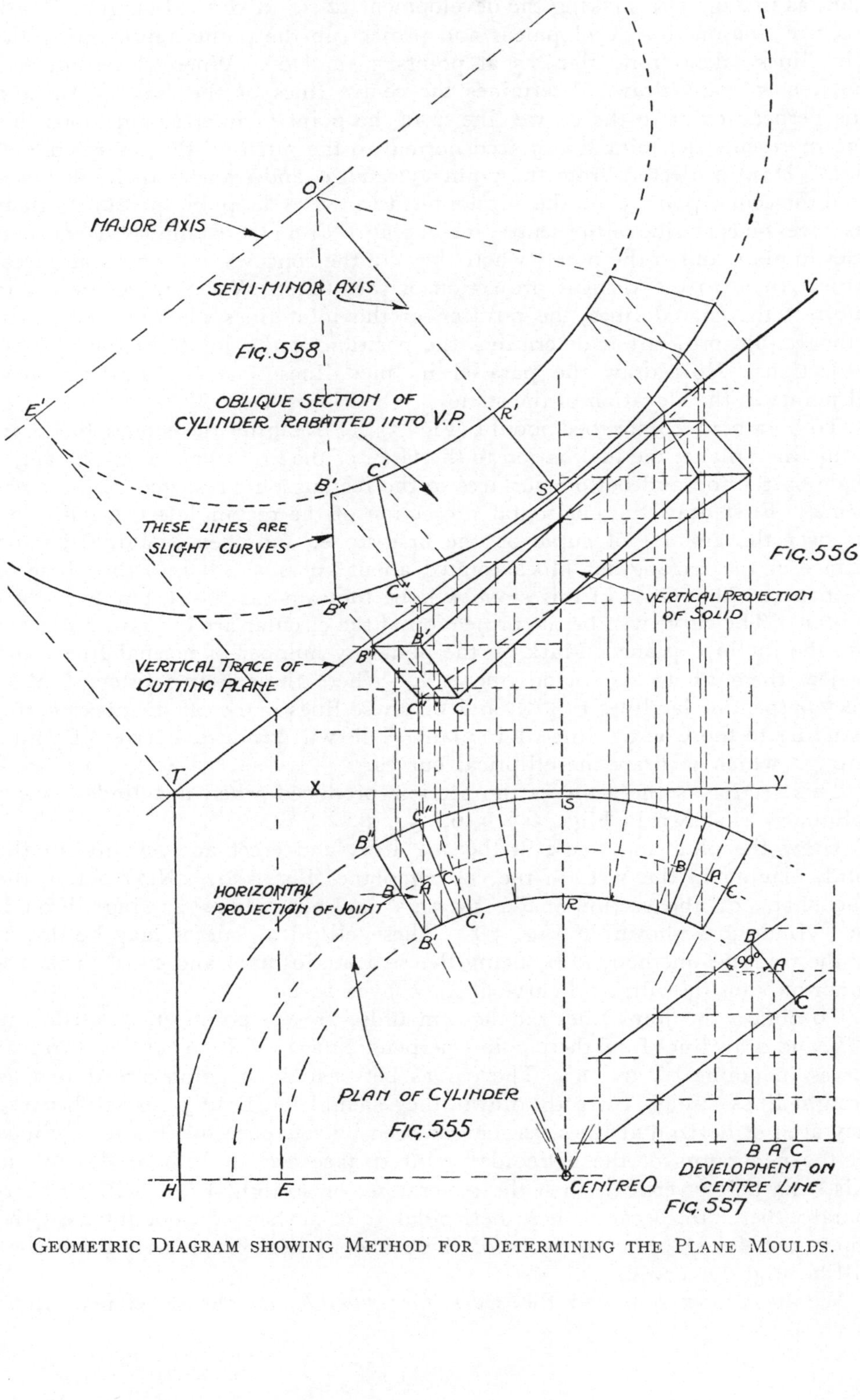

GEOMETRIC DIAGRAM SHOWING METHOD FOR DETERMINING THE PLANE MOULDS.

mould. Next develop the inner and outer face moulds, which will lie on the cylindrical surfaces.

Stretch out the points on the curve lines A C in plan, starting from point 1 on to the development lines A and C.

From these points erect perpendicular lines to cut horizontal lines drawn from the *height rod* representing the heights of the risers. Where these intersect determines the points through which curves may be drawn, thus producing the developed face moulds required, as shown in Figs. 559 and 560. The *boning lines* should be marked on these moulds, also the joint lines, which are obtained by measuring the distance of points 3 *y* and 3 *x* from the centre line of the joint in plan, and transferring these distances on either side of the point 3 on the development line A and point 3′ on the development line C. Project vertically from these points to intersect the top and bottom arrises of the developed face moulds in points 3 *y* 3 *x* and 3 *y*′ 3 *x*′. Lines drawn through these points determine the elevation of the joint lines in development and also the correct shape of the developed face moulds. Stones Nos. 2 and 3 are projected in a manner similar to that just described, their projection being given in Figs. 561 and 562. Stone No. 3 is shown bonded 4½ in. in the wall of the building, this portion of the stone being horizontal, as shown by the face moulds. If a capping course is required, the projections are the same as just described, and should not be difficult if the preceding methods of projection have been followed carefully.

Horizontal beds are sometimes required for the plinth stones. If this is the case, the setting out is simplified, the inner and outer developed face moulds being cut to the lines shown dotted on the development.

A developed plane mould would not be required, but a bed mould should be cut from the curve in plan.

A sketch of the second stone with a horizontal bed is shown in Fig. 563. The method of obtaining a developed joint mould to apply on these joint surfaces should a moulding be required is given in Fig. 432.

Setting Out Plinth or Capping for a Geometrical Stair Balustrade.—Set out the plan of the stairs to dimensions given, marking in the plan of the string, or baluster plinth, at points A B C. Now draw the treading line D, 18 in. from C, and mark off, along this line, distances equal to the tread of an ordinary flier—say 12 in.—and then draw the plan of the risers, numbering them as on the drawing (Fig. 564). Next set up a section through D. First put up a height rod, as on the extreme right-hand side of the drawing, and mark the heights of the risers—say 6 in. To draw the section on D, mark off along the line X X (Fig. 565) distances equal to the width of the tread at D. Draw perpendicular lines from these points, intersecting horizontal lines drawn from the height rod, thus giving the outline of the steps. The line of the soffit is determined on this section, and the joint in the rebate is drawn perpendicular to the soffit line, which is determined by marking 1½ in. in, on the tread, from the riser line, and drawing circles each of 1½-in. radius. Draw a tangent to these circles, which is the line of the soffit, and draw a perpendicular F G to the soffit line from the centre of the circle. This will give the section through the rebate at line D. Point G determines the

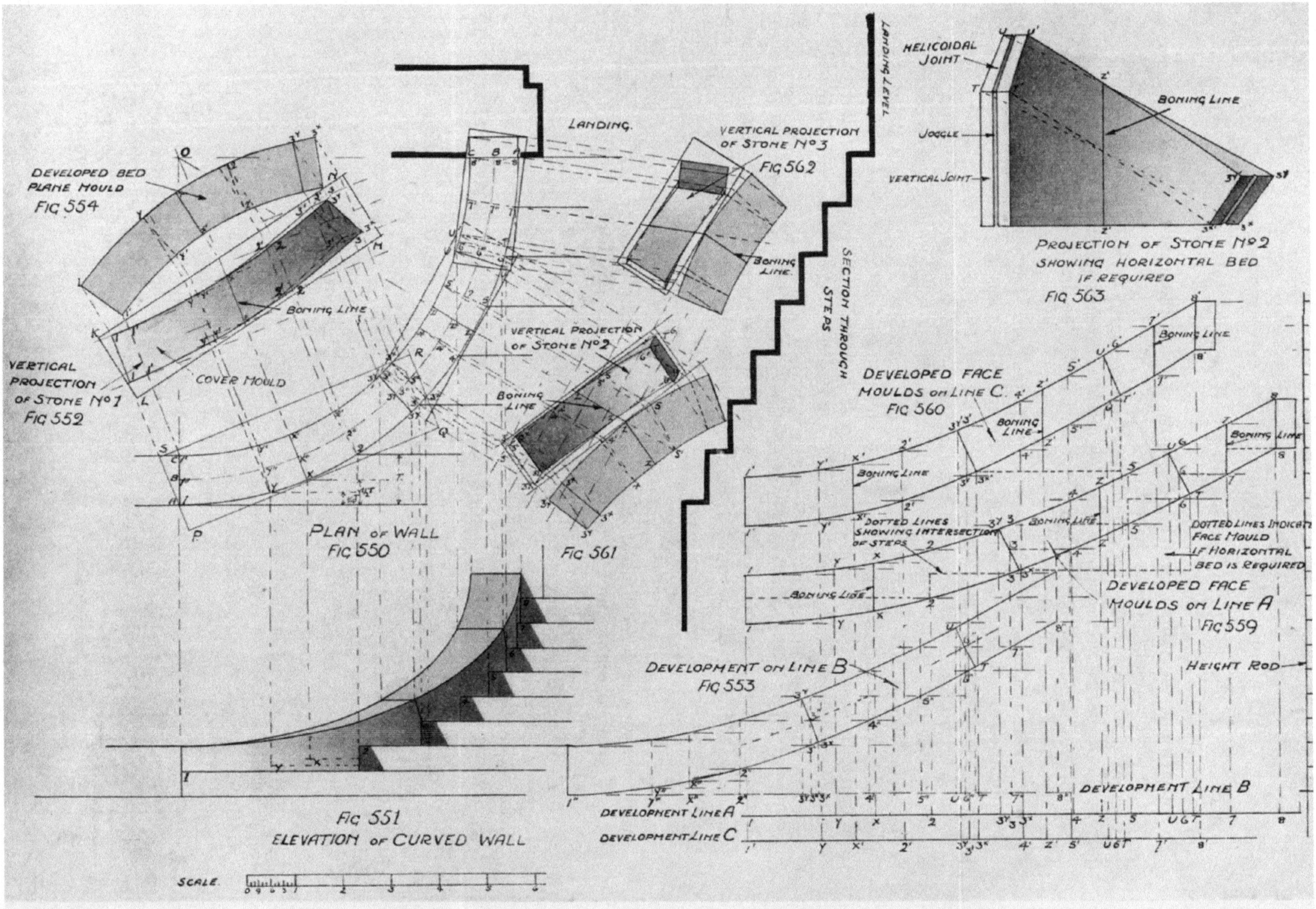
DEVELOPED BED PLANE MOULD
FIG 554
VERTICAL PROJECTION OF STONE No 1
FIG 552
COVER MOULD
BONING LINE
PLAN OF WALL
FIG 550
LANDING.
LANDING LEVEL
VERTICAL PROJECTION OF STONE No 3
FIG 562
VERTICAL PROJECTION OF STONE No 2
FIG 561
SECTION THROUGH STEPS
HELICOIDAL JOINT
JOGGLE
VERTICAL JOINT
PROJECTION OF STONE No 2 SHOWING HORIZONTAL BED IF REQUIRED
FIG 563
DEVELOPED FACE MOULDS ON LINE C.
FIG 560
DOTTED LINES SHOWING INTERSECTION OF STEPS
DOTTED LINES INDICATE FACE MOULD IF HORIZONTAL BED IS REQUIRED
DEVELOPED FACE MOULDS ON LINE A
FIG 559
HEIGHT ROD
DEVELOPMENT ON LINE B
FIG 553
DEVELOPMENT LINE B
DEVELOPMENT LINE A
DEVELOPMENT LINE C
FIG 551
ELEVATION OF CURVED WALL
SCALE.

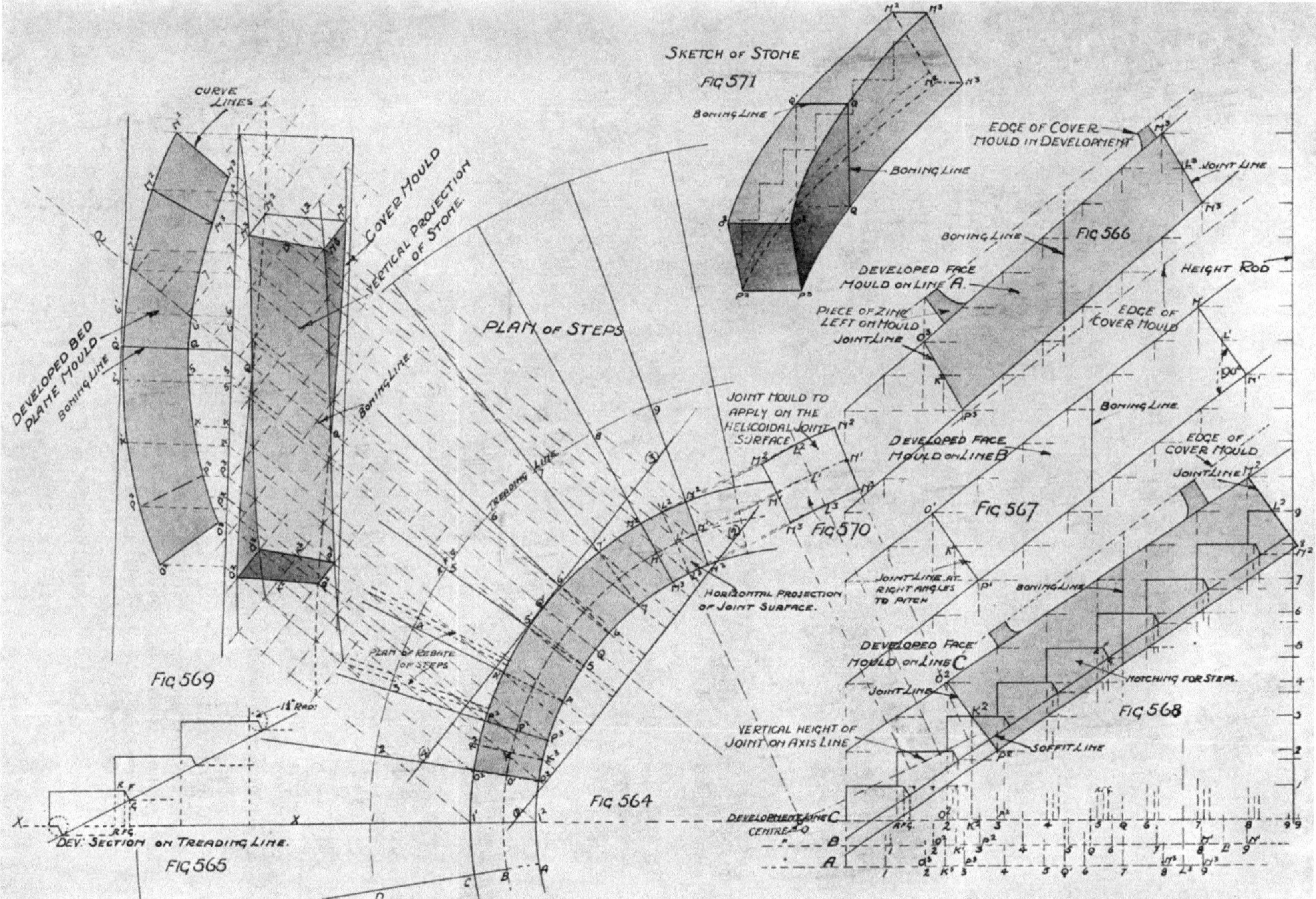

SETTING-OUT OF PLINTH FOR GEOMETRICAL STAIR BALUSTRADE

level for the horizontal joint lines on the soffit of the steps. Next erect developments of the lines A B C by stepping off the width of the step where the riser lines intersect the curved lines A B C, and stretching these distances along the development line for each curve, as in Figs. 566, 567, 568. From the points thus obtained, erect perpendiculars to intersect the horizontal lines drawn from the height rod. Next determine the height of the plinth. To do this it is necessary to determine the correct section of the steps on line C, so that a certain distance may be allowed for the plinth to project below the soffit line of the steps. To obtain the section, transfer point G from the development of the walking line (Fig. 565) to the developments just drawn, and draw a series of lines representing this distance below the tread lines.

Drop vertical lines from points F and G in the development of the walking line, and transfer the distances of points F and G from the riser line R in plan on line D. Through these points draw lines to the centre of plan, where these points intersect the curve line C, transfer their distances to the development of C (Fig. 568), and erect vertical lines which will give the section through the rebate and soffit line on line C.

Next determine the depth of the plinth, and mark the top and bottom bed lines. Note that the height is measured on a vertical line, and that the vertical height of the other developments will be the same as C. Now select the position of the joints in plan and obtain their horizontal projection. To do this, determine the positions of the joints on the centre line B, as at points K′ and L′, and transfer them to the development on line B. Project the points K′ L′ up to the developed elevation, as at K′ L′ (Fig. 567), and through these points draw lines perpendicular to the top and bottom lines of the development, as at O′ P′ M′ N′. Project the points M′ N′ O′ P′ down to the development line B, and transfer these distances from K′ L′ to M′ N′ O′ P′ in plan on the line B. Draw normals through these points intersecting the inner and outer curve lines A and C in points as shown.

Now transfer these distances from K^2 K^3 and L^2 L^3 to their development line, and project them up to intersect the top and bottom developed bed planes, as at M^2 N^2 M^3 N^3 and O^2 P^2 O^3 P^3. Lines drawn through these points determine the position of the joint line for the developed moulds.

It is now necessary to obtain a vertical projection of the stone, as in Fig. 569. It must be remembered that the pitch of this helical solid is the same as the going and rise of the steps. Therefore, select any point at or near the centre of the stone in plan, project a normal line Q R of an indefinite length, and erect perpendiculars to this line representing the height of the risers.

Project the points 1′ O^2 P^2 4′ 5′ 6′ 7′, etc., from the plan of the line C, and points 1^2 O^3 P^3, etc., from the plan of the line A to intersect their corresponding riser lines in the auxiliary elevation, thus obtaining a number of points through which to draw curves, representing the vertical projection of the helical surface. Add the vertical height of the plinth to these lines, thus obtaining curves which will give the outline of the helical solid. It is important to remember that all normal lines in plan are horizontal lines in elevation. Now project the joint lines from plan into elevation, proceeding as before, thus obtaining points O^2 O^3 and P^2 P^3 M^2 M^3 N^2 N^3. Join

O^2 P^2 and O^3 P^3 and M^2 N^2 and M^3 N^3, and this will determine the outline of the joint surfaces in elevation. These lines are actually curved lines. To assist in drawing these, a point on the curve may be obtained by projecting to elevation the centre line K^2 K^3 and L^2 L^3 from plan to intersect the horizontal riser line midway between the top and bottom arrises in points K^2 K^3 and L^2 L^3. When the elevation of the stone is complete, we are able to determine the minimum thickness of the slab of stone required to contain the finished stone. Draw a rectangle to enclose the figure in elevation, allowing sufficient material for practical purposes. The mould thus obtained is called a *cover mould*, and the vertical boning line should be marked on this mould, so that when applied to the surface of the stone, it determines the inclination of the top and bottom surfaces to the horizontal plane. The top and bottom surfaces of this rectangular solid are cutting planes passing obliquely through the cylinder containing the plan of the stair, as shown in Fig. 556. It is now necessary to obtain the projections of the figures lying on these planes. Let the top arris line of the cover mould be the ground line, or X Y. Turn the surface into the same plane as the cover mould, then, as the plane revolves, all the points on it will travel along lines at right angles to the arris of the cover mould. Project all lines at right angles to this arris line, and measure the distances for the points of the curves along these lines taken from plan. To do this, draw a line (1) (2) at right angles to the boning line Q R to clear the extreme arrises of the stone, and draw a line (3) (4) parallel to the line (1) (2).

These lines determine the full width of the stone, and are the plans or horizontal projections of the arrises forming the top and bottom inclined surfaces. Where the projectors from the various points in plan intersect the top arris line of the cover mould, draw perpendicular lines, and measure along these lines the points for the figure on the plane, from the line (1) (2) in plan. The true shape of the figure o1 curves will thus be obtained, determining the inclined mould to apply on the top and bottom inclined surfaces of the slab of stone. It is necessary to mark on all the moulds a *boning line* which is vertical in elevation and normal in plan.

Fig. 570 shows the developed mould to apply on the helicoidal joint surface if a moulding is required. The method of determining this mould is shown in Fig. 432.

It is often very difficult to set the developed face moulds correctly on the concave and convex surfaces of the stone, especially when the *vertical boning line* is very short compared with the length of the stone.

To assist in their correct application, the developed moulds should have portions of the zinc projecting to the top or bottom edges of the cover mould, so that the moulds may be adjusted to the top or bottom surfaces of the stone, instead of depending entirely upon their adjustment to the boning line. The method for obtaining these projections is shown in the drawing. Measure the vertical height of the arris line above points O^3 and M^3 in Fig. 569, and transfer these measurements to the development of face A (Fig. 566), thus obtaining the full height and position of the cover mould in relation to the face mould in development.

The zinc mould could then be cut as indicated in the drawing. The projection on the other developed face mould may be obtained in a similar manner.

Fig. 571 is a sketch of the stone showing the twisted surfaces.

Setting Out Marble Cut String for Geometrical Stair.—Set out the plan of the staircase, which is rectangular in this example, and plot the plan of the free end of the steps (Fig. 572).

A continuous flight of steps is arranged round a semicircular well-hole, landings being omitted. It is usual to plot the *walking line* 18 in. from the free end of the step, as this is found to be the usual line of ascent or descent. If the curve at the free end of these steps is small, and the winder steps converge to the centre, it causes the ends to be very narrow, also the going of the steps round the curve is very sharp and abrupt—a thing which should be avoided if possible. In the drawing is shown a method of varying the regularity of the steps, but improving the line of the steps at the free end, also easing the abrupt reduction of the first and last winders. The winders gradually diminish in width to the centre curve, and spread out again to pick up the full width of the fliers at the top part of the curve.

To form this arrangement for the steps, divide the curve in plan into a series of divisions, commencing at the centre point C, each division representing the width of the tread, should the steps be drawn normal to the centre O. Mark these points, including a few of the fliers above and below the curve on a stretch-out line, and erect a development of these points (Fig. 573). To do this, draw a vertical height rod and mark upon it a series of divisions representing the height of the risers. Where lines drawn from the height rod intersect the vertical lines drawn from the stretch-out line determines the string in development.

Draw a line touching the nosing points of the fliers at the lower and upper portion of the curve in plan. Where these lines are stopped at the development of the commencement of the curve, as at P P, a line drawn connecting points P P completes the line of going.

It is not advisable to have such an abrupt intersection with the curve portion, so that curves or *easings* are arranged, thus ensuring an easy curve throughout.

The points at the intersection of treads and risers in development should now be projected horizontally to meet these curves. Their intersections will represent the ends of the steps in development, as at points 3 4 5 6 7 C′, etc.

Drop vertical lines from these points to a development line, and step the various points on this line round the plan curve, commencing at the centre riser line C. Mark off the full width of the treads on the *walking line* and draw the outline of the steps in plan.

Next mark in the development the outline of the treads and risers and reduce the height already drawn by the thickness of the treads, thus obtaining the outline for the cut string.

A bird's-mouth joint is usually arranged at the mitre between the riser and string, as shown in Fig. 574. This varies the position of the riser face for the string about $\frac{3}{8}$ of an inch.

Now determine the depth of the string from the true section (Fig. 575).

Mark this depth on the development, on vertical lines. This will complete the developed face moulds for the string, with the exception of the joint lines, which may now be determined.

Mark on plan the thickness of the string. It is best to allow a portion of the straight string to be worked on the curved stone, because of the radiating joints. These joints are arranged at points C and 4 and C and 12, but more joints may be placed if desired. Now obtain the horizontal projection of the normal joints, as at C C^x, 4 4^x, and 12 12^x, by fixing their position in the development and dropping the points thus obtained into the development line, and transferring them to their correct position in plan. Join points C 12^x C^x 4. These lines form the face line of the stones out of which the curved stones are worked, as on line G H.

Now erect vertical projections of these stones viewed at right angles to the face lines, as shown in Figs. 576 and 577.

These elevations are obtained in a manner similar to that described in the previous example, by projecting all the points in plan.

Draw lines representing riser heights at right angles to these projectors. Now measure the vertical depth on the development and transfer it to the projected elevation, thus obtaining the elevation of the bottom edge of the string.

To obtain the developed plane mould, draw a line to clear the bottom points C^x 12^y, also another line parallel to it to clear point 10. Call this line G′ H′. Now rabat the top surface, formed by the line G′ H′, into the plane of the paper about G′ H′. Continue the vertical projectors from plan to intersect G′ H′, and from these points draw perpendicular lines to G′ H′. Measure the distance of the curve at various points in plan from the line G H, and transfer the distances on their corresponding projectors drawn at right angles to G′ H′. Through the points thus obtained draw the curve representing the plane mould for marking on the top and bottom inclined surfaces of the stone. These moulds are adjusted to the boning line marked on the stone from the cover mould. The construction of the steps is shown in Fig. 578, whilst a detail of the tread showing the notching for the string is given in Fig. 579. A geometrical construction for a curtail step is given in Fig. 580.

Setting Out Rib and Panel Vaulting.—The chief items to be considered when setting out this type of vaulting are: (1) the arrangement of the curves; (2) the setting up of the elevation of the several ribs so that the face moulds for the working of the voussoirs may be obtained, and the bevels for working the boss stones; (3) the determination of the bed moulds for the springers.

First decide upon the general form of the vault, and plot the centre line diagram of the ribs in plan (Fig. 581). Add the thickness for the various ribs, making the diagonal and transverse ribs wider than the intermediate and ridge ribs.

Draw in the plan of the boss stones at the intersection of the ribs as shown, and decide upon the position for the springing points of the ribs on the nosing line, as at E on the plan of the diagonal rib.

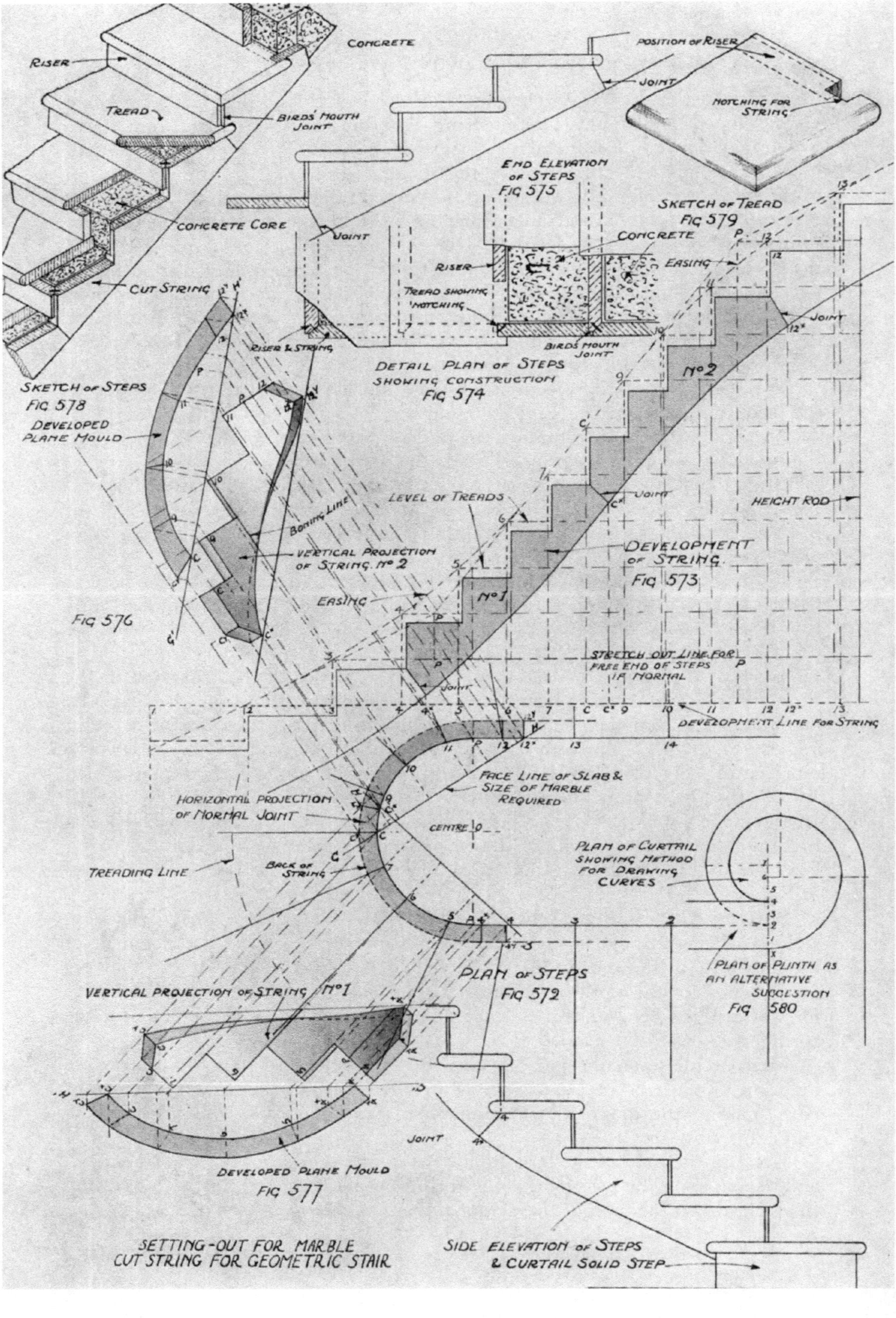

SETTING-OUT FOR MARBLE CUT STRING FOR GEOMETRIC STAIR

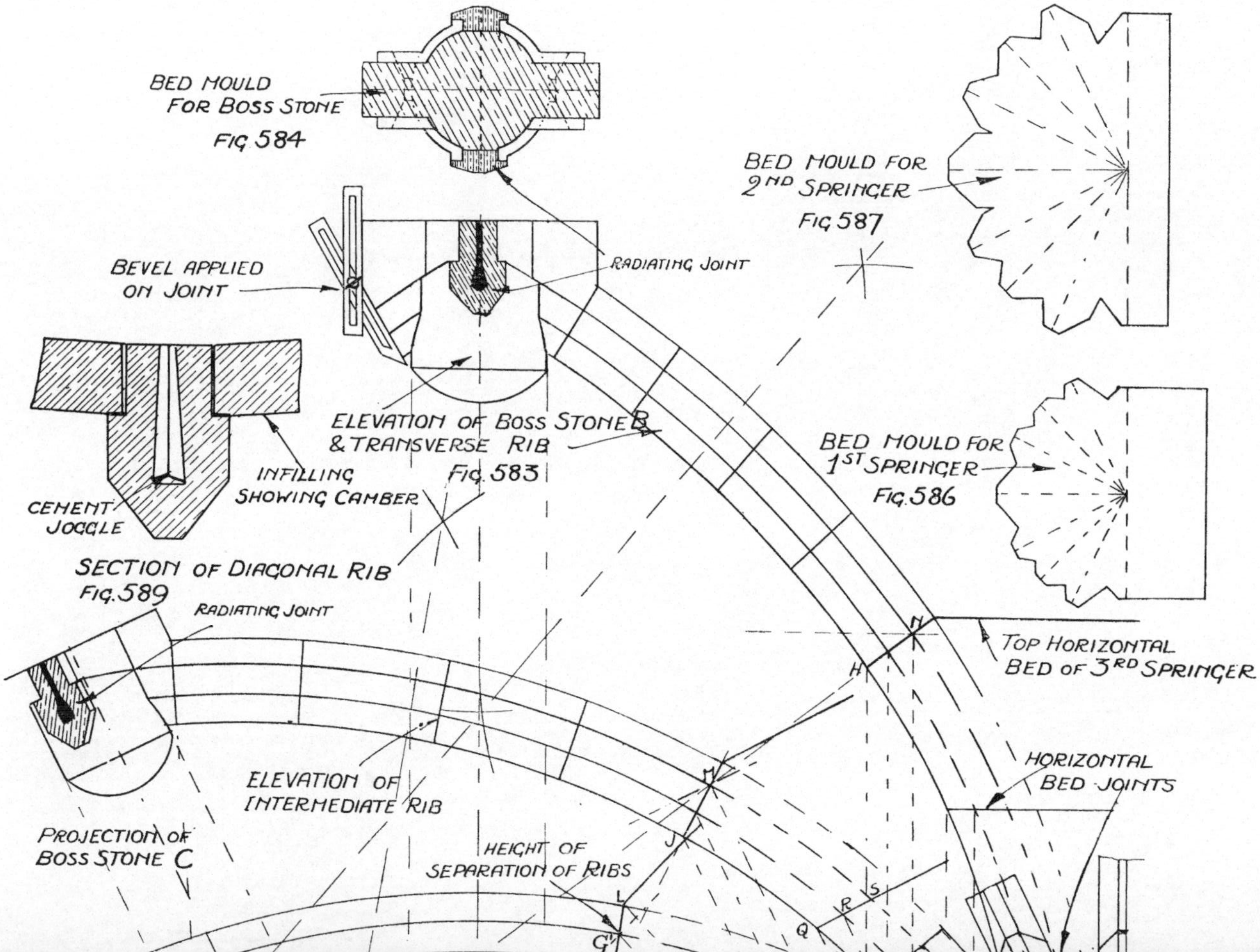

BED MOULD
FOR BOSS STONE
FIG. 584
BED MOULD FOR
2ND SPRINGER
FIG. 587
BEVEL APPLIED
ON JOINT
RADIATING JOINT
ELEVATION OF BOSS STONE B
& TRANSVERSE RIB
FIG. 583
BED MOULD FOR
1ST SPRINGER
FIG. 586
INFILLING
SHOWING CAMBER
CEMENT
JOGGLE
SECTION OF DIAGONAL RIB
FIG. 589
RADIATING JOINT
N
H
TOP HORIZONTAL
BED OF 3RD SPRINGER
M
ELEVATION OF
INTERMEDIATE RIB
HORIZONTAL
BED JOINTS
J
PROJECTION OF
BOSS STONE C
HEIGHT OF
SEPARATION OF RIBS
L
R
S
Q
G'

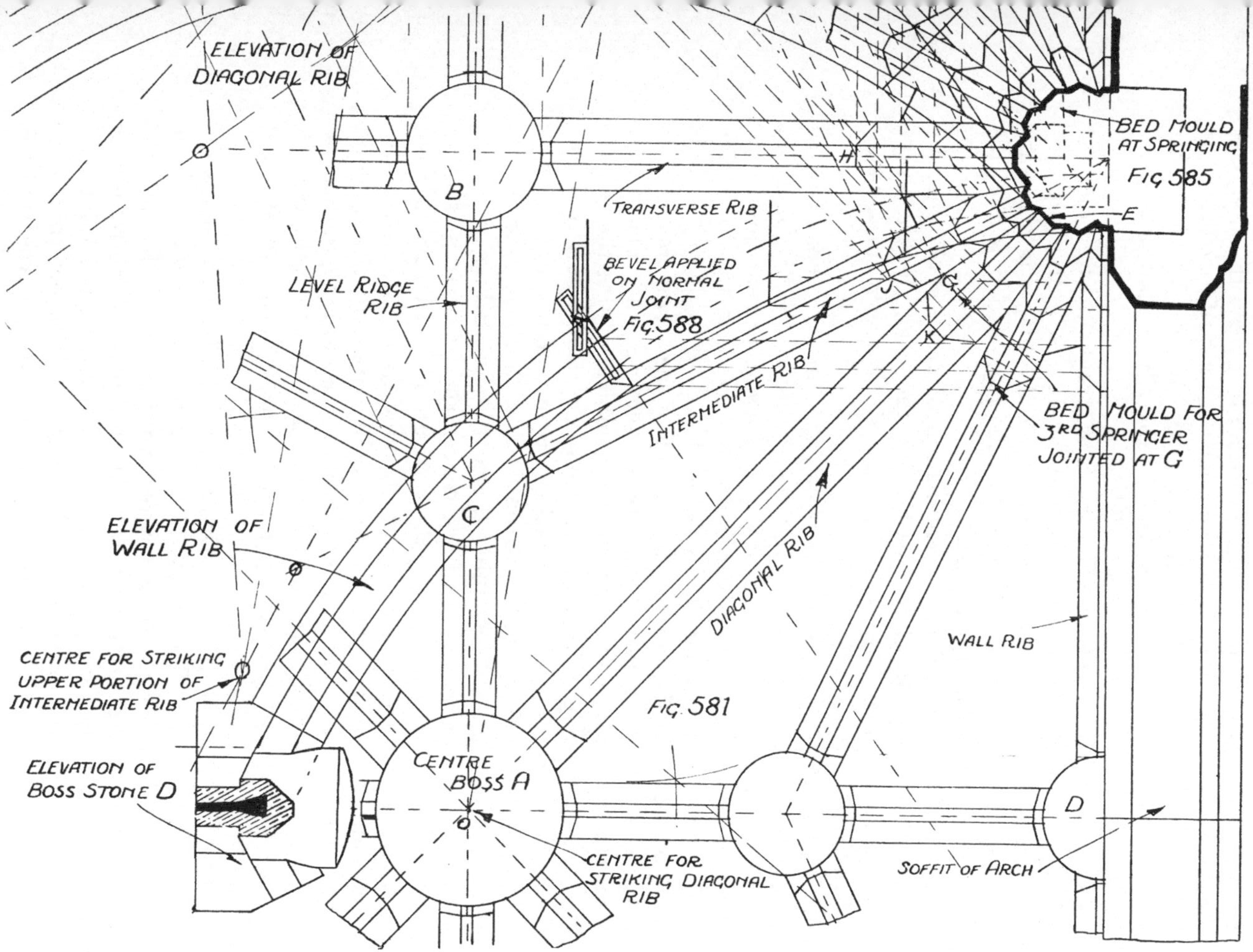

SETTING OUT OF GOTHIC VAULT.

Now determine the elevation of the ribs, commencing with the diagonal rib, this rib having the largest span. In the example given, the diagonal rib has been made semicircular in elevation, so that the centre for striking the curve of the rib is at the centre of the vault in point O. Draw a line from this striking centre perpendicular to the centre line of the rib in plan, and draw the curve from point E.

Where this curve cuts the perpendicular line from point O, determines the height of the vault, if the ridges are horizontal, as in the example given. Next proceed to erect elevations of the other ribs. To do this, it is necessary to decide upon the form required for the curve of the ribs, owing to the ribs being the same height, but of unequal span.

These curves may be composed of a combination of arcs, or obtained by ordinates, thereby producing elliptical curves. In the example, the curve of the ribs is struck from two centres, the radius of the lower curve being common to all the ribs.

Determine in plan the highest point at which the ribs are separated, as at G, and project this point up to point G′ in the elevation of the diagonal rib, on the soffit line of the in-filling. Through G′ draw a normal joint line cutting the nosing of the diagonal rib in point K. Produce the joint line through G′ to the curve representing the extrados of the rib in point L. This determines the top height for the horizontal bed line of the springers.

Now draw lines from the centre of each boss stone perpendicular to the centre line of the rib in plan, and mark on these perpendicular lines the vertical height of the vault, this height being taken from the centre height of the diagonal rib. Produce the centre line of the ribs in plan, and mark off on these lines, from the springing points of the rib nosings, a distance equal to the radius of the diagonal rib, thus determining the centres for striking the lower curve of the ribs in elevation. Now draw concentric curves representing the soffit line of the in-filling, and measure the vertical height of point G′ from the springing line, and transfer this height to the elevation of the other ribs, cutting them in points M and N. Through these points draw the normal joint lines cutting the nosing lines in points J and H, and place in the top horizontal bed line in each case. This determines the elevation of the springers. A diagram showing the method for determining the curvature of the ribs is given in Fig. 582.

Draw the centre line plan of the ribs. Determine the point D, which is the point where the ribs separate. This point in the setting out is taken on the soffit line of the in-filling. With O as centre, swing all the ribs into the vertical plane, which in the diagram is parallel to the plan of the transverse rib. From point C and radius C A, draw the elevation curve of the diagonal rib, to cut a perpendicular line drawn from C in point C^x. This determines the height of the vault. Now draw a horizontal line from C^x, representing the elevation of the horizontal ridge rib.

Erect perpendicular lines from the points E F G H, where the other ribs are rotated into the line O C, to cut the elevation line of the ridge rib in points E′ F′ G′ H′. Draw a normal line from point D through point C, producing it indefinitely beyond C. Now draw lines bisecting the points D E′, D F′,

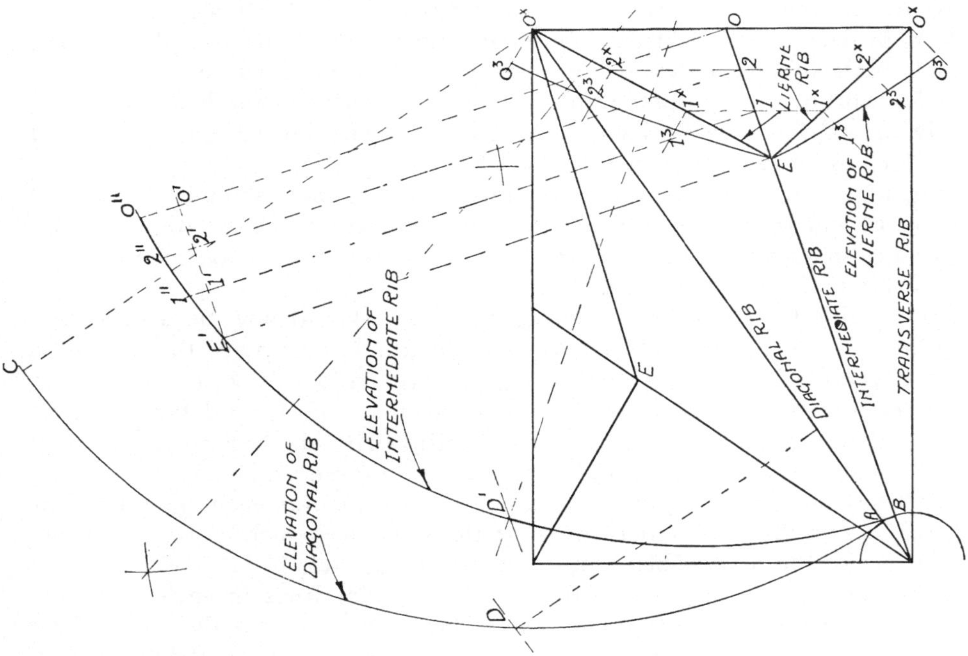

FIG. 590.

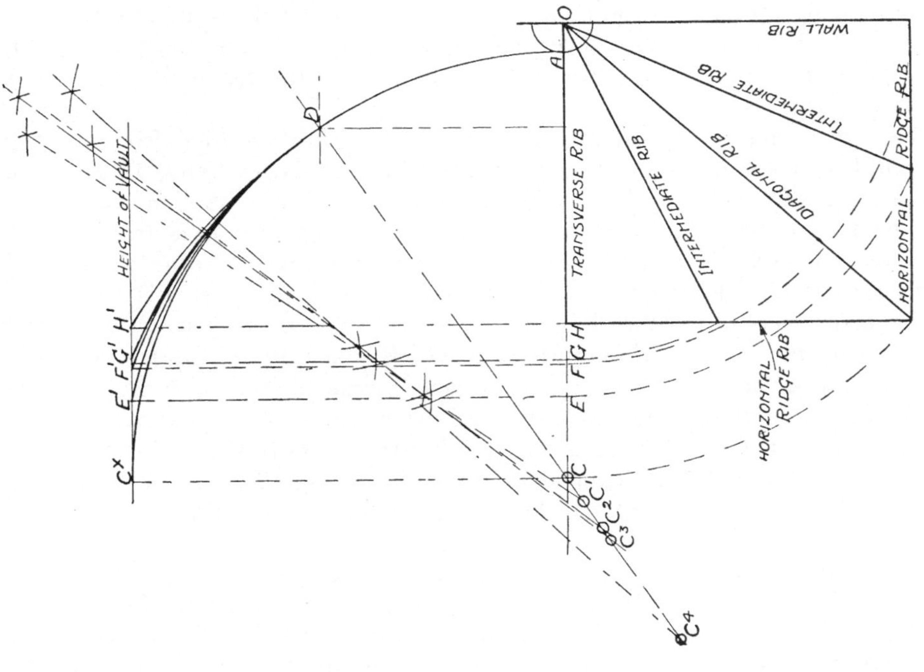

FIG. 582.

DIAGRAMS SHOWING METHOD FOR DETERMINING CURVATURE OF RIBS.

D G′, and D H′, to cut the normal line D C, produced in points C^1 C^2 C^3 C^4. These are the centres for drawing the upper portion of the ribs from point D.

Now project the plan of the boss stones up to elevation, and place in the normal joints in each elevation, as shown in Fig. 583, thus determining the bevel for working the normal joints. The top surfaces of the boss stones are usually arranged to be horizontal in order to facilitate working. The bed mould for boss stone B is shown in Fig. 584.

The ridge ribs being horizontal, radiating or key joints should be arranged so as to prevent the stones from falling out of position. Under these circumstances a raking section mould would be required for marking on the radiating joint surfaces of the ridge stones.

The moulds for working the springing stones should now be determined. These are shown divided into horizontal courses, so that, with the exception of the normal joints on the top springer, between points K and L, raking section moulds will be required for marking on the horizontal bed surfaces. To obtain the outline for these moulds, starting with the first springer, as at Fig. 585, cut a true section of each rib, and apply these to their corresponding nosing lines on the springing, as at point E, adjusting each mould to the centre line of each rib. Where these sections intersect each other determines the outline for the bottom bed mould at springing level.

Now mark in the position of the horizontal joint lines in each elevation, and project to the plan the points where these bed lines cut the lines of the ribs, as at Q and R, thus obtaining their outline in plan as shown. These bed moulds are shown projected out from the drawing in Figs. 586 and 587. The top springing course would be rather large if worked in one stone, therefore it would be best to place a vertical normal joint through point G in plan, as suggested in the drawing (Fig. 581), thus making the top springing course complete in three stones.

The top normal joints should be worked to a bevel applied from the top horizontal surface, the correct angle for setting the bevel being taken from the elevation of the ribs. As all the ribs are struck with the same radius up to point K, and the height of this point from the springing is the same in each case, the bevel which is shown applied on the normal joint of the wall rib, in Fig. 588, will be the same for the other normal joints at this level.

A detailed section of the diagonal rib is given in Fig. 589. This shows the rebate for the support of the in-filling, which is usually laid in thin courses. A slight camber should be arranged on each course, thereby forming a series of flat arches, so that when the joints of these stones are filled with mortar from the back surface it is impossible for the stones to drop or fall out.

The bed lines of the in-filling courses can be arranged inclined or at right angles to the main ribs in plan. The material for the in-filling should be as light as possible.

A diagram showing the method of obtaining the curves for lierne ribs is given in Fig. 590. Set out the centre line plan of the ribs and erect an elevation of the diagonal rib, thereby obtaining the height of the vault. Next determine the height of point D, where the ribs separate, and erect the elevation of the *intermediate rib* by drawing the lower curve of the rib up to the height D′ with the same radius as the *diagonal rib*, the centre for the upper portion of the curve being determined as already explained.

From point E on this rib, which is the junction of the *lierne ribs* in plan, erect a perpendicular to cut the curve line of the *intermediate rib* in point E′. From E′ draw a horizontal line to cut the perpendicular line from O in point

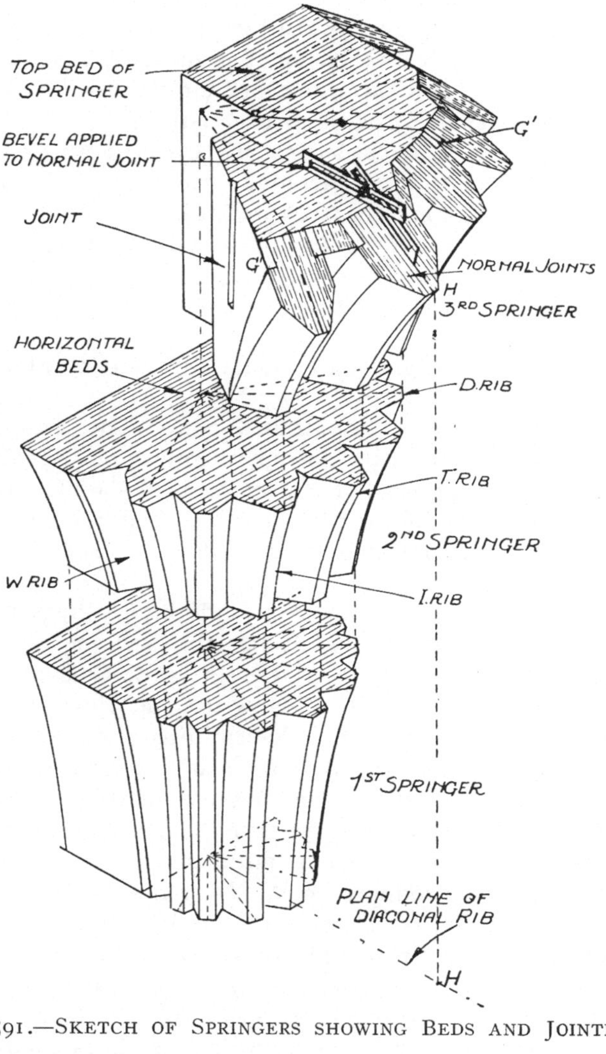

FIG. 591.—SKETCH OF SPRINGERS SHOWING BEDS AND JOINTING.

O′. Mark off on the line E′ O′ any number of points, as at 1′ 2′, and project these points to the plan of the *intermediate rib* in points 1 and 2. From these points draw lines parallel to the ridge rib O^x O^x, cutting the plan of the *lierne ribs* in points 1^x 2^x. From these points draw lines perpendicular to the plan of the *lierne ribs*, and mark on these lines the lengths of the ordinates O′, O″, 2′, 2″, 1′, 1″, as shown in points 1^3 2^3 O^3. A fair curve drawn through these points determines the curve for the *lierne ribs*.

A sketch of the springer stones showing the horizontal beds and jointing for the top springer is given in Fig. 591.

Setting Out Helicoidal Skew Arch.—This work would not be complete without an example of the setting out of a skew arch. It is not the author's intention to elaborate upon the mechanics involved, as numerous volumes have been published dealing almost exclusively with the subject. The example given is intended as an application of various geometrical principles which may be studied to advantage, but it is unlikely that the student would be called upon to carry out a piece of work similar to that shown, owing to the extensive use of concrete for work of this nature.

In the example given, the arch is constructed with a series of spiral courses, the beds and joints of which are helicoidal surfaces.

The joints extending along the arch are called *coursing joints*, and those across the arch *heading joints*, the plans of these lines being known as *coursing* and *heading spirals*.

Draw the plan of the abutments at C A B D, and place in the plan of the face lines (Fig. 592).

Now draw a section through the arch made by a cutting plane perpendicular to the axis, as in Fig. 593.

Divide the soffit line in section into a convenient number of divisions, and draw normal lines to the extrados curve in section. Project these points to the plan, and step the points A 1 2 3, etc., on the soffit, on to a stretch-out line on the right-hand side of the plan, and the points on the extrados on to a stretch-out line on the left-hand side. Erect perpendicular lines from all the points in the stretch-out lines.

Now use the plan line C C as an axis, and unfold the extrados surface in plan on the plane of the paper. In doing this, all the points in plan will travel along paths at right angles to the line C C, thus determining points D and D in development (Fig. 594). Now use the plan line B B as an axis, and unfold the soffit of the arch, thus determining the points A A in development (Fig. 595). Join points B and A, then the position of the *heading spiral* in development is determined. From point B in this development draw a line perpendicular to the *heading spiral* line B A, and produce it to meet the axis line in development, produced in point X. From X draw a horizontal line to meet the line A A produced in K. Now draw the development of the *coursing spiral* K K parallel to the line B X, and complete the *coursing spiral* lines for the other courses, from the points where the vertical lines, drawn from the stretch-out line, cut the *heading spiral* lines, as shown in Fig. 595.

To obtain the plan of these *coursing spirals*, divide the axis of the cylinder between points K and K^x into the same number of equal divisions as the section, and draw horizontal lines into the plan, to intersect their corresponding lines, projected into plan, from the section.

Through these intersections draw the curves as shown, as at A B C D. The surfaces between these lines, which cross on the centre line, are helicoidal surfaces.

To obtain the development of the oblique faces of the arches, project the points of intersection of the projectors, from the section, across to meet their corresponding *coursing spirals* in development, thus obtaining the curved lines in development, between points B and A.

Now place in the position of the *heading joints* in the intrados development as shown, thus dividing the courses into suitable-sized stones, and obtain their position in the plan by projecting the intersection of these *heading joints* with the vertical lines from the stretch-out line, to meet the lines in plan, drawn from the intrados in section as shown.

Now complete the development of the extrados in a manner similar to that described for the intrados.

First draw the development of the *coursing spiral* K K, and also the development lines for the *heading joints*, as at Q R. Where the vertical lines drawn from the divisions on the stretch-out line cut the line Q R determines the position for drawing the *coursing joints* parallel to the *coursing spiral* K K. Now obtain the development curves for the face lines of the arches between D C and D C, in a manner similar to that described for the intrados.

Next erect an elevation of the arch, viewed at right angles to the face, as in Fig. 596, by projecting the points, which are determined on the face line in plan, where the plans of the *coursing spirals* cut the face line in points S T P P M M, etc. Draw the elliptical curve representing the soffit in elevation, also the curve of the extrados. Where the points projected from S T P P M M on the face line in plan cut the elevation curves determines the points through which to draw the arch-joint lines in elevation. These lines are actually curves, because the face of the arch cuts the helical surfaces other than at right angles to the axis of the cylinder. In the drawing these joint lines are assumed as straight lines between the points. If the lines through these points are produced to the centre, they will meet at point G. The position for this point can be determined as shown in the diagram (Fig. 597).

From the centre O draw a line O F, making an angle with A O equal to the angle of obliquity θ, to cut a vertical line drawn from A. From A also draw a line A G, making an angle equal to the angle of the extrados ϕ. Draw a horizontal line from F, to cut this line in G, then F G will be the distance that point G will be below the springing line in elevation, and to which all the lines will converge, if drawn through the points of the arch joint in elevation, as in Fig. 596.

To Obtain the Moulds Necessary for Working one of these Stones.—The amount of twist in the surfaces of the stone is obtained from plan, and is determined by the plan of the *coursing spirals*, but in the drawing the stone selected has been redrawn for clearness in Fig. 598.

This stone may be worked from a cylindrical-shaped stone by projecting its extreme points down to the section (Fig. 597). The distance this stone travels round the cylinder is shown dotted in the section, whilst the length of the stone required is shown in dotted lines (Fig. 598). A mould cut to the portion of the shaded section within the dotted lines would be required for application to the stone, so that the cylindrical surfaces could be worked. The *boning line* should be marked on the cylindrical surfaces and the developed intrados and extrados moulds applied to these surfaces and adjusted to the boning line.

The following is given as an alternative method for working these stones :—

Draw a rectangle to clear the extreme points of the stone in plan, as shown in Fig. 598. Draw vertical projectors from the points 4 5 6 7, etc., in section (Fig. 597) to meet the top arris line of the rectangle X X in points 4 5 6 7, etc., and from these points draw lines perpendicular to X X. Now measure the vertical height of the points 4 4 5 5, etc., in section, from the springing line A B, and transfer these distances to their corresponding lines drawn perpendicular to line X X. This will give the outline of the section made by the cutting plane X X passing obliquely across the cylinder.

Now project the points O P N and M vertically, to cut the arris line X X. From these points in the line X X draw perpendicular lines to cut the curved section lines in O O and M M. Join O O and M M. These lines determine the extent of the plane mould to apply on the surface. Mark the *boning line* on the mould, as shown in Fig. 599.

To assist in placing these moulds in their correct position, small pieces of zinc may be left on the plane mould as shown, so that, when applying this mould to the top and bottom surfaces, the mould can be placed so that these projecting pieces of zinc are *flush* with the front face of the stone. This ensures that the moulds are out of twist.

Fig. 600 shows the projection of one of the stones taken from elevation.

Fig. 601 is a geometrical diagram illustrating the development and plans of the *heading* and *coursing spirals*.

The arch is shown apart from the abutments. In practice the skewbacks would be attached on a solid course below, such as a string-course, or impost cap, in order to avoid weakness at the springing, where the converging spirals meet the horizontal line.

BOOKS ON THE SUBJECT

Bates, E. L., and Charlesworth, F., "Practical Geometry and Graphics."
De la Rue, J. B., " Traité de la Coupe des Pierres " (1764).
Drury, F. E., " Geometry of Building Construction."
Dowsett, J. F., " Advanced Constructive Geometry."
Leroy, " Traité de Stéréotomie."
Morduit, Louis, " Traité Théorique et Pratique de la Stéréotomie."
Mowat, W. and A., " Treatise of Stair Building and Hand Railing."

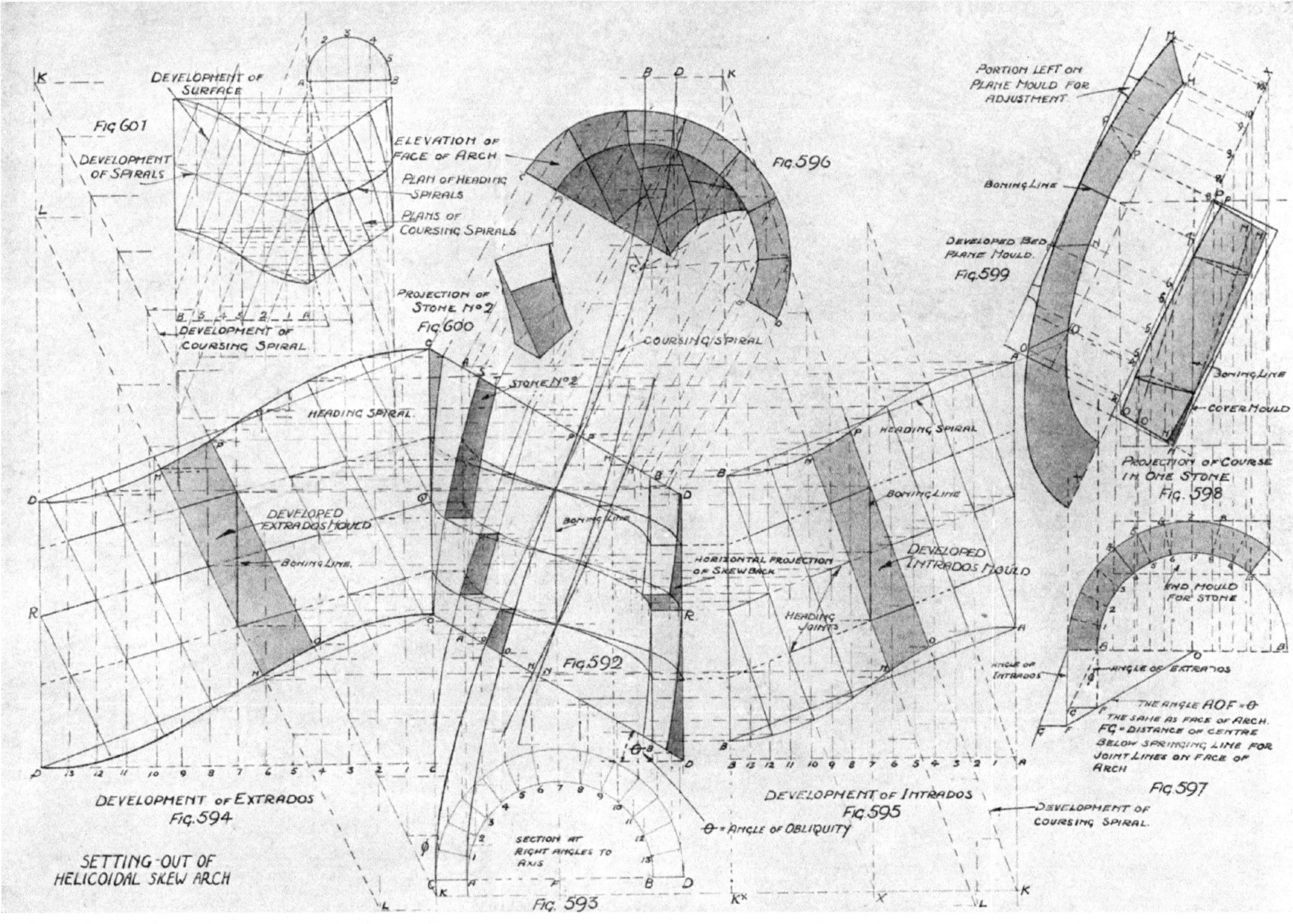

SETTING-OUT OF HELICOIDAL SKEW ARCH

SECTION III

COSTING AND ESTIMATING

CHAPTER I

MASONRY COSTING AND ESTIMATING

Outline of Costing System—Estimating—Estimate for Portland Stone Doorway—Estimate for Marble Wall Lining for Bathroom—Estimate for Marble Work to Conical Head over Window Opening.

IN considering the problems of costing and estimating, it is necessary that the student should be guided by a sound knowledge of the material under consideration, and of the processes or operations which are necessary to produce the finished work. He should also be equipped with a knowledge of the principles of good masonry and building construction, in order that he may be able not only to estimate the amount of material and labour involved, but also to guard against *vagueness*, *incomplete* or *inaccurate* information in "*drawings*," "*specifications*," or "*bills of quantities*" provided for the purpose of estimating.

The student must inform himself of the uses and capacities of machinery used in production, as well as of the hand labour still necessary.

This can be accomplished only by close observation and a careful tabulation of the results observed, to form a basis of the system known as *costing*. *Costing* has a twofold object, namely, the determination of the cost of any particular form of work, so as to provide the data for estimating the time and cost of proposed similar work; and also as a means of maintaining or improving the standard of efficiency in production, especially with regard to output. Without *accurate costing*, *good estimating* is impossible, so that its value needs no emphasis. Its adoption is absolutely essential, and as a means of comparing methods, or of comparing one type of machine with another, an indication is given in the direction of possible improvement.

There are many systems of costing, each doubtless having its own special merit, but the purpose of the writer is not to recommend any particular one, but to encourage the student to use his own powers of observation and to make a practice of recording, in a systematic way, information relating to this side of masonry production.

Owing to the varied character of masonry and the widely differing nature of the stones used, masonry firms have found it expedient to specialise in certain classes of stone only. We find that the general trade is divided up into three principal sections, namely, *stone*, *marble*, *granite work*. Although these branches have many methods of working in common, on the whole a different system of production and class of machinery has been developed or evolved in each section. We will consider them in the order given.

Stonework.—For the purpose of this section, stone may be defined as a sedimentary rock which is comparatively easy to work, the employment of which is confined mainly to the structural requirements of buildings. This includes the usual dressings ornamented by carving or otherwise, and comprises sandstones, and such limestones, as are not capable of receiving a high polish. There are many limestones approximating to marble which are partly crystalline and are capable of receiving a *fairly good* polish. When these stones are worked by *marble masons' methods*, they may, for the purpose of estimating, be considered as *marble*, but when used as ordinary building stones and dressed by the stone mason, they should be classified as *stone*.

The cheapest form of masonry is known as " *Rubble Walling*."

The stone for this class of work is taken from the rubble created in quarrying operations, this being considered of low value owing to the quantity produced when obtaining stones of larger dimensions and of better quality suitable for dressed stonework.

Stone walling of the rubble type is a branch of masonry executed by craftsmen who specialise in this class of work who are called *walling masons*. Owing to its simple and repetitive nature, the *costing* and *estimating* of rubble masonry presents little difficulty, so that only a passing reference to it need be made.

The rubble is usually sold at the quarries by the cubic yard, or by weight, according to local custom. The finished work is measured by the cubic yard, whilst the face of the work is measured separately, and priced at per foot super, according to the class of work.

The quoin stones are usually priced at per foot in height. Such work as tooled reveals and margins would be considered as labours coming under the heading of dressed stonework.

COSTING

Dressed Stonework.—Under this heading comes a large variety of work, ranging from the plainest to the most ornate forms in masonry. Since, as previously stated, *good estimating* can only be based on *correct costing*, we will first consider a costing system. It is to be presumed that the student possesses a good knowledge of the methods employed and the machinery used in present-day masonry works. In order to have some common ground of understanding, let us imagine the running of a medium-size masonry works, employing about twenty masons, with incidental labour, engaged in work of a varying character.

We will assume we are considering a bank front in Portland stone, whilst the equipment of the works will be assumed as follows :—

EQUIPMENT

Power	Electricity.
	No.
Machinery (stone-working)	3 Frame saws. 1 Diamond saw. 1 Carborundum jointer saw. 2 Moulding machines.
Lifting machinery	2 Travelling cranes on gantry.

Let us further assume that, to supervise and run these imaginary works, the following are also required :—

SUPERVISION.

General foreman or works manager.
Setter out, or draughtsman.
Time and costing clerk.
Office boy.
Shop foreman.
Head sawyer.
Fitter engineer.

MACHINISTS AND OTHER LABOUR.

3 Sawyers (frame).
1 Sawyer (diamond).
1 Sawyer (carborundum).
2 Machinists.
2 Cranemen (electric).
8 Labourers.

The distribution of the eight labourers could be as follows :—

2, Slingers for handling blocks and stones in course of process of manufacture.
2, Assisting machines, loading, etc.
3, Rubbing and general shop work.
1, General yard work.

As will be seen, the staff suggested comprises only those actually engaged on the productive side of the business, the financial, sales, and accountancy side (other than costing) being outside the scope of this section.

We will now consider the means of collecting information of manufacturing costs, upon which our estimating is to be based. To measure output, it is necessary to have some standard of measurement. Experience and custom have determined that the unit of measure should be one superficial or linear foot as the case may be. To record output, a form of cost sheet of such a simple character that no difficulty presents itself to the workman must be devised, since it is he who has to record the details of output from the machine under his control.

Frame Saws.—These are used for the conversion of large rough blocks into stones of sizes suitable for the work in hand.

Cost Sheet No. 1, page 236, is suggested as being suitable for this class of work, and would be a convenient medium for ascertaining output, the unit basis of cost, and a comparison of output with other machines.

The information collected on these sheets could be tabulated, and the average cost over a suitable period obtained, so as to form the basis of estimating. It will be noticed that the term "*effective sawing*" is used on the cost sheet. The need for this will be apparent to the practical mind, since the saw cuts on the outside of the blocks are often effective for one surface only.

Diamond and Carborundum Jointer Saws.—The use to which these saws may be put depends largely upon the type of work in hand; therefore, the cost sheet for recording output must be designed to cover possible variations. Usually the cost of the work of these machines can be directly allocated to a particular stone or order number, and, in such cases, this should be done. Further reference to this will be made when considering the collection of details to form the total costs. As compared with the *frame saw*, the entries for these machines will be more numerous; therefore, a larger size or greater number of sheets will be required over a given period.

COST SHEET No. 1

Frame Sawing

Machine No......... *Sawyer's Name and No...* *Week ending..............*

Date and Time.		No. of Hours.	Size of Blocks.			No. of Cuts Single Sawing.	Particulars of Effective Surface.				Kind of Stone, and Remarks, including No. or Identity Mark of Block.
Commenced.	Finished.		L.	W.	H.		No.	L.	W.	Super.	
											Space as desired. *Observations*— Records of break-down should be made in the remarks column.
Total hours .						Total feet super .					

Costing data arranged below could be kept separate.

For Office Use Only

	£	s.	d.
Power Charges . . @			
Cranage Charges . . @			
Stores Used			
Repairs and Renewals . .			
Insurance per cent. . . .			
Depreciation per cent. . .			
* Add Wages			
Total .			
Unit of Total Cost (per foot super)			

(Exclusive of Establishment Charges.)

Summary of Wages

	£	s.	d.
Sawyer (Feeder) . Hours @			
Sawyer (Foreman) . ,, @			
Engineer . . . ,, @			
Direct Labour Cost . . .			
Supervision per cent. . .			
*			
Unit of Labour Cost (per foot super)			

COST SHEET No. 2

Diamond or Carborundum Sawing

Machine No......... *Sawyer's Name and No..* *Week ending..............*

Date.	No. of Hours.	Order No.	Stone No.	Number and Sizes of Cuts.				Effective Surface Sawn.				Description and Remarks.
				No.	L.	W.	Super.	No.	L.	W.	Super.	
						←Space as required→						Full particulars of special work should be entered in this column (notchings, etc.).
Total hours		Total feet super. .						Total feet super				

For Office Use Only

	£	s.	d.
Power Charges . Hours @			
Cranage Charges . ,, @			
Stores Used			
Repairs and Renewals per cent. .			
Insurance per cent. . . .			
Depreciation per cent. . .			
* Add Wages			
Total .			
Unit of Total Cost (per foot super)			

(Exclusive of Establishment Charges.)

Summary of Wages

	£	s.	d.
Machinist . . . Hours @			
Machinist Assistant . ,, @			
Engineer . . . ,, @			
Direct Labour Cost . . .			
Supervision per cent. . .			
*			
Unit of Labour Cost (per foot super)			

Abstract of Costs

Order No.			
,, ...			
,, ...			

Cost Sheet No. 2 is a suggestion for general use. Reference to this sheet will show that columns have been provided to ascertain output, and to allocate costs direct to a particular piece of work, or order number.

Planing Machines.—The general purpose of these machines is the execution of plane, sunk, or moulded surfaces, the form of which may vary considerably upon stones which are usually but not necessarily sawn to size.

The practice of quantity surveyors is to measure the girth of mouldings or sinkings; therefore, the unit of measure should be based on this and the cost sheet designed accordingly.

Moulded work in masonry is seldom produced for stock; therefore, the output of the moulding machine can be allocated direct to a particular job, and usually to an individual piece of work.

The object of costing for this machine will be threefold:—

1. To find the whole cost of machine-moulded work for a given job or individual piece of work.

2. To discover the capacity of a particular machine.

3. To provide data as a basis for estimating similar work.

In Cost Sheet No. 3, page 239, arrangement has been made for the dissection of the various types of work executed on these machines.

Up to this point three different types of machines in general use have been considered, and although machines for other purposes, such as lathes, etc., are items of modern equipment, sufficient has been said to guide the student in recording output and ascertaining costs of work performed by any machine.

Banker Work.—As a rule, the use of a machine is limited to one kind of operation only. Thus, the *sawing machines* divide up the blocks, *moulding machines* plane out the mouldings, etc. The work of the banker mason, however, is of a varying nature; therefore, to make a record of his output must be more difficult than to do so for a machine designed for a single purpose.

It is the duty of the setting-out department to supply all working information for the purpose of building up a system of recording costs, not only the masons' work in particular, but of the collective work of the whole organisation. The draughtsman must supply complete details to the shop foreman of the work to be executed, or to those concerned. This should be done in such a manner that the time a mason has taken in the execution of a given job can be entered against the draughtsman's particulars, thus intimately associating the necessary work with the time taken.

For this purpose a *setting-out book* should be designed, having the following objects in view:—

1. To supply, in conjunction with the necessary moulds, working information for the foreman.

2. To afford particulars of the work performed and time involved for the costing staff.

3. To give a record for the draughtsman's own use and information.

Therefore, the book should be of a manifold type, having sheets in triplicate to the same folio.

Two of these should be detachable for passing on to the departments concerned, whilst the third is retained in the setting-out book.

COST SHEET No. 3

Planing and Moulding

Machine No....... *Machinist's Name and No..........................* *Week ending..............*

Date.	No. of Hours.	Order No.	Stone No.	State of Stone.	Sizes of Stone.				Dimensions of Work Executed.						Remarks and Sketches.
					No.	L.	W.	H.	No.	L.	W.	Plane.	Moulded.	Sunk or Notched.	
				←Space as required.→											
Total hours		Total of super feet .													

For Office Use Only

	£	s.	d.
Power Charges . @			
Cranage Charges @			
Stores Used . .			
Repairs and Renewals			
Insurance per cent. .			
Depreciation per cent.			
Total * .			

* This amount must be proportioned over the several classes of labour to ascertain the separate units of total cost.

Abstract of Wages

	Plane.			Moulded.			Sunk or Notched.		
	£	s.	d.	£	s.	d.	£	s.	d.
Machinist . . Hours @									
Machinist Assistant ,, @									
Engineer . . ,, @									
Direct Labour Costs . .									
Supervision per cent. . .									
Totals .									
Unit of Labour Cost (per foot)									

Allocation of Costs

	£	s.	d.
Order No. . . .			
,, . . .			
,, . . .			

Careful consideration should be given to the arrangement of the columns for collecting information, and a column should be devoted to every separate class of workman or machine employed in the production of the work, as well as to the incidental costs, such as supervision, power, stores, etc. Ample space should be allowed for the insertion of working information, including sketches.

When the work set out upon the sheet has been executed, the sheet should be passed to the costing department by the shop foreman, the latter being held responsible for the entries made in the time column.

A study of the example given will enable the student to see how the costs can be collected from the various sheets, and entered on the setting-out sheet. A complete record of the cost of an individual operation or a particular stone, or the total cost of a complete order, together with detailed information of the work and the amount of stone involved, is thus made.

Apart from the *cost sheet* already explained, *time sheets* should be made out weekly by all workmen, stating the nature of the work performed, and, in the case of the mason, the stone and order numbers.

These sheets are necessary for the computation of wages, and are also a check against the time book kept by the time clerk. The sheets are further useful as a partial check against the entries made in the setting-out sheet, and in the case of outside craftsmen, such as fixers, they form the principal means of ascertaining the cost and nature of the work performed. They are also useful for recording "extras over" a contract, in which case the time sheet should be endorsed by an authorised person.

The example shown on page 241 is termed a *mason's time sheet*, but the same type could be used for all workmen, in which case the heading would be "*General Time Sheet*."

Oncosts.—The previous remarks have been largely confined to recording output of machines and men engaged in operations which may be termed direct contributions to the actual production of masonry work.

Without the co-operation of other machinery and incidental labour, the work of these machines would be more or less ineffective. *Oncosts* may be defined as "*indirect costs*."

Under this heading would come the following: (1) Power; (2) lifting and conveying machinery; (3) supervision; (4) assistant labour to machines (other than operators); (5) general labourers in yard; (6) engine fitters; (7) blacksmith; (8) stores and renewals; (9) insurance.

As all these are necessary to production, the oncosts must be added to, or distributed over, those operations or direct labour in such a way as would fairly represent the costs which are necessary to supplement the work of the various machines and masons, so as to render their work effective.

The banker work would be subject to the following oncosts:—

(1) Shop foreman to direct; (2) labourers to clear away waste material and assist in the handling of the stones.

A proportionate charge would also be involved by the following: (1) Works manager; (2) draughtsman; (3) blacksmith; (4) crane; (5) office staff; (6) stores; (7) insurance.

MASON'S TIME SHEET

Name and No.. *Week ending*.......................

DAY.	Order No.	Stone No.	Description of Stone and Work to be Executed.	Labours Already Executed on Stone when Bankered.	Number of Hours on Job.	Hours Worked During Day.
FRIDAY						
SATURDAY						
MONDAY						
TUESDAY						
WEDNESDAY						
THURSDAY						

Total hours........................

Signature of Workman..

SPECI MEN OF SHEE

FOLIO NO. ______

PERFORATIONS FOR FILING.

SETTING OUT.

1	2	3	4		5	6		7			8	
Mason's Name.	Order No.	Mark.	Date and Time.		Particulars and Sketches.	Cubical Contents		Stone.			Frame Sawing	
			Com.	Fin.		Ft.	In.	£	s.	d.	£	s.

—Extent of O

←—Extent of Setter Out's and Shop Foreman's Sheet—→

OBSERVATIONS
Column 7 (S
,, 8, 9
,, 11 (
,, 12 (
,, 13 (
,, 14 (

Perforation line—→

↑ Perforation line

FOLIO NO.

COST COLLECTION.

9			10			11			12			13			14			15			16
Diamond and Carb. Saw.			Moulding Machine.			Mason.			Mason's Oncost.			Total.			General Oncost.			Total.			Remarks.
£	s.	d.	£	s.	d.	£	s.	d.	£	s.	d.	£	s.	d.	£	s.	d	£	s.	d.	
heet——→																					

—Should cost stacked in yard ready for use and including allowance for waste.
10.—Amounts obtained from the special Cost Sheets, which already provide for certain oncosts.
n).—Wages, with the addition of Health and Unemployment Insurance.
ns' Oncosts).—Should include cranage charge, shop foreman, and incidental labourers' wages, masons' stores and proportion of supervision charges.
).—Should be the total of columns 7 to 12.
al Oncosts).—Should provide for rent, rates, taxes, water, lighting, accident insurance, and office expenses (if any).

Now revert to the items under the heading of the oncosts, and apply them in the order given to the costing system of the imaginary works.

Power.—By this is meant the motive force used in the driving of the machines, which in this case is electricity. Each machine will require a certain amount of horse power for its work. The total amount of horse power will be determined by the number and requirements of the machines proposed to be used.

For illustration purposes we will assume the requirements of the machines mentioned on page 234 to be as follows :—

	No.	
Saw frames	3 at 8 H.P.	=24 H.P.
Diamond saw	1 ,, 8 H.P.	= 8 H.P.
Carborundum saw	1 ,, 3 H.P.	= 3 H.P.
Planing and moulding machines . . .	2 ,, 6 H.P.	=12 H.P.
Crane No. 1	1 ,, 21 H.P.	=21 H.P.
Crane No. 2	1 ,, 16 H.P.	=16 H.P.
The assumed total horse power is thus . . .		84 H.P.

Taking a short period of a week as a basis for computing the power cost, let us assume the following :—

	£ s. d.
Units of power used : at per unit . . .	
Proportion of engineer's time : add hours at per hour	
Allowance per cent. for depreciation of electric motors : based on value, divided by 50 weeks	
Proportion of supervision (based on engineer's wage) per cent.	
Stores	
Insurance against breakdown : per cent. of value of motor, divided by 50 weeks	
Total power cost .	£

This figure, divided by the total horse power, equals the cost per horse power. From this we are able to assess the power cost of an individual machine. Applying this figure to the cost of running a frame saw for power over a period of one week, we have 8 H.P. at per H.P., which, reduced to an hourly basis of 44 hours per week, would be divided by 44.

Having determined the power cost of the frame saw, let us proceed to find the unit cost for sawing by the frame saw. From the sawing cost sheet, page 236, we are able to obtain the amount of effective surface sawn in a given week. The costs to be placed against this figure would be roughly as follows :—

	£ s. d.
1. Sawyer's wages (feeder)	
2. Proportion of foreman sawyer's wage	
3. Proportion of engineer's wage	
4. Stores used (sand, sawplates, oil, etc.) (based on consumption)	
5. Supervision (based on direct wages)	
6. Power cost, 8 H.P., at	
7. Insurance against breakdown per cent. . . .	
8. Proportion of cranage cost (for loading and unloading) .	
9. Allowance for depreciation of machine per cent. . .	
10. Allowance for renewal and repairs per cent. . . .	
Total cost .	£

The figure thus obtained, divided by the total feet super sawn, would equal the cost per foot super.

In connection with sawing there is one important point which must not be overlooked, viz., wastage. Experienced masons know that it is impossible to utilise every part of the surface sawn, however economically used, so that an allowance must be made for the effective handling of the stones. This constitutes what is known as *waste sawing*, and, when estimating, this must be allowed for.

Lifting and Conveying Machinery (*Gantry Cranes*).—This item is an oncost which should be apportioned to machines depending upon its co-operation. To arrange this equitably, let us review the machines which it is called upon to serve. They are as follows:—

No. 3 Frame saws.
,, 1 Diamond saw.
,, 1 Carborundum saw.
,, 2 Planing machines and masons' shop.

It will be appropriate here to point out that it is usual for small hand cranes to be arranged in connection with certain machines where rapid feeding, or loading up, is necessary to supplement the work of the gantry crane. The cost of working these would be charged direct to the machines in question. Also, with regard to the masons' shop, a small supplementary system of overhead running tackle for handling the banker work is often used. This would be hand-worked machinery, the cost of which should be borne by the masons' work. To decide upon a fair distribution of the gantry cranes oncost, a careful observation and recording over a suitable period would have to be made of the calls by the various machines served upon the services of the gantry crane. This would have to be done on a time basis, for although some machines would require service more often than others, yet the time taken in handling heavy blocks is usually greater than that for less heavy stones. It will answer the purpose we have in view if we assess roughly the cranage requirements of the various departments. Let us assume them as follows :—

Frame saws: 3, at 6 hours per week each (including the stacking of the rough blocks)	18 hours.
Diamond saw: 1, at 4 hours per week	4 ,,
Carborundum saw: 1, at 4 hours per week	4 ,,
Planing machines: 2, at 3 hours per week each	6 ,,
Masons' shop, including loading for transport	12 ,,
Total	44 hours.

Having decided the proportion of charges to be made to the various departments, work out the total running cost of the two gantry cranes. This would be made up as follows :—

	£ s. d.
1. Drivers' wages: 2, at	
2. Labourers' wages: 2 (slinging), at	
3. Proportion of engineer's wage	
4. Supervision (based on wages)	
5. Power charges: assume 37 H.P.	
6. Insurance against breakdown: per cent. of value, divided by 50 weeks	
7. Depreciation of cranes: per cent. per annum, divided by 50 weeks	
8. Repairs and renewals	
9. Stores used	
Total charges	£

This amount, on the basis of proportionate charges assumed, would be divided out as follows :—

No.	£ s. d.
To 3 frame saws, $\frac{18}{44}$ of the cost determined	
,, 1 diamond saw, $\frac{4}{44}$ of the cost determined	
,, 1 carborundum saw, $\frac{4}{44}$ of the cost determined	
,, 2 planing machines, $\frac{6}{44}$ of the cost determined	
,, masons' shop, $\frac{12}{44}$ of the cost determined	
Total	£

We have now disposed of two of the chief items of oncost.

Great care must be taken in accounting for all the expenditure and in the correct allocation of costs, so that, when use is made of the costing system for estimating purposes, nothing has been overlooked which would lead to a loss when the work estimated for is produced. The proportionate distribution of the remaining oncosts must be made with this in mind.

Supervision, including salaries of works manager and office staffs, should be a charge upon the wages of all men who are under their control or direction, and for the purposes of costing, wages should be considered to include the amounts payable under the Health and Unemployment Acts.

Assuming the supervision salaries to amount to £20, and workmen's wages £160 per week, then the oncosts would be $\frac{1}{8}$, or $12\frac{1}{2}$ %, on wages, this being charged against the various departments in proportion to the wages paid.

The items of oncost still remaining to be dealt with are:—

(1) Assistant labour to machines; (2) general labourers in the yard and masons' shop; (3) blacksmith, stores, etc. These items would have to be charged to the various departments in the proportion as they are used or required. There are still a few matters which should be considered before concluding this part of the subject, viz., the provision for the inclusion of rent, water, rates, taxes, and stationery. These items are standing charges, and would have to be allowed for by the addition of a percentage to the total oncost, for which provision has already been made.

ESTIMATING

We now come to the second part of our subject, and if our system of costing is correct, we shall have little difficulty in applying the information gained thereby. In some parts of the country a "*bill of quantities*" is supplied for the purpose of obtaining estimates, and much depends upon the Quantity Surveyor as to the method of "*taking off*" the quantities, and the description of the labours involved. Some districts favour the supplying of detailed drawings, together with a specification of the work to be executed. From the mason's point of view, the latter is the best method, although the onus for the correct reading of the drawing would rest upon the mason. The most satisfactory method is for *specifications*, *quantities*, and *drawings* to be supplied, the responsibility thereby being shared and the possibility of error reduced.

Whatever mode of procedure for *taking off quantities* or *estimating costs* the student adopts, it is best that he should be guided by the orthodox methods, so that his work may harmonise with the accepted rules and conditions of the trade.

Let us assume that an estimate is required for the stonework in accordance with a drawing supplied. Let Figs. 191-194 represent the drawing. The first essential in estimating is to have a clear conception of what is required; the second, to formulate a plan of procedure for *taking off quantities*. By the latter is meant to arrange a system of identification of the various stones or parts comprising the whole, so that a ready means of reference is provided for checking, in order to avoid the overlooking of any particulars.

To prepare an estimate from a drawing, it is first necessary to prepare a "*bill of quantities*," which involves the following processes:—

1. *Taking off*, viz., measuring the quantity of stone required, also the labour involved, and recording the sizes and particulars in the order as read from the drawings.

2. *Abstracting*, viz., the abstraction of items of similar quality and their arrangement under headings, and finding their respective totals.

3. *Billing*, viz., the arrangement of the various items obtained by *abstracting*, in the form of a *bill*.

Applying this procedure to the case we have selected:—

Estimate of Portland Stone Doorway (Figs. 191-194).

TAKING OFF PORTLAND STONE DOORWAY (Figs. 191-194)

Best Whitbed, Finely Rubbed Finish

Identity Marks.	No. of Pieces or Times.	L.	W.	H.	Extension Column.	Extension Column.		Descriptive Particulars.	£	s.	d.	£	s.	d.
		Ft. In.	Ft. In.	Ft. In.	Ft. In.	Ft. In.								
1, 4	2	1 3	3 2	1 6		11 8	Cube	Stone, in quoins.						
	4	1 3	1 6		7 6									
	4	3 2	1 6		19 0									
	4	3 2	1 3		15 10									
						42 4	Super	Half sawing.						
	2	1 3	1 6		3 9									
	2	3 2	1 6		9 6									
						13 3	,,	Face (extra over half sawing).						
	2	3 2	1 3			7 11	,,	Bed (E.O. ½ S.).						
2, 3	2	2 0	2 0	1 6		12 0	Cube	Stone, in bases.						
	8	2 0	1 6		24 0									
	4	2 0	2 0		16 0									
						40 0	Super	½ S.						
	2	2 0	1 6			6 0	,,	F. (E.O. ½ S.).						
	2	2 0	2 0			8 0	,,	B. (E.O. ½ S.).						
			Girth.											
	2	1 6	1 5		4 3		,,	Sunk face in angles (girth measured).						
		Girth.												
	2	3 9	0 8		5 0									
						9 3	,,	Sunk face in attic base (measured over mould).						
			Girth.											
	2	0 6	0 7		0 7									
	No. 2 s	tops to	ditto, al	low	1 0									
						1 7	Super	Sunk rebate (stopped twice for steps).						
			Girth.											
	2	3 0	0 8		4 0									
	No. 2 s	tops to	ditto, al	low	1 4									
						5 4	,,	Circular mould to attic bases.						
5, 10	2	2 6	11 0	2 0		9 2	Cube	Stone, in plinth to pilaster returns.						
	4	2 6	2 0		20 0									
	4	2 6	0 11 }		16 6									
	4	2 0	0 11 }											
						36 6	Super	½ S.						
	2	2 6	1 2			5 10	,,	F. (E.O. ½ S.).						
	2	2 6	0 11			4 7	,,	B. (E.O. ½ S.).						
	2	2 0	0 11			3 8	,,	Joint (E.O. ½ S.).						
			Girth.											
	2	1 9	0 10		2 11									
	No. 2 s	tops to	ditto, al	low	1 8									
						4 7	,,	Sk. F.						
6, 9	2	1 10	1 5	2 0		10 5	Cube	Stone, plinth to pilaster.						
	4	1 10	2 0		14 8									
	4	1 5	2 0		11 4									
	4	1 10	1 5		10 5									
						36 5	,,	½ S.						
	2	1 10	1 2		4 3									
	2	1 5	2 0		5 8									
						9 11	,,	F. (E.O. ½ S.).						
	2	1 10	1 5			5 2	,,	B. (E.O. ½ S.).						
		Girth.												
	2	3 3	0 10			5 5		Sk. F.						
7, 8	2	1 5	2 1	2 0		11 10	Cube	Stone, engaged columns, shafts, and jambs.						
	2	1 5	2 0 }		14 0									
	2	2 1	2 0 }											
	2	1 5	2 1		5 11									
						19 11	Super	½ S.						

Note.—Money columns shown above are not necessary in this particular example, but they are sometimes useful in general practice.

TAKING OFF PORTLAND STONE DOORWAY—*Continued.*

Identity Marks.	No. of Pieces or Times.	L.	W.	H.	Extension Column.	Extension Column.		Descriptive Particulars.	£	s.	d.	£	s.	d.
		Ft. In.	Ft. In.	Ft. In.	Ft. In.	Ft. In.								
	2	2 1	2 0			8 4	Super	F. (prep.) E. O. ½.						
	2	2 1	1 5			5 11	,,	B.						
			Girth.											
	2	2 0	1 0			4 0	,,	Sk. rebate.						
	2	2 6	0 1½		5 0									
	No. 2 s	tops to	ditto, al	low	2 0									
7, 8						7 0	Lin.	Mould circular (cincture to column).						
	2	2 6	2 0			10 0	Super	Cir. F. Engaged columns, entasised and stopped for cinctures.						
	2	2 0	0 5		1 8									
	2	2 0	0 3		1 0									
						2 8	,,	Sk. F.						
	2	1 3	3 0	3 0		22 6	Cube	Stone. Pilasters with moulded bases.						
	8	1 3	3 0		30 0									
	4	3 0	3 0		36 0									
						66 0	Super	½ S.						
	2	2 5	3 0			14 6	,,	F. (E.O. ½ S.).						
	2	1 3	3 0			7 6	,,	B. (E.O. ½ S.).						
11, 14			Girth.											
	2	1 11	1 0		3 10									
	No. 2 m	itres to	ditto, al	low	2 0									
	No. 2 s	tops to	ditto, al	low	2 0									
						7 10	,,	Mould (mitres and stops measured in).						
			Girth.											
	2	1 10	2 5			8 10	,,	Sk. F. to pilaster.						
	2	1 11	1 8	3 0		19 2	Cube	Stone, engaged columns.						
	2	1 11	3 0		11 6									
	2	1 8	3 0		10 0									
	2	1 11	1 8		6 5									
						27 11	Super	½ S.						
	2	3 0	0 5			2 6	,,	F. (E.O. ½ S.).						
12, 13	2	1 11	0 8			2 5	,,	B. (E.O. ½ S.).						
	2	3 0	0 5		2 6									
	2	3 0	0 8		4 0									
						6 6	,,	Sk. F.						
	2	3 0	2 6			15 0	,,	Cir. F., column shafts.						
			Girth.											
	2	3 0	0 8			4 0	,,	Rebate.						
	2	2 6	0 9	3 0		11 3	Cube	Stone, ashlar returns.						
	4	2 6	3 0	}	39 0									
	4	0 9	3 0	}										
15, 20	4	2 6	0 9		7 6									
						46 6	Super	½ S.						
	2	2 6	3 0			15 0	,,	F. (E.O. ½ S.).						
	2	2 6	0 9			3 9	,,	Bed (E.O. ½ S.).						
	2	3 0	0 9			4 6	,,	Joint (E.O. ½ S.).						
	2	1 9	1 3	3 0		13 2	Cube	Stone. Pilasters.						
	4	1 9	3 0	}	36 0									
	4	1 3	3 0	}										
	4	1 9	1 3		8 9									
						44 9	Super	½ S.						
16, 19	2	3 0	0 6		3 0									
	2	3 0	0 8		4 0									
						7 0	,,	F. (E.O. ½ S.).						
	2	1 9	1 3			4 5	,,	B. (E.O. ½ S.).						
			Girth.											
	2	3 0	1 11			11 6	,,	Sk. F., to pilasters.						

TAKING OFF PORTLAND STONE DOORWAY—*Continued.*

Identity Marks.	No. of Pieces or Times.	L.	W.	H.	Extension Column.	Extension Column.		Descriptive Particulars.	£	s.	d.	£	s.	d.
		Ft. In.	Ft. In.	Ft. In.	Ft. In.	Ft. In.								
	2	1 5	2 0	3 0		17 0	Cube	Stone, engaged columns.						
	4	1 5	3 0											
	4	2 0	3 0		41 0									
	4	1 5	2 0		11 4									
						52 4	Super	½ S.						
17, 18	2	3 0	0 8			4 0	,,	F. (E.O. ½ S.).						
	2	1 5	2 0			5 8	,,	B. (E.O. ½ S.).						
	2	3 0	0 6		3 0									
	2	3 0	0 4		2 0									
						5 0	,,	Sk. F.						
	2	3 0	2 6			15 0	,,	Cir. F., engaged cols.						
	2	1 11	1 10	0 6		3 6	Cube	Stone, carved Ionic caps.						
	4	1 11	0 6											
	4	1 10	0 6		7 6									
	4	1 11	1 10		14 0									
21, 22						21 6	Super	½ S.						
	2	1 11	1 10			7 0	,,	B. (E.O. ½ S.).						
	No. 2	Masons' labour to caps, including Sunk Face and rebate, astragal, ovolo, abacus, and boasting.												
	No. 2	Carving to caps.												
	2	3 0	1 5	1 0		8 6	Cube	Stone, moulded head.						
	4	3 0	1 0											
	4	3 0	1 5		29 0									
	4	1 5	1 0		5 8									
						34 8	Super	½ S.						
	2	3 0	1 5			8 6	,,	B. (E.O. ½ S.).						
	2	1 0	1 5			2 10	,,	F. (E.O. ½ S.).						
23, 25			Girth.											
	2	3 6	1 5		9 11									
	6 mitres @		1 5		8 6									
	2 stops @		1 5		2 10									
						21 3	,,	Mould (mitres and stops included).						
	2	Recessed labour for secret key joints.												
	1	2 6	0 9	1 0		1 11	Cube	Stone, centre of head.						
	2	2 6	1 0		5 0									
	2	2 6	0 9		3 9									
	2	0 9	1 0		1 6									
						10 3	Super	½ S.						
24	1	2 6	1 0		2 6									
	1	2 6	0 9		1 11									
						4 5	,,	F. (E.O. ½ S.).						
			Girth.											
	1	6 6	0 2			6 6	Lin.	Fillet, round panel.						
	2	Labour to sunk joints to splayed secret tenons.												
	2	1 3	2 10	3 6		24 10	Cube	Stone, pilasters.						
	4	1 3	3 6											
	4	2 10	3 6		57 2									
	4	2 10	1 3		14 2									
26, 30						71 4	Super	½ S.						
	2	2 5	3 6			16 11	,,	F. (E.O. ½ S.).						
	2	2 10	1 3			7 1	,,	B. (E.O. ½ S.).						
			Girth.											
	2	3 6	1 8			11 8	,,	Sk. F. (pilaster).						
	2	2 1	2 7	2 0		21 6	Cube	Stone (springers to pediment).						
	4	2 1	2 0											
	4	2 7	2 0		37 4									
27, 29	4	2 1	2 7		21 6									
						58 10	Super	½ S.						
	2	2 1	2 0											
	2	2 7	2 0			18 8	,,	F. prep. (E.O. ½ S.).						

TAKING OFF PORTLAND STONE DOORWAY—*Continued.*

Identity Marks.	No. of Pieces or Times.	L.	W.	H.	Extension Column.	Extension Column.		Descriptive Particulars.	£	s.	d.	£	s.	d.
		Ft. In.	Ft. In.	Ft. In.	Ft. In.	Ft. In.								
27, 29	2	2 1	2 7			10 9	Super	B. prep. (E.O. ½ S.).						
	2	Labour to forming all moulding and sunk faces to pediment springer (see detail).												
28	1	4 1	1 3	1 3		6 4	Cube	Stone (moulded closers over head).						
	4	4 1	1 3		20 5									
	2	1 3	1 3		3 2									
						23 7	Super	½ S.						
	1	4 1	1 3			5 1	,,	B. (E.O. ½ S.).						
	1	1 3	1 3			1 7	,,	Jt. (E.O. ½ S.).						
	1	4 1	Girth. 1 3			5 1	,,	Sk. F.						
	1	4 3	Girth. 1 0		4 3									
	2	0 11	0 3		0 6									
	2 stops @		0 3		0 6									
						5 3	,,	Mould (stops measured in).						
31, 35	2	2 6	0 9	1 6		5 8	Cube	Stone (pilaster cap returns).						
	4	2 6	1 6		15 0									
	4	2 6	0 9		7 6									
	4	0 9	1 6		4 6									
						27 0	Super	½ S.						
	2	2 6	1 6			7 6	,,	F. (E.O. ½ S.).						
	2	2 6	0 9			3 9	,,	B. (E.O. ½ S.).						
	1	0 9	1 6			1 2	,,	Jt. (E.O. ½ S.).						
32, 34	2	3 1	1 4	1 6		12 4	Cube	Stone (carved pilaster caps).						
	2	3 1	1 6		9 3									
	2	3 1	1 3		7 9									
	2	1 3	1 6		3 9									
						20 9	Super	½ S.						
	2	1 9	1 6		5 3									
	2	0 7	1 6		1 9									
						7 0	,,	Sk. F. (in ashlar to clear abacus).						
	2	labours, forming additional Sk. F., mould, and boasting (see detail).												
	2	carvings to ditto.												
33	1	4 9	2 7	2 2		26 8	Cube	Stone (key to pediment).						
	2	4 9	2 2		20 7									
	2	4 9	2 7		24 6									
	2	2 7	2 2		11 2									
						56 3	Super	½ S.						
	1	4 9	2 2			10 4	,,	F. prep. (E.O. ½ S.).						
	1	4 9	2 7			12 3	,,	B. prep. (E.O. ½ S.).						
	1	1 6	1 3			1 11	,,	Jt. prep. (E.O. ½ S.).						
	2	2 7	0 10			4 4	,,	Sk. Joint.						
	1	4 9	1 2	average		5 6	,,	Sk. F.						
	1	5 0	1 5			7 1	,,	Cir. F. (weathering).						
	1	4 3	0 10			3 7	,,	Cir. Sk. F. (soffit).						
	1	4 1	0 4	average		1 4	,,	Sk. F. (deeply cut in tympanum).						
	1	5 0	Girth. 1 4			6 8	,,	Cir. mould to pediment.						
36, 39	2	1 3	2 11	1 9		12 9	Cube	Stone (pilaster architrave and frieze).						
	4	1 3	1 9		8 9									
	4	2 11	1 9		20 5									
	4	2 11	1 3		14 7									
						43 9	Super	½ S.						
	2	2 5	1 9			8 6	,,	F. (E.O. ½ S.).						

TAKING OFF PORTLAND STONE DOORWAY—*Continued.*

Identity Marks.	No. of Pieces or Times.	L.	W.	H.	Extension Column.	Extension Column.		Descriptive Particulars.	£	s.	d.	£	s.	d.
		Ft. In.	Ft. In.	Ft. In.	Ft. In.	Ft. In.								
	2	2 11	1 3			7 4	Super	B. (E.O. ½ S.).						
			Girth.											
	2	1 8	1 0			3 4	,,	Sk. F.						
36, 39			Girth.											
	2	1 7	1 0		3 2									
	2 m	itres @	1 0		4 0									
	2 sto	ps @	1 0											
						7 2	,,	Mould (mitres and stops measured in).						
	2	4 2	0 10	1 9		12 2	Cube	Stone.						
	4	4 2	1 9		29 2									
	4	4 2	0 10		13 11									
	4	0 10	1 9		5 10									
37, 38						48 11	Super	½ S.						
	2	4 2	1 9			17 6	,,	F. (E.O. ½ S.).						
	2	0 10	1 9											
	2	4 2	0 10			7 0	,,	B. (E.O. ½ S.).						
					Lin.									
	2	8 3	0 3		19 4	4 10	,,	Sk. F.						
	2	1 5	0 3											
	2	2 6	2 5	1 6										
	2	4 2	2 5	1 6		65 7	Cube	Stone (cornice).						
	1	4 9	2 5	1 6										
	4	2 6	1 6											
	4	4 2	1 6		54 3									
	2	4 9	1 6											
	4	2 6	2 5											
	4	4 2	2 5		87 5		Super							
	2	4 9	2 5											
	10	2 5	1 6		36 3									
						177 11	,,	½ S						
	1	13 0	1 9			22 9	,,	F. (E.O. ½ S.).						
	1	13 0	2 5			31 5	,,	B. (E.O. ½ S.).						
	5	2 5	1 9			21 2	,,	Jt. (E.O. ½ S.).						
	1	13 0	1 4			28 3	,,	Sk. F. to weathering lightly formed.						
	2	4 1	1 4											
40–44	No. 14	mitres	to ditto	allow		8 0		Sk. F. to weathering lightly formed.						
	No. 2 s	tops to	ditto											
	1	12 3	0 5		5 1									
	No. 2 m	itres @	0 5		1 8									
	No. 2 s	tops @	0 5											
						6 9	,,	Sk. F. (in chamfer) (mitres and stops measured in).						
			Girth.											
	1	13 0	2 3		47 8									
	2	4 1	2 3											
	No. 2 m	itres @	2 3		9 0									
	2 sto	ps @	2 3											
						56 8	,,	Mould (mitres and stops measured in).						
	2	1 11	0 4		1 3									
	2 mi	tres @	0 4		1 4									
	2 sto	ps @	0 4											
						2 7	,,	Mould.						
	No. 20	spaces 6	×5×2	in.				Between blocks (see detail).						
			Girth.											
	1	12 0	0 2			12 0	Lin.	Notch for asphalt.						
45–49	2	1 3	3 0	1 5	10 8									
	2	2 4	1 0	1 5	6 7									
	1	3 7	0 9	1 5	3 10									
						21 1	Cube	Stone (plinth to balustrade).						

TAKING OFF PORTLAND STONE DOORWAY—*Continued.*

Identity Marks.	No. of Pieces or Times.	L.	W.	H.	Extension Column.	Extension Column.		Descriptive Particulars.	£	s.	d.	£	s.	d.
		Ft. In.	Ft. In.	Ft. In.	Ft. In.	Ft. In.								
	4	1 3	1 5											
	4	3 0	1 5											
	4	2 4	1 5											
	4	1 0	1 5		55 3									
	2	3 7	1 5											
	2	0 9	1 5											
	4	3 0	1 3		15 0									
	4	2 4	1 0		9 4									
	2	3 7	0 9		5 5									
						85 0	Super	½ S.						
	2	2 9	1 0											
	2	1 0	1 0											
	4	2 4	1 0			28 0	,,	F. (E.O. ½ S.).						
	2	3 7	1 0											
	2	2 0	1 0											
	2	3 0	1 3		7 6									
	2	2 4	1 0		4 8									
	1	3 7	0 9		2 8									
						14 10	,,	B. (E.O. ½ S.).						
	6	0 9	1 5											
	2	1 0	1 5			9 3	,,	Jt. (E.O. ½ S.).						
			Girth.											
45–49	2	1 5	0 6		1 5									
			Girth.											
	2	1 5	0 7		1 8									
			Girth.											
	2	1 5	2 6		7 1									
						10 2	,,	Sk. F. (in angular checks).						
	1	10 11	0 6											
	6	0 3	0 6	Lin.	Sup.									
	2	3 0	0 6	28 1	14 0									
	1	5 6	0 6											
	2	2 1	0 6											
	10 stops to ditto @			0 6	5 0									
						19 0	,,	Sk. F. (in sub plinth, stops measured in).						
	1	18 5	0 2		18 5									
	14 mitres to ditto @			1 0	14 0									
						32 5	Lin.	Mould, cavetto (mitres measured in).						
	1	9 8	0 1		9 8									
	2 mitres, 4 stops				6 0									
						15 8	,,	Chamfer.						
	2	2 0	0 9	1 3	3 9									
	2	1 5	1 3	1 3	4 5									
	2	1 9	1 0	1 3	4 5									
	1	4 6	0 5	1 3	2 4									
						14 11	Cube	Stone (in dado to balustrade).						
	4	2 0	1 3											
	4	0 9	1 3											
	8	1 5	1 3											
50–56	4	1 3	1 3		49 0									
	4	1 9	1 3											
	2	0 5	1 3											
	4	1 0	1 3											
	4	2 0	0 9		6 0									
	4	1 9	1 0		7 0									
	4	4 6	0 5		7 6									
						69 6	Super	½ S.						
	4	1 9	0 11		6 5									

TAKING OFF PORTLAND STONE DOORWAY—*Continued.*

Identity Marks.	No. of Pieces or Times.	L.	W.	H.	Extension Column.	Extension Column.		Descriptive Particulars.	£	s.	d.	£	s.	d.
		Ft. In.	Ft. In.	Ft. In.	Ft. In.	Ft. In.								
	2	4 6	1 3											
	8	0 3	1 3											
	4	2 0	1 3		34 7	41 0	Super	F. (E.O. ½ S.).						
	4	1 2	1 3											
	4	1 0	1 3											
	2	2 0	0 9		3 0									
	2	1 5	1 3		3 7									
	2	1 9	1 0		3 6									
50–56	1	4 6	0 5		1 11									
						12 0	,,	B.						
	4	0 5	1 3			2 1	,,	Jt. (E.O. ½ S.).						
			Girth.											
	2	1 3	0 6		1 3									
	2	1 3	1 10		4 7									
						5 10	,,	Sk. F. in angles.						
	4	1 9	0 3			10 8	Lin.	Sk. F. fillet to raised panels.						
	4	0 11	0 3											
	2	1 3	3 2	0 5	3 4									
	2	2 3	1 4	0 5	2 6									
	1	4 0	0 9	0 5	1 3									
						7 1	Cube	Stone in capping.						
	4	1 3	0 5											
	4	3 2	0 5											
	4	2 3	0 5		21 8									
	4	1 4	0 5											
	4	4 0	0 5											
	4	1 3	3 2		15 10									
	4	2 3	1 4		12 0									
	2	4 0	0 9		6 0									
						55 6	Super	½ S.						
	4	2 3	0 5											
	2	4 0	0 5											
	2	2 11	0 5			11 5	,,	F. (E.O. ½ S.) (partly preparatory).						
	2	1 0	0 5											
	2	1 4	0 5											
	3	0 10	0 5			1 0	,,	Jt. (E.O. ½ S.).						
			Girth.											
57–61	2	0 5	0 6		0 5									
		Girth.												
	2	2 6	0 5		2 1									
						2 6	,,	Sk. F. to angles.						
	1	11 0	0 6		5 6									
	2	2 11	0 6		2 11									
	6	0 3	0 6		0 9									
	No. 14	mitres	@ 6		7 0									
						16 2	,,	Mould (mitres measured in).						
	2	3 2	1 3		7 11									
	2	2 3	1 4		6 0									
	1	4 0	0 10		3 4									
						17 3	,,	Sk. F. (to weathering lightly formed).						
	No. 14	mitres,	allow			7 0								
			Girth.											
	1	15 0	0 1			15 0	Lin.	Throating to capping.						
	No. 54							Lewis holes for lifting.						
	No. 4							Cramp mortises and copper cramps.						
	1	80 0	Lin.					Joggle, V-shaped, for cement grout (one joint measured only).						

ABSTRACT, PORTLAND STONE DOORWAY

Identity No.	Stone.		½ Sawing.		Face E.O. ½ S.		Bed E.O. ½ S.		Joint E.O. ½ S.		Sunk Face.		Sunk Joint.		Mould.		Circular Mould.		Circular Face.			
	Cube.		Super.		Super.		Super.		Super.		Super.		Super.		Super.		Super.		Super.			
	Ft.	In.	Ft.	In.	Ft.	In.	Ft.	In.	Ft.	In.	Ft.	In.	Ft.	In.	Ft.	In.	Ft.	In.	Ft.	In.		
1 and 4	11	8	42	4	13	3	7	11														
2 ,, 3	12	0	40	0	6	0	8	0			9	3	1	7			5	4				
5 ,, 10	9	2	36	6	5	10	4	7	3	8	4	7										
6 ,, 9	10	6	36	5	9	11	5	2			5	5										
7 ,, 8	11	10	19	11	8	4	5	11			2	8	4	0					10	0	Circular mould 7 ft. linear by 2 in.	
11 ,, 14	22	6	66	0	14	6	7	6			8	10			7	10						
12 ,, 13	19	2	27	11	2	6	2	5			6	6	4	0					15	0		
15 ,, 20	11	3	46	6	15	0	3	9	4	6												
16 ,, 19	13	2	44	9	7	0	4	5			11	6										
17 ,, 18	17	0	52	4	4	0	5	8			5	0							15	0		
21 ,, 22	3	6	21	6			7	0													Mason labour to caps No. 2.	Carving to caps No. 2.
23 ,, 25	8	6	34	8	2	10	8	6							21	3					Recessed labour to key No. 2. 6 ft. 6 in. by 2 in. fillet.	
24	1	11	10	3	4	5																Labour to key joints No. 21.
26 and 30	24	10	71	4	16	11	7	1			11	8										
27 ,, 29	21	6	58	10	18	8	10	9														
28	6	4	23	7			5	1	1	7	5	1			5	1						
31 ,, 35	5	8	27	0	7	6	3	9	1	2												
32 ,, 34	12	4	20	9							7	0									Add sunk labours and boasting No. 2.	Carved return No. 2.
33	26	8	56	3	10	4	12	3	1	11	5 1	6 4	4	4			6	8	7	1	Circular sunk face 3 ft. 7 in.	
36 and 39	12	9	43	9	8	6	7	4			3	4			7	2						
37 ,, 38	12	2	48	11	17	6	34	0			4	10										No. 4 cramps and mortises.
40 to 44	65	7	177	11	22	9	3	5	21	2	36 6	3 9			56 2	8 7					20 spaces between blocks, 6 by 5 by 2 in.	Notch 12 ft. lin. by 2 in.
45 ,, 49	21	1	85	0	28	0	14	10	9	3	10 19	2 0									Mould 32 ft. 5 in. by 2 in. lineal.	Chamfer 15 ft. 8 in. by 1 in. lineal.
50 ,, 56	14	11	69	6	41	0	12	0	2	1	5	10									11 ft. 8 in. by 3 in. fillet.	
57 ,, 61	7	1	55	6	11	5	1	0			2 17 7	6 0 0			16	2					Throating 15 ft. by 1 in. lineal. 54 lewis holes.	V joggles 80 ft. lineal.
TOTALS	383	1	1217	5	276	2	182	4	45	4	195	4	13	11	116	9	12	0	47	1		

BILLING, PORTLAND STONE DOORWAY

Ft.	In.					Rate.	£	s.	d.
383	0		Cube.	Stone (cost in yard and allow a percentage for waste) .	@				
1217	0		Super.	Half sawing (allow for waste)	@				
276	0		,,	Plane face (extra over half sawing)	@				
182	0		,,	Plane bed (E.O. ½ S.)	@				
45	0		,,	Plane joint (E.O. ½ S.)	@				
195	0		,,	Sunk face (assess at average cost)	@				
14	0		,,	Sunk joint (assess at average cost)	@				
117	0		,,	Mould (mitres and stops measured in)	@				
12	0		,,	Mould (circular)	@				
7	0		Lin.	Mould (circular), 2-in. girth	@				
47	0		Super.	Circular face in columns, etc.	@				
3	7		,,	Circular face sunk in soffit	@				
		No. 2		Mason labours to Ionic caps Nos. 21 and 22	@				
		,,		Carving labours to Ionic caps Nos. 21 and 22 . . .	@				
		,,		Recessed labour to secret radial joint stones Nos. 23 and 25	@				
		,,		Projecting labour to secret radial joint stone No. 24 .	@				
		,,		Labours to pediment springers Nos. 27 and 29 (E.O. ½ S. and prepared B. and F.)	@				
		,,		Additional sunk labours and boasting to pilaster caps Nos. 32 and 34 (over ½ S., etc.)	@				
		,,		Carved labours to caps Nos. 32 and 34	@				
		No. 20		Spaces, between cornice blocks 6 by 5 by 2 in. . . .	@				
12	0		Lin.	Notching for asphalt, 2-in. girth	@				
12	0		,,	Fillet to panel, 3-in. girth	@				
6	6		,,	Fillet to panel, 2-in. girth	@				
16	0		,,	Chamfer, baluster plinth, 1-in. wide	@				
32	0		,,	Mould, baluster plinth, 2-in. girth	@				
15	0		,,	Throating, in capping, 1-in. girth	@				
		No. 54		Lewis holes	@				
		No. 4		Cramp holes and copper cramps	@				
80	0		Lin.	Joggle in joint V-shaped (two half-joints measured as one)	@				
				TOTAL .	£				

ESTIMATE, PORTLAND STONE DOORWAY

		Rate.	£	s.	d.	£	s.	d.
Stone and labours (including all works oncosts)								
Establishment charges (rent, rates, etc., per cent.)								
Haulage charges (if any), tons	@							
Estimated cost of fixing (if any) 383 ft.	@							
Special charges								
Gross estimated profit per cent. (including expenses not previously provided for)								
Total amount of estimate .	£							

Observations.—In *taking off* the quantities for this estimate, the following assumptions have been made :—

1. That all surfaces of the rectangular blocks containing the finished stones have been sawn by the frame saw.

2. That such surfaces as one joint, one bed, and the back of each stone are in a condition suitable for the finished surface.

3. That all stones have been sawn slightly full of the finished sizes to allow of light tooling, or light tooling and rubbing, or rubbing only, to produce the required finished surface.

4. That joints or beds sawn to exact size by the diamond or carborundum saw are approximately equal in value to a surface sawn by the frame saw, and dressed as in assumption No. 3.

A priced *bill of quantities* is usually required by the architect upon acceptance of a *tender* or *estimate*. This provides for a basis of settlement in case of variations occurring after the acceptance, or during the progress of the work. In this case *haulage*, *fixing*, and *profit* would have to be included in the priced items of the *bill*. When *priced bills of quantities* are required, then the quantities should be supplied by the architect, and should form the basis of the contract.

In studying the foregoing example of an estimate, the student may find that it does not agree with some recognised standard, but standard rules are sometimes difficult to apply. *Masonry estimating* is beset with thorny problems, so that estimators seldom work along parallel lines. Although methods may differ, given the right groundwork, the results will prove satisfactory.

As the student progresses, he will develop his own method of procedure in *taking off quantities*, and by adopting short cuts will reduce the tedious details to a minimum. As his experience increases, so will his judgment to a corresponding degree. He will be able to assess at a glance the unit value of any class of work.

The ability to do this, however, can only be achieved as the result of close observation and a sound application of the knowledge thus obtained.

As a rule, masonry contractors prefer to quote a *lump sum* for a given quantity, stating the cubical contents of the work for which they have quoted. This is the equivalent of quoting at *per foot cube*. It must be borne in mind that this can only be done successfully by the estimator of experience, as the basis of the estimate must be an intimate knowledge of all the factors of the case in question.

MARBLEWORK

What has been said of stonework, from the point of view of costing and estimating, also applies to marblework. The basis of these systems must be a sound knowledge of both the processes or operations of manufacture, and the principles of good masonry and building construction. This given, *costing* resolves itself into a systematic recording of the results achieved by operations, whilst *estimating* resolves itself into the application of the facts obtained to the purpose of finding the probable cost of proposed similar work.

The term *marble* covers a large variety of stones, chiefly limestones capable of receiving a *high, glossy finish* or *polish*. The properties of marble have been dealt with in another section; it will therefore serve our purpose to restate the general uses to which it is put. These are:—

1. *Decorative*.
2. *Monumental*, including *sculpture*.
3. *Slab-work* for *shop fittings*, *wall linings*, and various utility purposes.
4. *Structural work*.

When used for decorative purposes, marble is usually selected in accordance with a colour scheme, sawn into thin slabs, jointed and polished as required, and fixed to prepared structural work, such as brick walls, etc. To prepare an estimate for this class of work, it is necessary to know what machinery and labour are involved.

ESTIMATE FOR LININGS FOR BATHROOM WALLS

Let us assume that an estimate is required for the supply of marble linings to the walls of a bathroom, the drawing of which is given in Fig. 131. Before we commence *taking off* the quantities, let us think out the processes which are necessary to produce the finished work. The "*raw material*" may be marble already sawn into slabs of *random size*, these being purchased in that state from the stock yard, or it may be in the form of *rough blocks*. We will assume the latter for the case in question. The first operation or process would be the sawing of the blocks into slabs of suitable thickness; this would be done by the agency of the "*frame saw*." The operation is termed "*slabbing*."

The next operation would be the jointing up of the slabs into pieces, the size and shape of which have been predetermined. This is called "*ripping*," and is accomplished by a carborundum sawing machine, similar to that shown in Fig. 60. The third operation would be the "*sanding*," *i.e.*, *rubbing down* to the required thickness and degree of smoothness by the use of a machine called a "*rubbing table or bed*," or by a machine designed for this purpose.

At this stage the marble slabs would be *fitted up*, and if coloured or of a weak nature, owing to natural flaws, would be treated accordingly. Therefore, the fourth process is that of "*fitting up*." We now come to the "*polishing*." In work of this class, a number of the pieces of marble are usually collected and arranged together in the form of a bed, set in plaster, and a polishing machine with sufficient capacity to polish a fairly large area is used. We will call this the fifth operation.

The sixth operation would be the "*hand polishing*" required to the edges, which completes the labours of production.

Machines are now being installed for *edge polishing*, but these are not assumed as being installed for this *estimate*.

ESTIMATE FOR LINING FOR BATHROOM WALLS

Ref.	Times.	Dimensions.				Rate.	£	s.	d.	£	s.	d.
		Material—Belgian black marble, per foot cube @ . . .										
		Ft. In. In. In.	Ft. In. Lin.	Ft. In.								
	2	9 11×6 ×1	19 10									
	2	2 10×6 ×1	5 8									
	1	9 9×6 ×1	9 9	17 8	Sup. foot skirting . @							
	2	0 5½×6 ×1¼		0 6	Sup. architrave blocks @							
	2	6 4 ×5 ×1	Lin. 12 8	5 3	Sup. architrave jambs @							
	1	4 1×5 ×1¾	Lin. 4 1	1 8	Sup. architrave head @							
	2	9 11×4 × ¾	Lin. 19 10									
	2	2 11×4 × ¾	5 10									
	1	9 9½×4 × ¾	9 10									
	2	0 4½×4 × ¾	0 9	12 0	Sup. dado rail . @							
	1	3 3 ×5½×1		1 8	Sup. window board @							
		Material—Dove-grey marble, per foot cube @										
		Ft. In. Ft. In. In.	Ft. In.	Ft. In.								
	2	9 11×2 8×¾	52 11									
	2	2 11×2 8×¾	15 7									
	1	9 9½×2 8×¾	26 3	94 9	Sup. dado filling . @							
		Material—Veined statuary marble, per foot cube @ . . .										
		Ft. In. Ft. In. In.	Ft. In.	Ft. In.								
	6	9 11×1 10×¾	109 1									
	4	2 11×1 10×¾	21 5									
	6	3 3×1 10×¾	35 9									
	1	9 9½×1 10×¾	18 0									
	1	3 8×0 10×¾	3 1									
	2	4 8×0 4½×¾	3 6									
	1	3 4½×0 4½×¾	1 4	192 2	Sup. . . . . @							
			Total .	325 8								
		Slab sawing (two sides measured as one).										
			Ft. In. 325 8									
					Carried forward							

ESTIMATE FOR LINING FOR BATHROOM WALLS—*Continued.*

Ripping (to rectangular sizes)—

1 in.		$1\frac{1}{4}$ in.	$\frac{3}{4}$ in.		$1\frac{3}{4}$ in.
Ft. In.		Ft.	Ft. In.		Ft. In.
39 8		4	39 8		8 10
11 4			11 8		
19 6			19 7		
24/6= 12 0			9 5		
25 4					
3 4			39 8		
7 5			11 5		
			19 7		
118 7			85 4	(32/2 ft. 8 in. vertical joints)	
			119 0		
			23 4		
			39 0		
			19 7		
			176 0	(96/1 ft. 10 in. vertical joints)	
			9 0		
			18 8		
			4 6		
			7 3		
			652 8		

Ref.	Item	Quantity	Description		Rate.	£	s.	d.	£	s.	d.
			Brought forward	.							
	Ripping	Ft. In. 4 0	lin., $1\frac{1}{4}$ in. thick	@							
		8 10	lin., $1\frac{3}{4}$ in. thick	@							
		Ft. In. 118 7	lin., 1 in. thick	@							
		Ft. In. 652 8	lin., $\frac{3}{4}$ in. thick	@							
		4 0	lin., $\frac{3}{4}$ in. extra rip to splays	@							
	Sanding to size	325 8	sup. (edges allowed for)	@							
	Fitting up	325 8	sup.	@							
	Mason— $2\frac{1}{4}$-in. girth rebate	4 1	lin. to doorhead	@							
	$\frac{3}{4}$-in. thick notching	4 4	lin., sunk joint over head	@							
	Polishing— Machine work	325 8	sup., plane face	@							
	Hand work	36 6	lin., $\frac{1}{4}$ in. top of skirting	@							
	Hand work	45 1	lin., $\frac{3}{4}$ in. edges to architrave and linings.	@							
			Packing for transport, allow	.							
	Supervision—Works oncosts per cent.			.							
	Office oncosts per cent.			.							
	Estimated profit per cent.			.							
	Selling price at works			£							

Note.—Add for transport, fixing, and special costs, if any.

Additional items to be allowed for are:—

Packing ready for transport, *transport* to site, and, finally, "*fixing*," including all expenses to be met in connection therewith.

With this statement of processes and the items involved, we are now in a position to *take off* the quantities and work out the estimate, provided, of course, we have at our disposal a *costing system* which will supply us with information regarding costs.

For a simple estimate of this kind, we can combine the three processes—"*taking off*," "*abstraction*," and "*billing*"—under one heading, the example given showing how this may be done.

The example given is of a very simple nature, including only a few of the labours which are to be met in the manufacture of marblework, but it will serve as a signpost to direct the student on the way to the more difficult problems of estimating.

ESTIMATE FOR MARBLE LINING TO CONICAL WINDOW HEADS WITH SPLAYED SOFFIT

As an intermediate example, we will take the drawing of the marble lining to a semicircular arch, for which an estimate is required. The setting out of this arch is shown in Figs. 409 and 410.

As in the case of the linings for the bathroom wall, we must think out what operations are entailed in our present example. Before proceeding with a study of the estimate, it will be advisable to examine the drawing carefully, as this will enable the student to visualise what labours are entailed. The first thing to consider is the material, which in this instance is to be marble. When measuring off the sizes of the stones, care must be taken to include the waste material which may be required to assist in the production of the work, but which ultimately may be cut away.

The necessary operations and processes are as follows:—

(1) *Slabbing*.

(2) *Ripping* (that is to the sizes which will contain the finished stones).

3) *Sanding* (that is to sizes ready for the application of the moulds).

(4) *Masoning* (that is hand work, or pneumatic tools).

(5) *Fitting up*.

(6) *Machine polishing* of plane surfaces.

(7) *Hand polishing* to *circular sunk face* and *moulding*.

(8) *Packing for transport*.

(9) *Transport*.

(10) *Fixing* (including an allowance for centering and other incidental expenses).

ESTIMATE, MARBLE LINING TO CONICAL WINDOW HEAD WITH SPLAYED SOFFIT

Ref.	Times.	Dimensions.	Extension Columns.		Super.	Particulars.	Rate.	£	s.	d.	£	s.	d.
		Ft. In. Ft. In. In.	Ft. In.	Ft. In.		*Material*—							
1	2	2 0×2 7½×2	10 6			Roman stone per foot cube . . @							
2	2	2 8×2 11 ×2	15 7										
3	2	2 10×1 6 ×2	8 6										
				34 7	Super.	Face stones . . . . . . @							
4	4	3 4×3 11 ×6		52 3	,,	Splayed soffit stones . . . @							
		Half sawing (slabbing)—				@							
		Ft. In.											
1, 2, 3	2	34 7		69 2	,,	Face stones (2 in. thick) . . . @							
4	2	52 3		104 6	,,	Soffit stones (6 in. thick) . . . @							
		Ripping (to rectangular sizes)—											
		Ft. In. In.											
1	4	4 7½×2	18 6										
2	4	5 7 ×2	22 4										
3	4	4 4 ×2	17 4										
				58 2	Lin.	Edges to face stones . . . @							
4	8	7 3 ×6		58 0	,,	Edges to soffit stones . . . @							
1	4	1 6 ×2	6 0										
2	4	1 10 ×2	7 4										
				13 4	,,	Extra ripping to arch joints . . @							
		Sanding (to sizes)—											
1, 2, 3 and 4				86 10	Super.	Face (measured one side) . . . @							
1		1 ft. 7 in., 2 ft. 8 in., 9 in., 1 ft. 6 in. =	6 6										
2		1 ft. 6 in., 2 ft. 3 in., 1 ft. 9 in., 1 ft. 9 in. =	7 3										
3		1 ft. 9 in., 2 ft. 10 in., 1 ft. 1 in. =	5 8										
				19 5	Lin.	Edges 2 in. thick . . . . @							
		Mason—											
		Ft. In. In.											
1, 2, 3	1	12 0×2		12 0	,,	Shaping 2-in. circular edge . . @							
4	1	12 0×3		12 0	,,	Mould, circular, 3-in. girth . . @							

4	4	2/3 ft. 11 in., 3 ft. 4 in., 2 ft.		52	8	,,	Splayed joint, 6 in. wide . .	@
4	4	Ft. In. Ft. In. 3 6×2 3 (average)		31	6	Super.	Circular sunk face (soffit). . .	@
4	4	2 6×1 8 (average)		16	8	,,	Add circular sunk face (panels) . .	@
4	4	3 11×2 11 (average)		45	8	,,	Rough circular back joint . .	@
4	8	2 6×0 ½		20	0	Lin.	Straight fillet to panels . . .	@
4	4	3 4×0 ½		13	4	,,	Circular fillet to panels . . .	@
		Selection and Fitting Up—						
1, 2, 3		All measured.		34	7	Super.	Straight face in arch . . .	@
4		,, ,,		52	3	,,	Circular face in soffit . . .	@
		Polishing—						
1, 2, 3		All measured.		34	7	,,	Straight face in arch . . .	@
4		,, ,,		52	3	,,	Circular face in soffit . . .	@
4		Extra over last item.		16	8	,,	Added value sunk panel . . .	@
1, 2, 3				12	0	Lin.	Mould 3-in. girth	@
1, 2, 3		Fillet.	12 0					
4		,,	13 4	25	4	,,	Fillet to arch and soffit panels . .	@
							Packing for transport, tons. .	@
							Total labour cost .	
							Works oncost (supervision) per cent. .	@
							Overhead charges (not already included in works oncost) per cent. .	
							Profit per cent.	
							Total of Estimate .	

GLOSSARY

A

Abutment.—The solid masonry supporting the arch on each side. That on which the arch abuts, or from which the arch springs.

Acute Angle.—A term used in geometry to denote an angle less than 90°—that is, less than a right angle.

Adhesion.—A term in physics denoting the tendency which different bodies have to remain attached to each other when their surfaces are brought into contact.

Alabaster.—A term commonly applied to a compact mass of gypsum or hydrated calcium sulphate. True Oriental alabaster, which is rarer and more expensive, is practically the same composition as a marble, and is carbonate of calcium, of stalagmitic formation.

Alomite.—A Canadian sodalite, a mineral containing a considerable amount of soda in its composition. Known as " Princess Blue."

Altitude of a Triangle.—The perpendicular distance from the vertex, or upper angular point, to the base.

Alumina.—A compound of oxygen and aluminium, and one of the constituents of clay.

Angle.—The inclination to each other of two straight lines meeting in a point, the point being called the vertex of the angle.

Annular Vault.—A vault springing from two walls each circular on the plan, the one being concentric with the other.

Apse.—The circular or multangular termination of a choir. In quarrying, a mass of rock locked in a " fault " is known as an apse.

Arc.—In geometry, a portion of the circumference of a circle or other curved line.

Arcade.—A range of arches supported on piers or columns, supporting a continuous wall.

Arch.—A mechanical arrangement of blocks of any hard material disposed in the line of some curve, and supporting one another by their mutual pressure. The arch itself is formed of voussoirs or arch stones.

Architrave.—The beam or lowest division of an entablature which extends from column to column. The term is also applied to the moulded frame which bounds the sides and head of a door or window opening.

Architrave Cornice.—An entablature consisting of an architrave and cornice only, without the frieze.

Archivolt.—The moulding round the face of an arch starting from the impost.

Area of a Figure.—The amount of space enclosed within its boundary.

Argillaceous.—Containing clay which is composed of silica, alumina, and combined water.

Arris.—The line or edge on which two surfaces, forming an exterior angle, meet each other.

Ashlar.—Wrought stone of uniform height, in contradistinction to rubble work; used chiefly for facing walls.

Attic Base.—The base of a column consisting of an upper and lower torus, with a scotia and fillets between them.

Axis.—The straight line in a plane figure round which it revolves to generate a solid.

B

Baluster.—A species of small column belonging to a balustrade.

Baluster Die-stone.—A half-baluster worked on the stone terminating a row of balusters. A die-stone in a balustrade.

Balustrade.—A parapet or dwarf wall comprising plinth, balusters, and capping.

BASE.—The lower portion of any structure or architectural feature.

BASE OF A COLUMN.—The part between the shaft and the pavement or pedestal.

BATTER.—An inward inclination of the exterior face of a wall.

BED JOINTS.—In cylindrical vaulting are the two surfaces intersecting the intrados of the vault in lines parallel to the axis of the cylinder, or the normal joints in an arch between the voussoirs.

BED MOULD OF CORNICE.—The lower and supporting portion of the cornice.

BEDS.—The lower surface upon which a block of stone rests, and the upper surface which supports the stone above.

BITUMINOUS LIMESTONES.—Those containing bitumen and having its characteristic odour when fractured.

BLOCK STONE.—Stone roughly sawn or quarry-axed to rectangular shape and in the condition in which it is received from the quarry.

BLOCKING COURSE.—A course of stone placed on the top of a cornice to give stability to the cornice by its weight, and to form the crowning feature to a wall.

BOASTING.—The act of chiselling the surface of a stone with a boasting chisel and mallet, the tool marks not being in uniform lines.

BOASTING FOR CARVING.—The roughly worked projection allowed for the carving details.

BOND.—The arrangement of stones or bricks so that the vertical joints of one course do not fall over the vertical joints of the courses above and below.

BOND STONES.—Those whose longest horizontal direction is placed in the thickness of the wall.

BONING.—The act of testing a surface to obtain a true plane.

BOSS.—A projecting ornament placed at the intersection of ribs in Gothic vaulting.

BRACKET.—A member projecting from a wall or pier for the purpose of supporting or assisting in the support of a superincumbent weight; a projecting ornament carrying the upper members of a cornice.

BREAK.—A recess or projection from the general surface of the wall.

BRECCIAS.—A rock made up of angular fragments of old rocks, re-cemented and re-crystallised in formation.

BROACH SPIRE.—An octagonal spire rising direct from a tower without a parapet.

BULL-NOSE STEP.—A step with a quadrant end.

BULL'S-EYE WINDOW.—A small circular window.

C

CALACATA.—Italian marble of white variety, having broad veinings of a slate-grey colour.

CALCAREOUS.—Containing lime. Thus calcareous springs may, under the action of air and sunlight, deposit their dissolved matter to form limestones.

CALCAREOUS SANDSTONES.—Those containing a large proportion of carbonate of lime.

CANOPY.—A roof-like covering over niches, etc.

CANTILEVER.—A projecting bracket to support cornices, balconies, etc.

CAPITAL.—The upper portion of a column or pilaster.

CARTOUCHE.—A tablet in the form of a scroll of paper for the reception of an inscription.

CATHETUS.—An imaginary line passing through the centre of a cylindrical body, or the eye of a volute.

CENTERING.—Wooden frames upon which the stones comprising arches and lintols are supported during erection.

CHANCEL.—That part of the eastern end of a church in which the altar is placed.

CHIMNEY-PIECE.—An ornamental finishing or framework for a fireplace.

CHORD.—A straight line joining the extremities of an arc.

CINCTURE.—The ring or fillet at the top and bottom of a column, which divides the shaft of the column from its capital and base.

CIPPOLINO.—A name given to laminated marble having a talcous stratification.

CIRCLE.—A plane figure enclosed by a curved line, every point of which is equidistant from a point within the figure, called a centre.

CLEAN BACK.—The inside vertical surface of a stone which extends through the thickness of the wall, forming a face on the inside.

CLEAVAGE.—The tendency of a rock to split in parallel layers, irrespective of its original structure.

CLOSER.—The last stone to be placed in a course to close or fill a gap.

COFFER.—A panel sunk in the surface of a vault or dome.

COLUMN.—A vertical support generally consisting of base, shaft, and capital.

CONCAVE.—The opposite to convex (*which see*).

CONE.—A solid described by the revolution of a right-angled triangle about one of the sides containing the right angle, which side remains fixed.

CONGLOMERATE.—A rock composed of rounded particles similar to pebbles. If composed of both angular and rounded particles, the rock is known as a conglomerate breccia, also known as pudding stone.

CONIC SECTIONS.—Obtained by the intersection of a cone by a plane.

CONVEX.—A form which swells or rounds itself externally.

COPE, TO.—The act of splitting or bursting a block of stone in a required direction.

COPING.—The capping or covering to a wall. Its function is to prevent rain-water from running down the surfaces of the wall.

CORBEL.—A stone projecting from a wall to support a weight.

CORNICE.—The projecting moulded course which crowns the part of the wall to which it is affixed.

CORONA.—The upper part of the cornice having a broad vertical face and a soffit recessed so as to form a drip to prevent rain-water from running down the face of a building.

COURSE.—The term applied to each horizontal layer of stone.

COVE.—A large hollow moulding.

CRINOIDAL MARBLE.—The term given to marble made up of fossilised fragments.

CROCKET.—A small ornament projecting from the sloping angles of pinnacles, spires, etc., in Gothic architecture.

CROSSETTE.—The horizontal seating worked on arch stones to fit over and rest on the adjacent stone.

CROWN OF AN ARCH.—The most elevated point on the arch curve.

CRYPTOCRYSTALLINE STRUCTURE.—A compact structure formed by an admixture of crystals and vitreous or glassy material determined by the conditions under which they have consolidated.

CUNEOID.—A wedge-shaped solid, partaking of the figure of a cone.

CUPOLA.—A hemispherical roof over a circular, square, or multangular building.

CURTAIL STEP.—A step with its free end semicircular in plan.

CURVE.—A line that may be cut by a straight line in more points than one.

CUTTING PLANE.—A plane cutting a solid into two parts in any direction.

CYLINDER.—A solid described by the revolution of a rectangle about one of its sides which remains fixed.

CYLINDRICAL WORK.—Any kind of work which partakes of the shape of a cylinder.

CYMA RECTA.—A moulding formed by a curve of contrary flexure, taking its name from its resemblance to a wave. Also known as an ogee moulding.

CYMA REVERSA.—Similar to the above, but the contour of the curves is reversed.

CYMATIUM.—The crowning member of a cornice.

D

DADO.—The wall surface between the skirting and dado rail.

DENTILS.—Tooth-like ornaments in the bed mould of a cornice.

DETAILS.—The term usually applied to drawings to a large scale.

DEVELOPMENT.—The laying out of the surface of a solid.

DIE OF A PEDESTAL.—The part included between the base and the cornice.

DIHEDRAL ANGLE.—The angle between two planes.

DOLOMITIC LIMESTONES.—A limestone containing a large percentage of magnesium carbonate.

DOME.—A spherical or polygonal roof over a building.

DRAFT.—A strip of surface or margin worked to the width of a drafting chisel either straight or to the curvature of a templet.

DRESSINGS.—All kinds of mouldings beyond the wall face.

DRIPSTONE.—Label or hood mould placed over the heads of doorways, windows, etc., for the purpose of throwing the rain-water clear of the aperture.

DWARF WALLS.—Walls of less height than the storey of the building.

E

ECHINUS.—An ornament in the form of an egg and dart, or anchor, peculiar to the ovolo moulding.

ELLIPSE.—A section produced by a plane cutting a cone or cylinder obliquely without intersecting the base of the cone. Considered as a plane curve, an ellipse is the path traced by a point moving so that the sum of its distances from two given points, called foci, is constant.

ENTABLATURE.—The part of an order above the columns including the frieze, architrave, and cornice.

ENTASIS.—The delicate swelling on the shaft of a column which prevents a hollow appearance in outline.

EXTRADOS.—The exterior curve of an arch, or the upper curved surface of a vault.

EYE OF A DOME.—The horizontal aperture at the top.

EYE OF A VOLUTE.—The circle at the centre, from the circumference of which the spiral line commences.

F

FAÇADE.—The front elevation or face of a building.

FACE MOULD.—The templet for application to the face of a stone.

FACETTE.—The fillet between the flutes of a column.

FAN VAULT.—A system of vaulting peculiar to the Perpendicular period, all the ribs having the same curve, and resembling the framework of a fan.

FASCIA.—A band or fillet; often used to denote a plain vertical course over a large opening.

FAULT.—A dislocation in geological strata, or a break preventing the continuance of the strata.

FELSPAR.—A widely distributed mineral consisting of silica and alumina, with potash, soda, or lime. An important constituent of granite. Clay is produced by the decay of felspar.

FILLET.—A narrow, flat band used for the separation of one moulding from another.

FINIAL.—The top or finishing portion of a pinnacle, etc.

FLIERS.—Steps of parallel width.

FLUSH.—A term applied to surfaces which are in the same plane. The arrises of wrought stones are said to be flushed when damaged by chipping.

FLUTINGS.—The upright channels on the shafts of columns.

FOUNDATION.—That part of the construction in direct contact with the ground.

FREESTONE.—Any stone composed of sand or grit that works freely. In the trade, limestones in general are considered as freestones.

FRIEZE.—The member in the entablature of an order between the architrave and the cornice. If curved in section, it is said to be cushioned or pulvinated.

G

GABLE.—The vertical piece of wall at the end of a roof.

GABLET.—A gable-shaped termination to a buttress.

GARGOYLE.—A spout for throwing the rain-water clear of the wall of a building.

GNEISS.—A stone differing from granite only in that the materials of which it is composed (quartz, felspar, and mica) are segregated in layers.

GRANITIC TEXTURE.—When the constituents of a holocrystalline rock, such as granite, are intimately interlocked with each other, suggesting that they have crystallised out from fused material.

GRANULAR TEXTURE.—Holocrystalline rock, the consolidation of which has apparently been interrupted.

GRILLAGE.—A framework of rolled steel joists in a foundation.

GRIT-STONE.—A stone (which may be of various degrees of hardness) consisting of loose particles of silica.

GROIN.—The arris formed by the intersection of vaulting surfaces.

GROUT.—Cement or mortar made into a thick liquid for pouring into the joint cavities.

H

HEAD OR LINTOL.—The stone which spans in one piece the top of an aperture.

HELIX.—A curved line generated by a point which moves along the surface of a cylinder in such a way that a constant ratio is maintained between its travel round the surface of the cylinder and the distance travelled along its axis.

HOLOCRYSTALLINE STRUCTURE.—A complete mass of crystals without vitreous or glassy material.

HORIZONTAL PLANE.—A plane parallel to the ground, or, more correctly speaking, parallel to the surface of still water.

HYPERBOLA.—A curve formed by the section of a cone made by a plane parallel to the axis.

I

IMPOST.—A horizontal moulded course on which an arch immediately rests.

INTRADOS.—The inner curved outline of an arch, or the curve forming the inside surface of a vault or dome.

J

JAMBS.—Stones forming the vertical surfaces at the sides of a door or window opening.

JAMB LININGS.—Thin slabs covering the rough jambs on the inside of a window opening.

JOGGLE.—An indentation cut in the joint surfaces of stones.

JOINT.—The surface of contact between two adjacent blocks of stone.

K

KERF.—A saw-cut in stone.

KEYSTONE.—The central stone of an arch which locks the units comprising the arch.

L

LABEL.—The drip or hood moulding over an aperture.

LAGGINGS.—Battens connecting the ribs of centering from which the stones are wedged to their correct position.

LIASSIC LIMESTONES.—More or less argillaceous limestones, used chiefly for the manufacture of hydraulic limes.

LIERNE RIBS.—Those which merely cross from rib to rib.

LINE.—The path or locus of a moving point.

LOGGIA.—A covered portico for a building.

LUNE.—Of the surface of a sphere is the portion lying between two planes which contain an axis of the sphere.

LUNETTE.—An opening in a vaulted ceiling; the space formed by the intersection of a wall and a vault, sometimes pierced by a window.

M

MADREPORIC MARBLES.—Those containing fossil remains of Madrepores or similar organisms.

MAGNESIA.—A compound of oxygen and magnesium, nearly always found where lime is present. Magnesian or dolomitic limestones contain a large amount of magnesia.

MICA.—A mineral consisting of thin flexible laminæ or scales having a shining and almost metallic lustre. The scales are easily separated. A constituent of granite, but also found in many sedimentary rocks, enabling them to be split easily. Has a complex composition, mainly silica and alumina. Where the alumina is almost entirely replaced by magnesia, talcs are formed.

MICACEOUS SANDSTONES.—Those containing a large proportion of mica distributed over the bedding planes.

MODILLIONS.—The projections under the corona of a cornice, resembling brackets.

MORTISE.—A sinking in a stone to receive a corresponding projection.

MULLIONS.—Stone piers dividing a window into two or more lights.

N

NATURAL BED OF STONE.—The surface upon which the stone was originally deposited.

NEWEL.—A post at the end of a flight of steps to support a handrail.

NICHE.—A recess in a wall for the reception of a statue or ornament.

NOSING.—The vertical surface at the extreme projection of a cornice. The rounded portion of a tread projecting over a riser.

O

OOLITIC LIMESTONES.—Consisting of egg-shaped grains; forming a structure similar to the roe of a fish.

ORDER IN ARCHITECTURE.—Signifies a column with its base, shaft, and capital, and the entablature which it supports.

ORIEL.—A window projecting from the face of the wall, and supported on corbellings.

OUT OF WINDING.—A surface in a true plane.

P

PANEL.—An area sunk or raised from the general face of the surrounding work.

PARABOLA.—The section of a cone made by a plane parallel to one generator.

PENDENTIVE.—The portion of a dome-shaped vault which descends into a corner of an angular compartment.

PILASTER.—A square pillar slightly projecting from a pier or from a wall face.

PINNACLE.—A small turret-like termination placed on top of buttresses, etc.

PISOLITE.—A coarse oolitic limestone with grains about the size of a pea.

PLAIN ASHLARS.—Stones with rubbed, dragged, or polished plane surfaces.

PLANE.—A surface coinciding in every direction with a straight line.

PLINTH.—The lower vertical member at the base of a column or pedestal. Also applied to the projecting base of a building.

POLISHED WORK.—The surfaces of stones of crystalline texture gradually reduced to a degree of fineness constituting a polish.

POLYPHANT.—A close-grained serpentinous rock.

PORPHYRITIC.—A rock in which some of the crystals are developed on a larger scale than the rest.

PROFILE.—The contour outline of mouldings taken at right angles to their length.

Q

QUADRILATERAL.—A plane figure bounded by four straight lines.

QUIRK.—A narrow groove or sinking.

QUOIN.—A term applied to corner stones, but originally applied to the angle itself.

R

RAKING MOULDINGS.—Mouldings which are inclined to the horizontal.

RAMP.—A surface which is curved in elevation and straight in plan.

RAMP AND TWIST.—Surfaces which rise and curve at the same time.

RECTANGLE.—A plane figure having its opposite sides equal and all its angles right angles.

RESPOND.—A half-pillar or corbel attached to, or projecting from, a wall, for the support of an arch at the end of an arcade.

REVEAL.—The wall surfaces at the side of an opening.

ROUGE.—A distinctive name given to Belgian marbles of red variety.

ROUGH BACK.—A term applied to the vertical surface of stone that is entirely hidden in the thickness of a wall.

S

SACCHAROIDAL MARBLES.—Those with texture similar to loaf-sugar.

SCRIBE.—To make a hard line on the surface of a stone with a sharp-pointed tool, thereby transferring to the stone the profile of the mould or templet.

SCROLL STEP.—Having a spiral free end.

SECTION.—The geometrical representation of the figure produced by a cutting plane passing through a solid.

SERPENTINE.—A soft rock with shades and markings similar to those on a serpent's skin. Composed chiefly of silica, magnesia, and water. Takes a high polish.

SETTING.—The placing or fixing stones in walls, etc.

SHAFT OF A COLUMN.—That part between the base and the capital.

SHELL MARBLES.—Those containing only a few shells.

SICILIAN.—A variety of white Italian marble having a bluish tinge.

SILICA.—The most abundant substance in the earth's crust. A compound of silicon and oxygen. Sand, quartz, and flints are forms of silica.

SILICEOUS LIMESTONES.—Those containing a large percentage of silica, such as Kentish ragstone.

SILL.—The lower horizontal mêmber of a door or window opening.

SITE.—The position of a building; the place whereon it stands.

SKEW ARCH.—An arch whose face is not at right angles to its axis.

SKEWBACK.—A sloping joint surface, forming an abutment for the voussoirs of a segmental or flat arch.

SKIDS.—Short lengths of timber used for packing stones to the height required.

SOFFIT.—The under surface of a lintol or arch, or the lower surface of a vault.

SPAN.—The clear horizontal distance between the supports of an arch.

SPHERE.—A solid bound by a closed curved surface, such that all straight lines drawn from the centre to the surface are equal.

SPHEROID.—A body or figure similar to a sphere, but not perfectly spherical.

SQUINCH ARCH.—The small arch formed across the angle of a square or polygonal apartment, to support the side of a superimposed polygonal structure of greater number of sides.

STAIR.—A series of steps enabling a person to ascend or descend from one floor level to another.

STAIRCASE.—The apartment containing the stairs.

STATUARY MARBLE.—The name given in a general way to pure white marble free from marking.

STEP.—A ledge on which the foot may be securely placed in the process of ascending or descending.

STEREOTOMY.—The art of cutting solids into sections to suit certain conditions.

STILTED ARCH.—The springing of the arch curve raised for some distance above the impost, as upon stilts.

STOOLINGS.—The horizontal surfaces at the ends of a sill, forming seatings for the jambs.

STRING-COURSE.—A moulded or plain projecting course continued horizontally along the face of a building.
STRINGS.—The slabs at the free end of the steps, covering the concrete core.
SUB-PLINTH.—A second or lower plinth placed under the principal one.
SURFACE OF OPERATION.—A prepared plane surface for the application of the square or templets, from which the stone is worked to its correct shape.

T

TALC.—A magnesian mineral, oily to the touch, of shining lustre, translucent, and sometimes transparent when in thin plates. Also known as French chalk.
TANGENTS.—Lines or planes touching a curved line or curved surface, but not cutting it.
TAS-DE-CHARGE.—Springing stones of a Gothic vault, having horizontal bed joints.
TEMPLATE.—The stone placed under a beam to distribute the weight.
TEMPLET.—A mould or pattern giving the contour to which stones are to be cut.
TENON.—A projection formed for insertion in a mortise.
THROAT.—A groove worked in the under surface of projecting stonework to form a drip.
TILE STONES.—Thin slabs of stone suitable for roof coverings and made from an easily cleaved rock, cleft along its laminations.
TRACERY.—The geometrical and other forms in the head of Gothic windows.
TRANSOM.—A horizontal bar dividing a window into two or more lights in height.
TRAVERTINE.—A limestone deposited by chemical precipitation from calcareous springs. A large proportion of the buildings of ancient and modern Rome are built of travertine
TRIGLYPH.—A channelled ornament in the frieze of the Doric Order.
TURRETS.—Small towers usually containing stairs.
TYMPANUM.—The part of a pediment included between the horizontal and raking moulding.

U

UNICOLOURED MARBLES.—Those either black or white, or of uniform colour throughout.

V

VARIEGATED MARBLES.—Those having irregular spots and veins.
VENT.—A natural flaw in a block of stone.
VERTICAL PLANE.--A plane at right angles to a horizontal plane.
VOLUTE.—A spiral curve forming the principal feature of the Ionic capital.
VOUSSOIR.—A wedge-shaped stone forming a unit of an arch.

W

WEATHERING.—The slope given to the upper projecting surfaces of cornices and mouldings for the quick discharge of rain-water.
WELCH GROIN.—A groin formed by the intersection of two cylindrical vaults, of which one is of greater height than the other.
WELL-HOLE.—The space in the centre of a staircase.
WINDER.—A step in the circular part of a stair.
WINDING.—A twisted surface not in a true plane.
WREATHED COLUMNS.—Those twisted in the form of a screw.

Z

ZONE.—Of the surface of a sphere is the portion lying between two planes which are perpendicular to an axis of the sphere.

INDEX

A

B

C

D

E

F

G

H

I

J

K

L

M

N

O

P

Q

R

S

T

V

W